SEXUAL COERCION IN PRIMATES AND HUMANS

SEXUAL COERCION IN PRIMATES AND HUMANS

An Evolutionary Perspective on
Male Aggression against Females

EDITED BY

MARTIN N. MULLER

RICHARD W. WRANGHAM

HARVARD UNIVERSITY PRESS

Cambridge, Massachusetts, and London, England

2009

Library of Congress Cataloging-in-Publication Data

Sexual coercion in primates and humans: an evolutionary perspective on male aggression against females / edited by Martin N. Muller and Richard W. Wrangham.

   p. cm.

  Includes bibliographical references.

  ISBN 978-0-674-03324-5 (alk. paper)

  1. Primates—Sexual behavior. 2. Aggressive behavior in animals. I. Muller, Martin N., 1971– II. Wrangham, Richard W., 1948–

  QL737.P9S442 2009

  156'.5—dc22   2008042366

For Sherry and Elizabeth

Contents

IV FEMALE COUNTERSTRATEGIES

V SUMMARY AND CONCLUSIONS

SEXUAL COERCION IN PRIMATES AND HUMANS

I

INTRODUCTION AND THEORY

Male Aggression and Sexual Coercion of Females in Primates

Martin N. Muller, Sonya M. Kahlenberg,
and Richard W. Wrangham

Primates have long played a prominent role in the study of sexual selection. Darwin (1871) drew particular attention to the thick manes of male baboons, the bright facial coloration of male mandrills, and the large canines of male gorillas as sexually selected traits, and his first scientific paper on sexual selection famously concerned the "brightly-coloured hinder ends and adjoining parts of certain monkeys" (Darwin 1876). Subsequently, primates have provided model systems for the study of sexual selection in a wide range of contexts, including male infanticide (van Schaik and Janson 2000), sperm competition (Harcourt 1995; Birkhead and Kappeler 2004), sexual signaling (Nunn 1999; Zinner et al. 2004), and body size dimorphism (Mitani et al. 1996).

Although Darwin originally formulated the theory of sexual selection to account for exaggerated traits that appeared to decrease survival, such as the peacock's tail, he viewed sexual selection largely as a positive force working alongside natural selection to promote an adaptive fit between animals and their environments (Arnqvist and Rowe 2005). "Amongst many animals," he wrote, "sexual selection will give its aid to ordinary selection, by assuring to the most vigorous and best adapted males the greatest number of offspring" (Darwin 1859). This led Darwin to view reproductive interactions between the sexes as essentially cooperative ventures that inevitably produced improvements in the species as a whole (Arnqvist and Rowe 2005).

More recently, it has been recognized that males and females routinely have conflicting reproductive interests that result from fundamental asymmetries in levels of parental investment (Trivers 1972) and potential reproductive rates

(Clutton-Brock and Vincent 1991). Because in most cases females invest more time and energy in offspring than do males, male reproductive success (1) is limited primarily by access to temporally rare, fecund females and (2) is potentially much higher than that of females. Consequently, males often benefit by being more eager to mate, and less choosy about their mating partners, than are females, and selection can favor male traits that override female preferences. The result is an evolutionary arms race between the sexes, in which strategies and counterstrategies are selected to minimize reproductive costs imposed by the opposite sex (Arnqvist and Rowe 2005; Chapman 2006; Parker 2006). In some cases, these arms races result in reduced fitness for the population as a whole (Arnqvist and Rowe 2005).

Sexual conflict has been shown to have important evolutionary effects at multiple levels, from the morphological and physiological to the behavioral and life-historical. One of the most striking behavioral manifestations of sexual conflict was first recognized in studies of wild primates. Smuts and Smuts (1993) observed that males in a number of primate species appear to use force, or the threat of force, to coerce unwilling females to mate with them. They proposed that such coercion could productively be viewed as a third mechanism of sexual selection, distinct from intrasexual competition and intersexual choice (see also Clutton-Brock and Parker 1995). Although the utility of this distinction has been disputed (e.g., Pizzari and Snook 2003), there is no doubt that sexual coercion is a potentially important mechanism of mating bias within the broad framework of sexual conflict theory (Watson-Capps 2005). Jana Watson-Capps examines the relationship between studies of sexual coercion and the broader field of sexual conflict theory in Chapter 2 of this volume.

Since the publication of Smuts and Smuts's seminal review, few systematic attempts have been made to assess the frequency of male aggression against females in wild primates, or to test the idea that such aggression often functions as sexual coercion—this despite a burgeoning interest in sexual conflict theory that began in the early 1990s (Arnqvist and Rowe 2005). The purpose of this volume, which originated in a special session at the 2006 meetings of the International Primatological Society in Entebbe, Uganda, is to remedy this deficiency. The rationale is twofold. First, field observations of a range of primates have now been carried out for sufficient periods to permit researchers to address questions about male coercion empirically. Second, recent theoretical work by van Schaik et al. (2004) and Clarke et al. (Chapter 3 in this volume) has clarified the relationship between sexual coercion, female promiscuity, and male infanticide. These developments set the stage for a detailed exploration of the evolutionary forces that have shaped male aggression against females in primates, both human and nonhuman.

Male Aggression against Females: Noncoercive Functions

Male aggression against females shows considerable variation across primates. In some species such aggression appears to be rare, or limited to mild threat behaviors; in others it is an almost daily occurrence and can lead to serious injury. Interspecific comparisons are complicated, however, by both the paucity of quantitative data and variation in the methods used to sample behavior.

Although male-male aggression generally receives greater attention in the primate literature, in a number of species males direct aggression toward females as frequently or more frequently than they do toward other males (e.g., semi-free-ranging populations of rhesus, long-tailed and Tonkean macaques: Thierry 1985). In 1998 an adult anestrous female chimpanzee traveling with five adult males in Kanyawara, Kibale National Park could expect to receive male aggression, on average, once per day (Muller et al. 2007). This rate is statistically indistinguishable from the rate of male aggression toward males in the same year. For estrous females, the rate of received male aggression was substantially higher.

Similar associations between female fecundity and received male aggression are widespread in primates, a fact that Smuts and Smuts cited in support of the coercion hypothesis. In this chapter, we discuss the potential coercive functions of male aggression. It is important to note from the outset, however, that male aggression against females is not a unitary phenomenon and that coercion is unlikely to account for all instances of such aggression. In some species, for example, females do not experience heightened rates of male aggression when they are most likely to conceive (e.g., black-handed spider monkeys: Campbell 2003). Furthermore, within species aggression sometimes occurs in contexts that are not related to mating (Smuts and Smuts 1993). We therefore begin by reviewing four noncoercive explanations for the occurrence of male aggression against females: feeding competition, status competition, male policing, and redirected aggression. Although these represent alternatives to the coercion hypothesis, they do not exclude it, nor do they exclude each other.

Feeding Competition

Access to resources generally represents a stronger constraint on female than male reproduction (Wrangham 1979). However, males still require adequate resources to maintain health and survive. Consequently, within groups, male-female aggression may reflect intersexual contest competition for food. In species in which males are socially dominant to females, this often takes the form of agonistic supplants from shared resources. Intersexual feeding competition has not been well studied in primates, but the regular occurrence of male-female aggression in the context of feeding supports this idea (e.g., wedge-capped capuchins: 63% of male-female aggression, O'Brien 1991; olive baboons: 20%, Smuts 1985; chimpanzees: ~18%, Goodall 1986, Figure 12.3). If contest competition explains why males are aggressive toward females, then male-female aggression should occur more frequently in species or populations that depend on monopolizable foods, since these make contests feasible (van Schaik 1989). Male-female aggression would also be expected to occur more often inside food patches and during food shortages. However, these predictions have not been systematically tested.

Disputes over resources may also explain why males of some species are aggressive toward females from other social groups. Chimpanzee males, for example, regularly patrol the boundaries of their territories and attack strangers of both sexes (Wilson and Wrangham 2004; Watts et al. 2006). Such attacks can be fatal, although this outcome appears more likely for male than female victims (Muller and Mitani 2005). If attacks on strange females represented sexual coercion, then males would be expected to target either estrous females (to obtain copulations) or nulliparous females (to recruit them into their communities). However, long-term observations of the Gombe chimpanzees in Tanzania showed that males most frequently attack older, anestrous females during intergroup encounters (Williams et al. 2004).

Chimpanzee females maintain distinct but overlapping foraging areas (Wrangham and Smuts 1980), and when community boundaries shift, they typically adjust their ranges to remain within male-protected zones (Williams et al. 2002). Because older females have firmly established foraging areas, the Gombe researchers concluded that males probably benefit by eliminating such females or by forcing them to contract their ranges. This results in expansion of their own community's territory to include more food sources. The fact that larger home ranges at Gombe were associated with more mixed-sex parties (a correlate of increased food availability) and shorter interbirth intervals for resident females supports the idea that male aggression against extragroup females is related to intergroup feeding competition in chimpanzees (Williams et al. 2004). In black-and-white colobus monkeys, males do not kill extragroup females, but they sometimes chase them and their male associates during intergroup encounters. In support of a feeding competition explanation for this type of male-female aggression, intergroup aggression increased when high-quality resources were being contested, and groups that won conflicts often supplanted losers from food sources (Fashing 2001; Harris 2006).

Status Competition

In species in which males dominate females, male-female aggression might also reflect intersexual struggle for dominance. This explanation predicts that young males should be highly aggressive toward females when initially entering the dominance hierarchy and that aggression should diminish once females respond submissively to male challenges. Such patterns have been reported for both chimpanzees and mountain gorillas, indicating that male-female dominance relations explain some aspects of male-female aggression in these species (Pusey 1990; Watts and Pusey 1993; Nishida 2003; Muller et al., Chapter 8 in this volume). In fission-fusion species, such as chimpanzees, establishing dominance over females may even extend into adulthood, since some male-female dyads associate infrequently, leaving males with few opportunities to establish dominance during adolescence. However, in chimpanzees at Kanyawara, more frequent female companions received the highest rates of aggression from adult males (Muller et al., Chapter 8), suggesting that status competition is a relatively unimportant cause of male-female aggression.

It has been proposed that even after dominance relationships are decided, males may continue to direct aggression toward females in order to reconfirm their dominance relationships. In support of this hypothesis, Goodall (1986, Figure 12.3) found that in Gombe chimpanzees a large proportion of male aggression against females occurred when a male and female first met after an extended separation. This hypothesis could be further tested by examining the relationship between rates of female submissive behavior and male-female aggression within dyads, predicting a negative correlation if dominance reconfirmation is important. However, the dominance reconfirmation explanation for male-female aggression is problematic in that it is unclear why males would have to repeatedly enforce their dominance over females, particularly given that females never or very rarely challenge adult males.

Male Policing

Another possible function of male-female aggression is policing. This behavior involves males directing aggression at female group members to break up female-female conflicts. Males may intervene to defend a particular female ("partial interventions") or may direct aggression at both participants to halt the conflict impartially ("controlling interventions"). Partial interventions are expected when a male has a more valuable relationship with the target than with the aggressor or when the target is the male's close kin (Watts et al. 2000). In both cases, this behavior may function as male mating effort. For example, in baboons and macaques, some males form special relationships ("friendships") with particular females, and males direct aggression toward females who threaten their friends or their friends' offspring (Smuts 1985; Manson 1994; Palombit, Chapter 15 in this volume). By providing agonistic support to female friends, males help maintain friendships and thereby increase the probability of future reproductive or social payoffs. In gelada baboons, a species with one-male breeding units in which females are philopatric and form close alliances with female kin, males intervene in female conflicts mainly on behalf of females lacking kin allies (Dunbar 1980). Interventions of this kind may help persuade females to stay with the male during harem fissions and takeover attempts by new males (Watts et al. 2000). In species such as mountain gorillas, in which females transfer between groups, males commonly intervene in female conflicts, mainly through control interventions, but sometimes by supporting new immigrants, who receive high rates of aggression from resident females (Watts 1991, 1997). Interventions among mountain gorillas therefore diminish inequities among females and may deter low-ranking females from emigrating (Watts et al. 2000). A similar argument has been made for chimpanzees (Pusey 1980; Nishida 1989; Kahlenberg et al. 2008). If male-female aggression is explained by male policing behavior, then it is expected to increase during periods of high female-female aggression. Females should also benefit by experiencing lower rates of aggression from other females, maintaining greater access to resources, or having reduced stress. Data from mountain gorillas support these predictions (reviewed in Watts et al. 2000). Specifically, rates of male display toward females increased with rates of female-female harassment (Watts 1992), and male interventions led to decreased aggression from resident females toward immigrants, with targets being less likely to lose contests (Watts 1991).

Redirected Aggression

Finally, it is possible that some cases of male-female aggression have more to do with relationships among males than they do with intersexual relationships.

For example, male victims of male aggression sometimes redirect aggression to nearby females (olive baboons: Smuts 1985; chimpanzees: Goodall 1986), whether to reduce aggression-induced stress, deflect attention away from themselves to avoid further aggression, or facilitate reconciliation with aggressors (Watts et al. 2000). Males also attack close female associates of rival males, seemingly to instigate conflicts with these males (olive baboons: Smuts 1985; captive chimpanzees: de Waal 1982). If male-male competition explains male-female aggression, then these behaviors should occur in parallel and should be more common when male dominance hierarchies are unstable. These predictions have not been fully evaluated. However, in a number of primates, male aggression toward females increases in reproductive contexts (Smuts and Smuts 1993). Ostensibly, this pattern offers support for the redirected aggression hypothesis because male-male competition commonly increases in reproductive contexts, and unsuccessful males might redirect aggression toward vulnerable females. However, the redirected aggression hypothesis would predict increases in aggression toward any low-ranking target, including noncycling females and juveniles. The fact that estrous females are singled out as victims suggests that redirected aggression cannot account for most male-female aggression.

Male Aggression as a Sexually Selected Strategy

Definitions

Across primates the predominant functional explanation for male aggression against females is sexual coercion. Specifically, males are hypothesized to use aggression strategically to control female sexuality and increase their chances of fathering offspring. Sexual coercion can take a variety of forms, but no clear terminological consensus has emerged in the literature. Different investigators sometimes use the same word to label discrete behaviors (cf. "sexual harassment" *sensu*: van Schaik et al. 2004 and Clutton-Brock and Parker 1995) or different words to label the same behavior. We thus begin by establishing a basic taxonomy of coercion.

Smuts and Smuts (1993: 2–3) provided the essential definition of sexual coercion: "use by a male of force, or threat of force, that functions to increase the chances that a female will mate with him at a time when she is likely to be fertile, and to decrease the chances that she will mate with other males, at some cost to the female." This definition remains useful because it suggests more or less straightforward criteria that must be met for a behavior to be considered coercive. It also suggests an important, but generally neglected, distinction between direct and indirect forms of sexual coercion.

Direct coercion (that which "functions to increase the chances that a female will mate" with a male) involves the use of force to overcome female resistance to mating. This is the simplest, and probably the most taxonomically widespread, form of coercion. It is the primary form recognized by Clutton-Brock and Parker (1995), who suggested three distinct categories of direct coercion: *forced copulation, harassment,* and *intimidation.* These three strategies are differentiated primarily by the temporal proximity of their effects. Forced copulation ("rape" in the human literature) involves violent restraint, resulting in immediate mating. Harassment involves repeated attempts to mate that impose costs on females, inducing eventual female submission. Intimidation involves physical retribution against female refusals to mate, increasing the likelihood of submission to future advances. Because they presume female resistance, all three forms of direct coercion are expected to primarily involve nonpreferred males.

The second part of Smuts and Smuts's definition involves indirect coercion: the use of force "to decrease the chances that [a female] will mate with other males." We refer to this class of behavior as *coercive mate guarding.* Coercive mate guarding represents a distinct form of sexual coercion because it functions to constrain female promiscuity rather than to overcome female resistance (Smuts and Smuts 1993; van Schaik et al. 2004). Consequently, both preferred and nonpreferred males may practice coercive mate guarding. In many species, however, it is primarily a strategy of high-ranking males. It is expected to be prominent in species with high rates of multimale mating and, as we will see, can often be viewed as a counterstrategy to female attempts at paternity confusion (van Schaik et al. 2004; Clarke et al., Chapter 3 in this volume).

As with direct coercion, discrete strategies can be identified under the rubric of coercive mate guarding. The temporal distinction that proved useful in categorizing direct coercion is more problematic here. Nevertheless, we recognize three basic types of coercive mate guarding: *herding, punishment,* and *sequestration.* Herding involves short-term aggression directed toward females to induce immediate separation from rival males and to restore proximity to the guarding male. Punishment involves physical retribution against female association or copulation with other males, decreasing the likelihood of these behaviors in the future. Sequestration involves forceful separation of females from a social group, particularly during periods of maximal fecundity, to prevent copulations with rival males. Herding and punishment are widespread in primates, to separate females from rivals both within a group and in neighboring groups (Smuts and Smuts 1993; Clutton-Brock and Parker 1995). Sequestration shows a more restricted distribution but finds clear expression in hamadryas baboons (Swedell and Schreier, Chapter 10 in this volume) and chimpanzees, and extraordinary elaboration in human societies (Wilson and Daly, Chapter 11 in this volume).

We use the terms *herding* and *punishment* to refer solely to male aggression against females. However, these behaviors have parallels in male-male competition. Males sometimes exhibit direct aggression against male competitors, apparently to separate them from fecund females or to punish them for mating with females. It is useful to consider such aggression separately from male-female aggression, however, because this can help to discriminate female strategies. If a mate-guarding male directs aggression at both male competitors and the estrous female, it suggests that the female is resistant to being guarded. If aggression is directed solely at male competitors, then it is more likely that a female is amenable to being guarded. This distinction is important because in many mammals, aggression in the context of mate guarding is directed primarily at male competitors, and females actively try to escape the attentions of nonguarding males. Bighorn sheep provide a notable example. When a subordinate ram evades the mate defense of a dominant and approaches a cycling ewe, the ewe often flees to a position where she cannot be mounted (e.g., on a steep ledge or under a rock overhang or a deadfall) or enters thick vegetation (such as an aspen grove), where pursuing males have difficulty maneuvering (Hogg and Forbes 1997). The fact that, among primates, males frequently direct aggression at both male suitors and cycling females strongly suggests an alternative social dynamic in which females are often resistant to males' mate-guarding efforts.

Infanticide is a final form of sexual coercion involving aggression directed not at a female but at her infant. A large body of evidence suggests that infanticide by male primates commonly represents a sexually selected strategy by which newly immigrated or newly dominant males shorten a female's period of lactational amenorrhea (Hrdy 1979; van Schaik and Janson 2000). Infanticide does not constitute intimidation in the usual sense of the word. However, it should probably be considered a form of direct coercion under the Smuts and

Smuts definition because it can increase the probability that a female will be fecundable before the killer is dead or expelled from her group. We classify infanticide as a third category of coercion because, although it accelerates mating opportunities, it does not inherently determine which males take advantage of those opportunities.

Because previous publications offer comprehensive treatments of infanticide as a sexually selected strategy (e.g., van Schaik and Janson 2000), in this volume it is addressed primarily as a basis for female promiscuity. In many mammals multimale mating appears to have evolved to confuse paternity and decrease the risk of male infanticide (Hrdy 1979; van Schaik and Janson 2000; Wolff and Macdonald 2004). Coercive mate guarding is thus intimately related to infanticide because it represents the male response to the female counter-strategy of paternity confusion (van Schaik et al. 2004; Clarke et al., Chapter 3 in this volume).

Assessing Sexual Coercion

Because Smuts and Smuts's definition of sexual coercion is functional rather than behavioral, identifying coercion in the wild can be complicated. Several specific conditions must be satisfied for male aggression to be interpreted as sexual coercion. These criteria are addressed in detail in the following chapters, which employ long-term field data to test predictions of the coercion hypothesis in specific populations. However, we provide a brief introduction here.

If male aggression is employed strategically to control female sexuality, then the most fecund females (i.e., those with the highest probability of conception) are expected to receive the most aggression from males. This prediction is supported by data from a range of species showing that cycling adult females receive higher rates of male aggression than either adolescent females or adult females undergoing lactational amenorrhea. Among hamadryas baboons, for example, Kummer (1968) observed that juvenile females and mothers with new infants enjoyed greater freedom of movement than other females, wandering further from their harem leaders before incurring a neck-bite. Swedell and Schreier (Chapter 10) similarly found that female hamadryas were least likely to receive aggression when they were lactating. This pattern is also found in mountain gorillas (Robbins Chapter 5). Among chimpanzees, nulliparous females receive significantly less aggression from males than do more fecund, parous females (Muller et al. 2007, Chapter 8). And in a rare observational study of a human population, Flinn (1988) showed that in a Caribbean village, agonistic interactions between co-resident mates were more likely when the woman was fecund than when she was pregnant or nursing a young infant.

Further support for an association between male aggression and female fecundity comes from the fact that cycling females in a range of species suffer increased rates of male aggression in mating contexts. Cycling chimpanzee, chacma baboon, and hamadryas baboon females, for example, all receive elevated rates of male aggression when exhibiting maximally tumescent sexual swellings (Muller et al. 2007, Chapter 8; Kitchen et al. Chapter 6; Swedell and Schreier, Chapter 10, all in this volume).

If male aggression against females functions as sexual coercion, then it should also be associated with increased mating activity. As noted by Smuts and Smuts (1993), this prediction is problematic because the relevant question, "Does a male who is aggressive toward a female increase his probability of siring offspring with her above the probability had he not been aggressive?" is unanswerable. For this reason researchers have generally focused on asking whether there is a correlation between rates of male aggression against females and mating success. Specifically, do individual males show higher copulation rates with the females toward whom they are relatively more aggressive than with those toward whom they are less aggressive? This appears to be the case for chimpanzees (Muller et al. 2007, Chapter 8), hamadryas baboons (Swedell and Schreier, Chapter 10 in this volume), and Japanese macaques (Soltis et al. 1997), but empirical data are absent for most species.

A positive correlation between mating and aggression per se is not sufficient to establish coercion because alternative processes could theoretically produce such an association. Female choice is one confounding variable. If, for reasons of their own, females prefer to mate with high-ranking males, and high-ranking males are more aggressive, then a noncoercive association between mating and aggression could result. An additional complication is that females necessarily maintain proximity with males in order to copulate. If males are generally aggressive toward all females in direct proximity with them, then a spurious correlation could result (Soltis et al. 1997). Any robust attempt to relate mating access to aggression must control for such confounds.

A final criterion for male aggression to qualify as sexual coercion is that it must impose fitness costs on females. Although receiving aggression may appear inherently detrimental, it could potentially benefit females by, for example, allowing them to test the quality of potential mates (Szykman et al. 2003).

If the fitness benefits of such testing outweigh the costs of the aggression, then there would be no evolutionary conflict between males and females, and thus no coercion under the Smuts and Smuts (1993) definition (cf. Watson-Capps Chapter 2). For this reason, inferring sexual coercion from observations of behavior alone is problematic (Pizzari and Snook 2003).

Male mountain gorillas illustrate this difficulty (Robbins, Chapter 5 in this volume). Female gorillas receive increased rates of male aggression during mating periods, but this aggression primarily takes the form of mild displays of threat. The prolonged attacks and beatings seen in chimpanzees do not occur in gorillas. In the absence of information on how costly such moderate aggression is for female gorillas, it is difficult to exclude the hypothesis that male aggression in this species represents courtship rather than coercion (Robbins, Chapter 5 in this volume).

Despite the difficulties of quantifying fitness outcomes in long-lived species like primates, it is useful to ask whether male aggression imposes proximate costs on females because selection should favor female strategies to reduce such costs. If, for example, female resistance is designed to select "more persistent and thus better quality partners" (Pizzari and Snook 2003), and the costs of such testing are significant, then selection should favor females who utilize information from previous interactions in their mating decisions. This might be impossible for the invertebrate species studied in most sexual conflict research, but it is well within the capacity of primate species that show complex social cognition (e.g., recognition of conspecifics as individuals, memory of prior social interactions). In orangutans, for example, females consistently exhibit resistance toward particular males and show strong preferences for other males without engaging in initial resistance (Fox 1998; Knott, Chapter 4 in this volume). This suggests that in orangutans female resistance is due not to mate assessment but to negative preference, supporting the idea that forced copulations in orangutans represent sexual conflict (Knott and Kahlenberg 2007; Knott, Chapter 4).

Male aggression can inflict a variety of costs on females, ranging from physical injury and physiological stress to reduced feeding time. The severity of these impacts differs by species. Chimpanzee females experience relatively brutal aggression that can lead to severe wounding and stress (Muller et al., Chapter 8, Novak and Hatch, Chapter 13, both in this volume). Chacma baboon females experience increased stress and direct fitness losses from male infanticide, but appear to experience lower costs of sexual aggression (Kitchen et al., Chapter 6 in this volume). Orangutan females are frequent victims of forced copulation, but they almost never experience violence that results in physical injury (Knott, Chapter 4 in this volume).

Interspecific Differences in Male–Female Aggression

Indirect Coercion

As the succeeding chapters illustrate, male aggression against females varies dramatically across primates, occurring regularly in some species and virtually never in others. Recently, van Schaik et al. (2004) and Clarke et al. (Chapter 3 in this volume) have proposed that much of this variation can be explained by differences in the risk of infanticide by males. Specifically, they note that in species where infanticide represents a sexually selected male strategy to renew ovarian cycling in lactating mothers, females often adopt multimale mating as a counterstrategy. Increased female promiscuity functions to confuse paternity, and thereby decreases the risk of male infanticide (Hrdy 1979; van Schaik and Janson 2000; Wolff and Macdonald 2004), but simultaneously drives selection for coercive mate guarding in males.

Coercive mate guarding requires that males keep track of female movements, intervene in female social interactions, and intimidate females with physical force. Thus, van Schaik et al. (2004) argue that coercion is most feasible in species that are highly sexually dimorphic, terrestrial, and diurnal. Accordingly, they note that most reports of successful coercion come from Old World monkeys and apes, many of which exhibit these traits. In Chapter 7 of the present volume, however, Link et al. evaluate the evidence for coercion in an arboreal New World primate, the spider monkey.

The van Schaik model makes further predictions about the distribution of male-female aggression among species in which males kill infants. Multimale mating offers females some protection from infanticide by altering potentially antagonistic males' perceptions of paternity. However, in primates that maintain cohesive foraging groups and strong linear dominance hierarchies, females who subtly bias paternity toward high-ranking males gain additional protection for their offspring by inducing those males to defend their infants against potential aggressors (van Schaik et al. 2000). Females in these species are thus expected to mate promiscuously early in the follicular phase in order to give all males a nonzero probability of conception, but to adopt a preference for high-ranking males during the periovulatory period (POP), when conception is most likely to occur (reviewed in Clarke et al., Chapter 3 in this volume). In this scenario, the interests of high-ranking males in monopolizing females during the fertile period coincide with the interests of females in concentrating paternity in high-ranking males, and a decrease in coercive mate guarding is expected as the POP approaches.

In fission-fusion species such as the chimpanzee, on the other hand, high-ranking males are not predictably present to defend infants from infanticide attempts, and females are thus expected to be less interested in concentrating paternity and more likely to maintain a conflict of interest with high-ranking males during the POP (Clarke et al., Chapter 3). In these species, coercive aggression by high-ranking males is expected to continue or even intensify around ovulation, a hypothesis that is tentatively addressed in Chapter 8 of the present volume.

Clarke et al. (Chapter 3) see paternity confusion as the primary benefit driving multimale mating across primates. However, coercive mate guarding can arise in any system in which males and females disagree on the optimal number of female mating partners. This includes pair-bonded species like humans, in which females are not generally promiscuous, but can potentially benefit from selective extra-pair mating (e.g., Gangestad and Thornhill 2004).

Humans represent an interesting exception to the association among infanticide, female promiscuity, and male sexual coercion. Men rarely commit infanticide, and women are less promiscuous than many primate females, yet men exhibit an elaborate repertoire of coercive mate guarding that includes extreme forms of sequestration (Smuts 1992; Wilson and Daly, Chapter 11 in this volume). These behaviors regularly inflict severe costs on women (Wilson and Daly 1988; Wilson and Daly, Chapter 11, and Novak and Hatch, Chapter 13, in this volume). This exceptional human pattern likely results from the fact that, unlike most primate males, men invest heavily in direct paternal care and thus face potentially high costs to female promiscuity (Daly and Wilson 1988; Smuts 1992; Wilson and Daly, Chapter 11). In addition, humans form pair-bonds in the context of multimale, multifemale social groups, even though males and females typically spend most of their days in separate locations. This combination of features provides near-constant opportunities for extra-pair copulation. The closest nonhuman primate approximation of the human social system is maintained by hamadryas baboons, with males and females forming pair-bonds in the context of a broader multimale, multifemale society, but there the mated pair is never apart. As with humans, male coercion, and specifically sequestration of females, is taken to an extreme in hamadryas (Swedell and Schreier, Chapter 10 in this volume).

As was the case with nonhuman primates, however, not all forms of male violence against women necessarily reflect sexual coercion. In Chapter 12, Rodseth and Novak explore the idea that intersexual violence sometimes represents political signaling among men. We further explore the political use of violence against women in Chapter 18.

Direct Coercion

Direct coercion, at least in its most severe form of forced copulation, is uncommon in primates. In species with high levels of female promiscuity forced copulation is rare, arguably because females are less likely to reject male suitors. In chimpanzees, for example, females do sometimes refuse male solicitations but predictably mate with every male in their community during a given reproductive cycle (Stumpf and Boesch 2006). Perhaps as a result, male chimpanzees almost never attempt to force copulations (Muller et al., Chapter 8 in this volume). Rare instances generally involve close male relatives, such as brothers or sons, for whom females are expected to maintain strong mating aversions (Goodall 1986).

Among primates with more selective females, direct coercion is sometimes minimized because male conspecifics intervene to prevent it (e.g., Smuts and Smuts 1993). Accordingly, the two primate species exhibiting the highest apparent rates of forced copulation, orangutans and humans, maintain grouping patterns that often leave females solitary, and thus vulnerable to male coercion (Smuts 1992; Knott, Chapter 4 in this volume). Emery Thompson (Chapter 14, this volume), for example, notes that the incidence of rape in the United States is highest among college-age women, who are more likely to be unmarried and living alone, away from natal kin.

The idea that human rape represents a male reproductive strategy that has been shaped by sexual selection is highly controversial, in large part because some scholars contend that to treat rape as a natural phenomenon subtly legitimizes it (e.g., Dusek 1984). In Chapter 14, Emery Thompson shows that an evolutionary account of male coercion neither offers support for violence against women nor suggests that sexual violence is genetically determined or inevitable. She presents evidence against the hypothesis that rape represents an adaptation through which low-quality males gain immediate opportunities to fertilize fecund females. Emery Thompson also stresses the functional importance of the distinction between stranger rape and acquaintance rape, and explores the idea that acquaintance rape represents a component of a long-term male reproductive strategy.

The Evolution of Sexual Coercion.

In sum, developments in understanding sexual coercion give us the first opportunity to explore and elaborate a general theory through the perspectives of different species. Beginning with general principles, the theory seeks to account for variations in the occurrence, nature, and patterning of male aggression toward females according to the evolutionary ecology and social system of each species. This book includes a sufficiently wide array of social systems to offer robust challenges to the theory, and the result gives considerable support to its essential idea: that male behavior has been selected to include a wide variety of coercive tactics, which in many primates reflect long-term relationships rather than merely short-term copulation attempts. The application of this idea to different contexts, whether to different populations or different species, represents an exciting opportunity for understanding coercion in both humans and nonhuman primates.

References

Arnqvist, G., and L. Rowe. *Sexual Conflict.* Princeton, N.J.: Princeton University Press, 2005.

Birkhead, T. R., and P. M. Kappeler. "Post-copulatory Sexual Selection in Birds and Primates." In *Sexual Selection in Primates: New and Comparative Perspectives,* eds. P. Kappeler and C. P. van Schaik, pp. 151–171. Cambridge: Cambridge University Press, 2004.

Campbell, C. J. "Female-Directed Aggression in Free-ranging *Ateles geoffroyi." International Journal of Primatology* 24 (2003): 223–237.

Chapman, T. "Evolutionary Conflicts of Interest between Males and Females." *Current Biology* 16 (2006): 744–754.

Clutton-Brock, T. H., and G. A. Parker. "Sexual Coercion in Animal Societies." *Animal Behaviour* 49: 1345–1365.

Clutton-Brock, T. H., and A. C. J. Vincent. "Sexual Selection and the Potential Reproductive Rates of Males and Females." *Nature* 351 (1991): 58–60.

Daly, M., and M. Wilson. *Homicide.* New York: Aldine de Gruyter, 1988.

Darwin, C. *The Origin of Species.* London: John Murray, 1859.

————. *The Descent of Man and Selection in Relation to Sex.* London: John Murray, 1871.

————. "Sexual Selection in Relation to Monkeys." *Nature* 15 (1876): 18–19.

de Waal, F. B. M. *Chimpanzee Politics.* Baltimore, Md.: Johns Hopkins University Press, 1982.

Dunbar, R. I. M. "Determinants and Evolutionary Consequences of Dominance among Female Gelada Baboons." *Behavioral Ecology and Sociobiology* 7 (1980): 254–265.

Dusek, V. "Sociobiology and Rape." *Science for the People* 16 (1984): 10–16.

Fashing, P. J. "Male and Female Strategies during Intergroup Encounters in Guerezas *(Colobus guereza):* Evidence for Resource Defense Mediated through Males and a Comparison with Other Primates." *Behavioral Ecology and Sociobiology* 50 (2001): 219–230.

Flinn, M. V. "Mate Guarding in a Caribbean Village." *Ethology and Sociobiology* 9 (1988): 1–28.

Fox, E. A. "The Function of Female Mate Choice in the Sumatran Orangutan *(Pongo pygameus abelii)*." Ph.D. Dissertation, Duke University, Durham, N.C., 1998.

Gangestad, S. W., and R. Thornhill. "Female Multiple Mating and Genetic Benefits in Humans: Investigations of Design." In *Sexual Selection in Primates: New and Comparative Perspectives,* eds. P. Kappeler and C. P. van Schaik, pp. 90–116. Cambridge: Cambridge University Press, 2004.

Goodall, J. *The Chimpanzees of Gombe: Patterns of Behavior.* Cambridge, Mass.: Harvard University Press, 1986.

Harcourt, A. H. "Sexual Selection and Sperm Competition in Primates: What Are Male Genitalia Good For?" *Evolutionary Anthropology* 4 (1995): 121–129.

Harris, T. R. "Between-Group Contest Competition for Food in a Highly Folivorous Population of Black and White Colobus Monkeys *(Colobus guereza)*." *Behavioral Ecology and Sociobiology* 61 (2006): 317–329.

Hogg, J. T., and S. H. Forbes. "Mating in Bighorn Sheep: Frequent Male Reproduction Via a High-Risk 'Unconventional' Tactic." *Behavioral Ecology and Sociobiology* 41 (1997): 33–48.

Hrdy, S. B. "Infanticide among Animals: A Review, Classification, and Examination of the Implications for the Reproductive Strategies of Females." *Ethology and Sociobiology* 1 (1979): 13–40.

Kahlenberg, S. M., M. Emery Thompson, M. N. Muller, and R. W. Wrangham. "Immigration Costs for Female Chimpanzees and Male Protection as an Immigrant Counterstrategy to Intrasexual Aggression." *Animal Behaviour* 76 (2008): 1497–1509.

Knott, C. D., and S. M. Kahlenberg. "Orangutans in Perspective: Forced Copulations and Female Mating Resistance." In *Primates in Perspective,* eds. C. J. Campbell, A. Fuentes, K. C. MacKinnon, M. Panger, and S. Bearder, pp. 290–305. Oxford: Oxford University Press, 2007.

Kummer, H. *Social Organisation of Hamadryas Baboons. A Field Study.* Chicago: University of Chicago Press, 1968.

Manson, J. H. "Mating Patterns, Mate Choice, and Birth Season Heterosexual Relationships in Free-Ranging Rhesus Macaques." *Primates* 35 (1994): 417–433.

Mitani, J. C., J. GrosLouis, and A. F. Richards. "Sexual Dimorphism, the Operational Sex Ratio, and the Intensity of Male Competition in Polygynous Primates." *American Naturalist* 147 (1996): 966–980.

Muller, M. N., S. M. Kahlenberg, M. Emery Thompson, and R. W. Wrangham. "Male Coercion and the Costs of Promiscuous Mating for Female Chimpanzees." *Proceedings of the Royal Society B: Biological Sciences.* 274 (2007): 1009–1014.

Muller, M. N., and J. C. Mitani. "Conflict and Cooperation in Wild Chimpanzees." *Advances in the Study of Behavior* 35 (2005): 275–331.

Nishida, T. "Social Interactions between Resident and Immigrant Female Chimpanzees." In *Understanding Chimpanzees,* eds. P. G. Heltne and L. A. Marquardt, pp. 68–89. Cambridge, Mass.: Harvard University Press, 1989.

———. "Harassment of Mature Female Chimpanzees by Young Males in the Mahale Mountains." *International Journal of Primatology* 24 (2003): 503–514.

Nunn, C. L. "The Evolution of Exaggerated Sexual Swellings in Primates and the Graded-Signal Hypothesis." *Animal Behaviour* 58 (1999): 229–246.

O'Brien, T. G. "Female-Male Social Interactions in Wedge-Capped Capuchin Monkeys: Costs and Benefits of Group Living." *Animal Behaviour* 41 (1991): 555–567.

Parker, G. "Sexual Conflict over Mating and Fertilization: An Overview." *Philosophical Transactions of the Royal Society B* 361 (2006): 235–259.

Pizzari, T., and R. R. Snook. "Sexual Conflict and Sexual Selection: Chasing away Paradigm Shifts." *Evolution* 57 (2003): 1223–1236.

Pusey, A. E. "Behavioural Changes at Adolescence in Chimpanzees." *Behaviour* 115 (1990): 203–246.

Smuts, B. B. *Sex and Friendship in Baboons.* New York: Aldine de Gruyter, 1985.

———. "Male Aggression against Women: An Evolutionary Perspective." *Human Nature* 3 (1992): 1–44.

Smuts, B. B., and R. W. Smuts. "Male Aggression and Sexual Coercion of Females in Nonhuman Primates and Other Mammals: Evidence and Theoretical Implications." *Advances in the Study of Behavior* 22 (1993): 1–63.

Soltis, J., F. Mitsunaga, K. Shimizu, Y. Yanagihara, and M. Nozaki. "Sexual Selection in Japanese Macaques I: Female Mate Choice or Male Sexual Coercion?" *Animal Behaviour* 54 (1997): 725–736.

Stumpf, R., and C. Boesch. "The Efficacy of Female Choice in Chimpanzees of the Tai Forest, Cote d'Ivoire." *Behavioral Ecology and Sociobiology* 60 (2006): 749–765.

Szykman, M., A. L. Engh, R. C. Van Horn, E. E. Boydston, K. T. Scribner, and K. E. Holekamp. "Rare Male Aggression Directed toward Females in a Female-Dominated Society: Baiting Behavior in the Spotted Hyena." *Aggressive Behavior* 29 (2003): 457–474.

Thierry, B. "Patterns of Agonistic Interactions in Three Species of Macaque *(Macaca mulatta, M fascicularis, M tonkeana)."* *Aggressive Behavior* 11 (1985): 223–233.

Trivers, R. L. "Parental Investment and Sexual Selection." In *Sexual Selection and the Descent of Man, 1871–1971,* ed. B. Campbell, pp. 136–179. Chicago: Aldine, 1972.

van Schaik, C. P. "The Ecology of Social Relationships amongst Female Primates." In *Comparative Socioecology: The Behavioural Ecology of Humans and Other Mammals,* eds. V. Standen and R. A. Foley, pp. 195–218. Oxford: Blackwell, 1989.

van Schaik, C. P., and C. H. Janson, eds. *Infanticide by Males and Its Implications.* Cambridge: Cambridge University Press, 2000.

van Schaik, C. P., G. R. Pradhan, and M. A. van Noordwijk. "Mating Conflict in Primates: Infanticide, Sexual Harassment and Female Sexuality." In *Sexual Selection in Primates: New and Comparative Perspectives,* eds. P. Kappeler and C. P. van Schaik, pp. 131–150. Cambridge: Cambridge University Press, 2004.

Watson-Capps, J. J. "Female Mating Behavior in the Context of Sexual Coercion and Female Ranging Behavior of Bottlenose Dolphins (*Tursiops Sp.*) in Shark Bay, Western Australia." Ph.D. Dissertation, Georgetown University, 2005.

Watts, D. P. "Harassment of Immigrant Female Mountain Gorillas by Resident Females." *Ethology* 89 (1991): 135–153.

———. "Social Relationships of Immigrant and Resident Female Mountain Gorillas. I. Male-Female Relationships." *American Journal of Primatology* 28 (1992): 159–181.

———."Agonistic Interventions in Wild Mountain Gorilla Groups." *Behaviour* 134 (1997): 23–57.

Watts, D. P., F. Colmenares, and K. Arnold. "Redirection, Consolation, and Male Policing: How Targets of Aggression Interact with Bystanders." In *Natural Conflict Resolution,* eds. F. Aureli and F. B. M. de Waal, pp. 281–301. Berkeley: University of California Press, 2000.

Watts, D. P., M. N. Muller, S. J. Amsler, G. Mbabazi, and J. C. Mitani. "Lethal Intergroup Aggression by Chimpanzees in Kibale National Park, Uganda." *American Journal of Primatology* 68 (2006): 161–180.

Watts, D. P., and A. E. Pusey. "Behavior of Juvenile and Adolescent Great Apes." In *Juvenile Primates: Life History, Development, and Behavior,* eds. M. E. Pereira and L. A. Fairbanks, pp. 148–167. Oxford: Oxford University Press, 1993.

Williams, J. M., G. W. Oehlert, J. V. Carlis, and A. E. Pusey "Why Do Male Chimpanzees Defend a Group Range?" *Animal Behaviour* 68 (2004): 523–532.

Williams, J. M., A. E. Pusey, J. V. Carlis, B. P. Farm, and J. Goodall. "Female Competition and Male Territorial Behaviour Influence Female Chimpanzees' Ranging Patterns." *Animal Behaviour* 63 (2002): 347–360.

Wilson, M. L., and R. W. Wrangham. "Intergroup Relations in Chimpanzees." *Annual Review of Anthropology* 32 (2003): 363–392.

Wolff, J. O., and D. W. Macdonald. "Promiscuous Females Protect Their Offspring." *Trends in Ecology and Evolution* 19 (2004): 127–134.

Wrangham, R. W. "On the Evolution of Ape Social Systems." *Social Science Information* 18 (1979): 335–368.

Wrangham, R. W., and B. B. Smuts. "Sex Differences in the Behavioural Ecology of Chimpanzees in the Gombe National Park, Tanzania." *Journal of Reproduction and Fertility Supplement* 28 (1980): 13–31.

Zinner, D. P., C. L. Nunn, C. P. van Schaik, and P. M. Kappeler. "Sexual Selection and Exaggerated Sexual Swellings of Female Primates." In *Sexual Selection in Primates: New and Comparative Perspectives*, eds. P. Kappeler and C. P. van Schaik, pp. 71–89. Cambridge: Cambridge University Press, 2004.

2

Evolution of Sexual Coercion with Respect to Sexual Selection and Sexual Conflict Theory

Jana J. Watson-Capps

In this chapter I discuss how the biological explanation of sexual coercion depends on understanding the interplay between female and male interests regarding sex. Sexual selection theory contains a well-established body of principles, but the importance of sexual coercion as an independent arena of conflict has only been appreciated in recent years. As a result, theories describing the evolution of sexual conflict and sexual coercion are still developing.

Sexual coercion theory and broader sexual conflict theory are rarely discussed simultaneously, even though they are interrelated. Different terms associated with sexual conflict or sexual coercion are often invoked to describe identical scenarios. For example, in seaweed flies *(Coelopa ursine)*, there is a "pre-mating struggle" between males and females owing to sexual conflict over when to mate (Crean and Gilburn 1998); in red-sided garter snakes *(Thamnophis sirtalis parietalis)*, males also struggle with females to achieve mating, but it is referred to as male "harassment" of females (Shine et al. 2000), a type of sexual coercion. Integrating sexual conflict and sexual coercion will allow for recent advances in one area to further our understanding in the other. To accomplish this goal, I will present an account of the dynamic development of sexual coercion theory as it pertains to broader sexual conflict and sexual selection theories.

Sexual Selection Theory

Darwin's original definition of sexual selection remains valid. It is "the advantage which certain individuals have over others of the same sex and species, in exclusive relation to reproduction" (Darwin 1871). In essence, selection favors individuals who obtain a relatively high number of mating opportunities or superior mate quality, or both.

Even though individuals may employ both devices to increase their reproductive success, usually one sex is choosier about whom they mate with, while the other invests more in competing for more mating opportunities. Determining which role a sex will play has been the subject of much theoretical discussion, but the overall pattern is clear: in most species, females tend to be choosier than males. Bateman (1948) provided experimental support for this tendency in fruit flies *(Drosophila melanogaster)*. He found that males who mated more females (up to five) had higher reproductive success, whereas females who mated three or more males had no more reproduction than those who mated two males. In many other species, the relationship between the number of mating partners and reproductive success has also been positive and steep for males, whereas for females, the relationship reaches its peak very quickly and then may even become negative with additional matings (Andersson and Iwasa 1996).

In order to mate with more females, males compete with other males over access to females (intrasexual selection). Males may interact directly through fighting and endurance tests, or indirectly, such as when males scramble to find females before other males do or when their sperm must compete with other males' sperm in the female reproductive tract (for a review of how female behavior affects male competition, see Luttbeg 2004).

Females discriminate between males in order to select certain preferred traits in males (mate choice, a form of intersexual selection). Female preference for certain male traits may evolve through several methods: females may have a preexisting bias toward certain traits unrelated to mate quality (sensory bias); females may gain direct benefits, such as food or protection, by mating with certain males (direct benefits); females may produce superior offspring by mating with males who display a trait indicative of their genetic quality (good genes/indicator traits); or, if the female preference for a trait and the trait itself are genetically linked, both traits may increase in a population (Fisher runwaway process) (Andersson and Iwasa 1996; Cunningham and Birkhead 1998; Ryan 1998; Welch et al. 1998; Fisher 1999; Møller and Alatalo 1999; Kokko et al. 2003).

When a male and female are in conflict over whether or not to mate, one sex, typically male, may use force to overcome resistance to mating. The suggestion that this sexual coercion represents a third, separate mechanism of sexual selection came in the mid-1990s (Smuts and Smuts 1993). Recent reorganizations of mate choice, competition for mates, and sexual coercion have high-lighted how all three of these mechanisms are interrelated.

During the 1990s biologists reassessed the separation of sexual selection into separate components. Some took the view that all sexual selection was the result of mate choice, either directly or indirectly (Wiley and Poston 1996; Cunningham and Birkhead 1998). Direct mate choice requires discrimination between potential mates, whereas indirect mate choice includes traits that do not involve discrimination of mates by the individual but still result in mating biases. For example, males may compete with each other over resources attractive to females, or the females themselves, but female traits select the context for the competition by determining the resource (e.g., breeding site) or by mating with the winner. In an analogous fashion, other scientists proposed that all sexual selection was the result of competition between members of the same sex for matings (Andersson and Iwasa 1996). For example, males have to compete with each other for access to females and also to be preferred by them. In a sense, both are true, but where does sexual coercion fit into these new frameworks of sexual selection theory?

Sexual Coercion as a Form of Mating Bias

There are already many well-developed models of how mate choice and competition for mates can explain the evolution of some traits that would not otherwise be selected for naturally. It is useful to compare sexual coercion to these other, well-established methods of sexual selection to determine how traits can evolve from sexual coercion and how coercion itself evolves. Sexual coercion is most similar to mate choice and should be classified as a type of mating bias, alongside traditional female mate choice. Mating bias, defined as female (for most species) traits that increase mating success for a subset of males, is no longer restricted to traditional female preference for males (mate choice), but includes female resistance to sexual advances (coercion). Coercion, therefore, encompasses all instances of males using aggression to overcome female "preference" and females using "resistance" as a method of assessing male coercive ability (i.e., a method of mate choice).

Female Mate Preference versus Resistance

The difference between female preference and resistance seems clear at first glance but becomes murky upon further inspection. Some authors do not distinguish between female preference and resistance because they both result in mating bias (Gavrilets et al. 2001). Others consider them distinct but related phenomena, as resistance can be viewed as a negative preference (Kokko 2005). Here I discuss them separately because ethologists typically classify these behaviors based on function: females who seek out and allow mating to occur with males of a certain trait are said to prefer that trait, whereas females who avoid and try to prevent mating with males of a certain trait are said to be resisting that male and therefore are biasing against that trait.

Female preference for mates, also called traditional female choice, clearly results in biased mating because females only mate with males who have certain preferred traits or a certain level of that trait. For example, females may bias their mating toward larger males by preferring to mate with the largest male they encounter.

Female resistance to mating can bias male mating success in a few different ways. If resistance is successful, female resistance will act as a negative preference, resulting in more matings for males with the preferred trait (see Kokko 2005 for a theoretical model on this process). If resistance can be overcome or if resistance is used as a means of assessing male quality, it may bias mating toward coercive males (e.g., in seaweed flies, *Coelopa ursine:* Crean and Gilburn 1998; and in house flies, *Musca domestica*: Andrés and Arnqvist 2001). Resistance may promote male contest or sperm competition by advertising the attempted copulation to other males in the area (e.g., in northern elephant seals, *Mirounga angustirostris*: Cox and Le Boeuf 1977). In addition,, females may use resistance as a conditional strategy depending on the costs of mating and the female's energetic reserves, such as in convenience polyandry (Watson et al. 1998). In convenience polyandry, females adjust their mating rate to balance the costs imposed by male persistence. For example, female water striders, *Aquarius remigis,* who carry males on their back, and incur a cost doing so, may try to dislodge an unpreferred male only if they have the energy reserves to do so (Watson et al. 1998).

Even though sexual coercion is partly demonstrated by the female's resistance, the processes whereby male coercive traits and female resistance traits evolve are very similar to those involved in traditional female mate preference. Considering these two forms of mating bias simultaneously, rather than as distinct mechanisms, will help us understand how sexual coercion works to produce sexually selected traits.

Evolution of Resistance

Males use coercion to mate with (or try to mate with) unwilling females; if males are faced with the option of not mating or sexually coercing a female, selection would favor sexual coercion. But coercion would not be possible without a reluctant female, which posits the question: why do females resist sexual advances by (some) males? The reasoning is similar to why females prefer males with certain qualities. With a little modification, we can apply well-developed models of preference evolution to the evolution of resistance.

As mentioned earlier, females who resist mating with males of certain traits are exerting a negative preference for those traits, thereby increasing their chances of mating with preferred males. However, resisting males, like searching for preferred mates, comes with some cost to the female.

If resisting exacts a cost from the female greater than the benefit she would receive from mating with a preferred mate instead, there would be no net direct gain for resisting (Cordero and Eberhard 2003; Kokko et al. 2003). In this case, selection would favor females who do not resist mating, resulting in convenience polyandry, when females mate with multiple males to reduce sexual harassment by males. However, if coerciveness can be passed on from father to son, the female may benefit indirectly because her sons will also be coercive and able to overcome female resistance. This is a version of the "sexy son" hypothesis, in which the cost of coercion exacted on the female is outweighed by the reproductive benefit to her sons (Parker 1979). Recent theoretical models suggest that, in some cases, these indirect benefits are large enough that a female would be able to pass on more genes if she mated with a coercive male (Kokko 2005). If these indirect benefits were larger than the direct costs for females, there would be no selective pressure to avoid coercion (Cordero and Eberhard 2003), but as yet there is no empirical data to support this idea.

I have talked broadly about "male" and "female" mating strategies, but it is important to remember that both sexes, even coercive males, can exert mate preferences. If coercion of a female is an expensive exercise for males (e.g., energetically, increased mortality risk), males may only attempt coercion on preferred female mates (Trivers 1972; Kokko and Monaghan 2001). For example, in seaweed flies, 6% of mating struggles were ended by the male giving up (Crean and Gilburn 1998); the authors hypothesize that these instances may reflect struggles that have become too costly or not worthwhile for the males. A further study on seaweed flies, *Gluma musgravei*, addressed this issue directly.

Females who were rejected by males had higher mortality rates than those with whom the males copulated (Dunn et al. 2001). The authors therefore suggest that "decisions" to mate with the female occur during the mating struggle. Larger males were also found to be more discriminatory (Dunn et al. 2001). Perhaps the indirect benefits of mating with coercive males were so great that males only mated with resistant females to ensure that their female offspring would have the ability to select for coercive males (i.e., "resistant daughters"). The finding that male seaweed flies often dismounted passive females without copulating (Dunn et al. 2001) is consistent with the idea that males may prefer resistant females.

Forms of Sexual Coercion

The primary example of sexual coercion is forced copulation. However, several forms of coercion fit our behavioral definition of increasing male mating opportunity over other males through aggression. In addition to forced copulation, sexual coercion includes intimidation, harassment, sequestering, and infanticide (Hrdy and Hausfater 1984; Smuts and Smuts 1993; Clutton-Brock and Parker 1995; Andersson and Iwasa 1996; Kokko 2005). Clutton-Brock and Parker (1995: 1345) define three types of sexual coercion: forced copulation—when "males gain matings by using superior speed or strength to restrain a resisting female"; harassment—when "males' repeated mating attempts cause a female to incur such a high cost that she immediately mates with the male"; and intimidation—when "males punish a female who refuses a mating attempt, thereby increasing his chance of mating with her in the future." Infanticide is also classified as sexual coercion if males kill infants to increase their chances of fathering an offspring with the mother (Hrdy and Hausfater 1984; Smuts and Smuts 1993). Sequestering is included when males use force to keep females away from mating opportunities with other males, such as with coercive mate guarding (Connor et al. 1992; Andersson and Iwasa 1996; Watts 1998; Kokko 2005).

Examples of sexual coercion in different forms have been documented in all classes of vertebrates and in many of the better known invertebrates (Kummer 1968; Thornhill 1980; McKinney et al. 1983; Mitani 1985; Goodall 1986; Mesnick and Le Boeuf 1991; Smuts 1992; Bercovitch 1995; Olsson 1995; Stone 1995; Wrangham and Peterson 1996; Crean and Gilburn 1998; Michiels and Newman 1998; Watts 1998; Linklater et al. 1999; Shine et al. 2000; Bisazza et al. 2001; Stutt and Siva-Jothy 2001; Cunningham 2003). The mating biases resulting from that coercion are similar to those derived from mate choice, although the method (preference vs. resistance) differs. Mating, though beneficial to both males and females, is often the battleground for differing agendas between the sexes.

Sexual Conflict Theory

The view that males and females are living at odds with each other rather than amicably has received much attention lately and has sparked a resurgence of interest in intersexual conflict theory (see reviews: Snook 2001; Chapman et al. 2003; Cordero and Eberhard 2003; Parker 2006; Lessells 2006; Tregenza et al. 2006). Sexual conflict refers to any evolutionary conflict of interest between the sexes and includes any situation in which the optimal outcome of a trait or an interaction is different for each sex (Parker 1979, 2006). Because there can only be one outcome, there is necessarily a fitness cost to one sex. Gowaty and Buschhaus (1998: 209) take a slightly more ecological approach by defining sexual dialectics as "sexual coevolution that occurs when opposite sexes create ecological problems that the other sex must solve in order to survive and reproduce." Sexual conflict is not restricted to mating behaviors. Other interactions between males and females, such as parenting, can also exhibit sexual conflict (Lessells 2006). Sexual conflict is not limited to animals; it has also been investigated in plants that alternate between haploid and diploid generations (Haig and Wilczek 2006).

Sexual conflict is common in mating situations because of the necessary difference in size between eggs and sperm (anisogamy), which leads to opposing optimal mating tactics for males and females. Sexually conflicted interactions result in at least one sex being pushed away from its optimum. For example, males and females may be in conflict over how often to mate: the optimal number of mates for males may be five partners, while the optimal number for females could be one. As a result of this conflict, the female, for example, may be pushed from her optimal level (one partner) in the direction of the male optimum (five partners). In this example, the male benefits by mating with an optimal number of females, whereas the female incurs a cost by mating with too many males. This cost is measured in terms of her evolutionary fitness. However, not every aspect of mating is ruled by sexual conflict and fitness costs; mating provides obvious fitness benefits to both sexes in the form of shared offspring.

Sexual Conflict Effecting Evolution

How can the presence of this conflict between males and females result in genetic changes in a population over time? The evolutionary response to sexual conflict depends on whether the antagonism is the result of conflicting optima on a specific trait (e.g., body size), or over a certain scenario, which is solved by different traits in each sex (e.g., female mate preference/resistance and male coercion). Conflict involving a specific trait can be addressed through intralocus contest evolution, while conflict over a certain scenario can be accomplished through interlocus contest evolution (Rice and Holland 1997).

Contests between two alleles at the same loci (intralocus contest evolution) may eventually result in sex-limited expression of those traits (Lande 1980; but see Michiels and Newman 1998 for an interesting example in hermaphrodites). Sex-linked genes can spread in a population even if the cost to one sex is greater than the benefit to the other (Rice 1992).

If the contest is instead mediated by gene products at two different loci (interlocus contest evolution), then the two genes may continually evolve in response to one another. Sexual selection could select for male tactics that increase their mating success but imposes direct costs to females; natural selection may then select for countertraits that reduce those costs for females, and/or sexual selection may select for traits that increase her reproductive benefits through biases in paternity (Rowe 1994; Partridge and Hurst 1998; Cordero and Eberhard 2003). Once this pattern of trait and countertrait is set up, the traits in both sexes may continue to coevolve with each other antagonistically. Sexually antagonistic coevolution has been well demonstrated in fruit flies (*Drosophila melanogaster:* Holland and Rice 1999; Pitnick et al. 2001), house flies (*Musca domestica:* Andrés and Arnqvist 2001), and water striders (*Aquarius remigis:* Rowe et al. 1994; Arnqvist et al. 2000; Arnqvist and Rowe 2002).

In a study of fruit flies, Holland and Rice (1999) removed sexual selection and the possibility for antagonistic coevolution between the sexes for 47 generations by artificially enforcing lifetime monogamy in mating pairs of fruit flies, which normally mate promiscuously. Because both the male and female had no access to other mates, there was no sexual conflict, and therefore no opportunity arose for mate choice or competition for mates, and there was no need for coercion. In normal fruit fly mating, where sexual selection can occur, males produce seminal fluid that can reduce the probability a female will subsequently mate with another male, but this tactic comes at a cost to the female because proteins in the seminal fluid are harmful to the female's health. However, when monogamy was enforced, they found that males evolved to be less harmful to females (presumably by lowering seminal fluid toxicity) and females evolved to be less resistant to this harm (Holland and Rice 1999). Therefore, the properties of seminal fluid that reduce the chance a female will mate with subsequent males is thought to evolve due to sperm competition between males, but because it also increases female mortality (presumably as a by-product), females then evolved a resistance to the toxicity (Holland and Rice 1999). As females increase their resistance to the seminal fluid, males may intensify the potency in response, thereby continuing the coevolutionary "arms race". Sexual conflict has been implicated in speciation due to the rapid nature of evolution with antagonistically coevolving traits (Arnqvist et al. 2000; Gavrilets 2000).

Sexually Antagonistic Coevolution or Palliative Adaptation?

Evolutionary conflict between the sexes may lead to sexually antagonistic coevolution, but this is not necessarily the case. To help determine whether sexual conflict would lead to antagonistic coevolution, Lessells (2006) has argued the importance of differentiating between adaptive costs (or harm; also called conflict load) and collateral costs. When there is sexual conflict over a trait or an interaction's outcome, there is unavoidably a cost to the sex that is not able to achieve the optimal trait or outcome (i.e., that sex becomes the "loser"). Both sexes may even incur fitness costs as a result of the conflict. In order to understand how traits evolve in response to sexual conflict, and how these traits may perpetuate the conflict, we must first be able to differentiate between the two types of harm.

Adaptive harm refers to direct costs resulting from changes in the optimal value of the trait in conflict. For example, if a male and female are in conflict over whether to mate, and the male physically punishes the female for not mating, the harm inflicted from that punishment is the means of pushing the interaction in the male's favor, mating. Adaptive harm establishes a selective advantage for the opposite sex to develop a counteradaptation to bring the trait back to their optima. In our example, the female may be pressured to develop a strategy to avoid or withstand the punishment, so that she can avoid mating with that male. This pattern of trait followed by countertrait can lead to continuous sexually antagonistic coevolution (Lessells 2006; Parker 2006).

The second type of harm, *collateral harm*, is a side effect of manipulating the trait in conflict. Returning to our example of conflict between a male and a female over whether to mate, if a male forces a copulation with a female and the female is harmed during a forced copulation, that injury is not necessary for the male to achieve that mating, but it is a by-product of male aggression during the forced copulation. In other words, it is not adaptive for the male to harm the female in this example if his optimal outcome is to mate with her; the harm is collateral (although if he is setting a precedent of aggression to discourage future resistance, then it would qualify as adaptive harm). The occurrence of collateral harm establishes a selective pressure for palliative adaptations to reduce that harm (but not the outcome itself). Palliative adaptations are in the interests of both sexes and not likely to lead to sexually antagonistic coevolution (Lessells 2006; Parker 2006). Assessing the type of harm to females, and also to males, is important in determining whether the behaviors are in the process of antagonistic coevolution between the sexes or the result of palliative solutions.

It may be difficult to detect a cost if an antagonistic or a palliative adaptation has already occurred. However, sexual conflict is defined by the cost one or both sexes bear, so it is necessary to demonstrate a cost in order to establish the presence of sexual conflict. There are ways to quantify costs due to sexual conflict, if that conflict can be removed. In experimental studies, such as in fruit flies and in yellow dung flies, *Scathophaga stercoraria,* cost was measured by removing sexual conflict through artificially enforced monogamy (Holland and Rice 1999; Hosken et al. 2001). The degree of conflict can also be manipulated by artificially changing the sex ratio (Wigby and Chapman 2004). In other studies, it may still be possible to measure a cost because conflict is not necessarily balanced for every trait (e.g., if somatic costs are balanced out by reproductive benefits, you can still measure somatic costs) or equilibrium may not be achieved. This last point is particularly important; because of the nature of antagonistic coevolution, there is not likely to be an equilibrium end point. Many studies found costs to females associated with mating, and some have quantified costs to both sexes (Mesnick and Le Boeuf 1991; Smuts and Smuts 1993; Rowe 1994; Rowe et al. 1994; Chapman et al. 1995; Stone 1995; Michiels and Newman 1998; Watson et al. 1998; Linklater et al. 1999; Civetta and Clark 2000; Shine et al. 2000; Pitnick et al. 2001; Stutt and Siva-Jothy 2001).

Sexual Coercion within Sexual Conflict Theory

Sexual coercion is one mechanism of sexual selection, brought about by sexual conflict over when to mate and whom to mate with. Sexual coercion may also perpetuate the conflict between the sexes if antagonistic coevolution occurs. Because sexual coercion is only a small piece of sexual conflict theory, it is not usually referenced in discussions of sexual conflict (but see Kokko 2005; Parker 2006). Sometimes studies of sexual conflict are also studies of sexual coercion, but they are not addressed as such. For example, in seaweed flies, large males have a reproductive advantage because of success in pre-mating struggles, which the authors interpret as a side effect of sexual conflict (Crean and Gilburn 1998). These cases involving pre-mating struggles can be categorized as sexual coercion because the use of force by a male has been linked directly to his copulation success. Another, more extreme, example can be found in bed bugs, where a male pierces a female's abdominal wall with his genitalia, reducing her longevity (Stutt and Siva-Jothy 2001). I would classify this behavior as forced copulation, a type of sexual coercion. Furthermore, even though conflict between the sexes is fundamental to the evolution of sexual coercion, sexual conflict is rarely mentioned in studies of sexual coercion (but see Shine et al. 2000; Kokko 2005).

This separation between the sexual conflict and sexual coercion literatures may have formed because studies of sexual coercion tend to be rooted in ethology, while general studies of sexual conflict are not. Two different sets of literature, sexual coercion in behavior journals and sexual conflict in broader journals, have led to a lack of communication between the two growing fields. The differences in terminology often illustrate the differences in focus between the two subjects: coercion deals primarily with obvious male behaviors, while sexual conflict is explicitly interested in the interaction between males and females. Sexual coercion terminology often describes only the male behavior indicating that it is happening to the female (i.e., coercion, forced copulation, harassment, sequestering), whereas the sexual conflict literature often uses terms indicating both sexes are active participants (i.e., struggle, conflict). Correspondingly, studies of sexual coercion usually quantify only the cost the female incurs, but sexual conflict studies often look at costs to both sexes. Because these topics are so closely interrelated, it is necessary to connect common concepts so that students of sexual coercion can benefit from the recent advances in sexual conflict theory, and vice versa.

In the following sections, I discuss how the concepts of adaptive versus palliative harm (from sexual conflict theory) are useful when making predictions about the evolution of sexual coercion and female responses to coercion.

Sexual Coercion and Adaptive Harm

Adaptive harm refers to costs imposed on one sex by the other in order to achieve their optimal trait or outcome. In the case of sexual coercion, this would include any example where the harm itself is responsible for increasing a male's mating opportunity, such as costs incurred by females from harassment, intimidation, infanticide, loss of control over mate choice, and sequestering. Even though they were not classified as adaptive at the time, examples of adaptive harm, such as decreased time resting, decreased time foraging, and increased predation risk due to harassment, have been observed in a variety of animals (olive baboons, *Papio anubis:* Bercovitch 1983; solitary bees: Stone 1995; feral horses: Linklater et al. 1999; southern elephant seals, *Mirounga leonine:* Galimberti et al. 2000; red-sided garter snakes: Shine et al. 2000).

Antagonistic adaptations to adaptive harm would alter the result of the mating interaction itself. There are several potential ways to accomplish antagonistic adaptation. Females may spatially avoid males, as documented for solitary bees (Stone 1995). Females may breed synchronously with other females to reduce the male's ability to monopolize estrous females or to reduce the risk of infanticide by confusing paternity, as proposed for some primates (Hrdy and Hausfater 1984; Smuts 1992). Female-bonded societies (kin or otherwise) may use coalitions to combat males directly (Smuts 1992; Perry 1997; but see Hohmann and Fruth 2003). Obtaining a male bodyguard could protect females from coercion from other males, such as is seen in gorillas *(Gorilla gorilla),* rhesus macaques *(Macaca mulatta),* chacma baboons *(Papio cynocephalus ursinus),* and Sumatran orangutans *(Pongo pygmaeus abelii)* (Manson 1994; Mesnick 1997; Harcourt et al. 2001; Palombit et al. 2001; Fox 2002).

Sexual Coercion and Collateral Harm

Collateral harm refers to costs incurred during an interaction in which conflict is present, but these costs are not responsible for any change in the outcome; they are simply a by-product of participating in that interaction. Several types of harm sustained during sexual coercion would fit into this category: harm resulting from forced copulations, and efforts to protect an offspring from infanticide, as well as harm incurred as a by-product of being sequestered, besides the force used to keep the female present). Female mortality due to mating attempts would also fall under collateral harm because this is obviously not in the male's interest; this situation has been observed in feral sheep, *Ovis canadensis*, (Réale et al. 1996), northern elephant seals, (Le Boeuf and Mesnick 1990), and red-sided garter snakes (Shine et al. 2000). Strategies to counter these costs may develop over time and can be either palliative or antagonistic.

Palliative adaptations would alleviate the collateral harm associated with coercion without interfering with the coercive result. There are several ideas about how females may reduce these costs. Living with other females to reduce the risks associated with infanticide and predation has been hypothesized as a main force behind the evolution of primate sociality (Brereton 1995; Sterck et al. 1997). Females may mate with unpreferred males if the costs of resisting are higher than any benefit gained by resisting (convenience polyandry: Thornhill and Alcock 1983; Cordero and Andrés 2002). In a similar vein, female red-sided garter snakes that are too young to reproduce may mate with males anyway because this renders them unattractive to other males and lets them leave without further harassment (Shine et al. 2000). To avoid collateral costs associated with infanticide, females may reabsorb fetuses when new males enter a group, that is, exhibit the Bruce effect (alpine marmots, *Marmota marmota:* Hackländer and Arnold 1999).

Males and females are not in conflict about all aspects of sexual coercion. For example, the immediate health of the female is of concern to both sexes. It is possible that males will coevolve with females to develop palliative adaptations that decrease harm to the female. For example, an extreme form of forced copulation is observed in the bed bug, where traumatic insemination occurs. Because a male pierces the female's abdominal wall with his genitalia, females have less control over mating rate; female mating frequency is over 20 times their optimal mating frequency, and mated females have reduced longevity (Stutt and Siva-Jothy 2001). Stutt and Siva-Jothy (2001) hypothesize that the design of the males' genitalia has evolved to reduce the female's healing costs and risk of infection, probably to increase both male and female reproductive success. There may be other instances of this more cooperative coevolution between males and females.

Conclusion

Understanding sexual coercion becomes easier when we realize that males and females are in frequent conflict over sex. Sexual conflict is not a new area of study, but a recent surge of interest on the topic has altered how we think about sexual selection in general. In addition to traditional images of mating, involving males vying for a female's sexual attentions and females choosing particularly pleasing males, sex can be hostile. Females can resist males they do not want to mate with, and males can attempt to overcome this resistance through coercion.

Even when mating is viewed through the lens of sexual conflict, models of evolution used to explain female preference can still be used (once slightly modified) to investigate the evolution of female resistance behaviors and male coercion. Sexual coercion in response to female resistance is just one mechanism of intersexual sexual selection (similar to mate choice).

Taking advantage of recent advances in sexual conflict theory, we can predict when sexual coercion may lead to escalating coevolution between the sexes and when to expect palliative adaptations in the interest of both sexes.

References

Andersson, M., and Y. Iwasa. "Sexual Selection." *Trends in Ecology and Evolution* 11 (1996): 53–58.

Andrés, J. A., and G. Arnqvist. "Genetic Divergence of the Seminal Signal-Receptor System in Houseflies: The Footprints of Sexually Antagonistic Coevolution?" *Proceedings of the Royal Society of London, B* 268 (2001): 399–405.

Arnqvist, G., M. Edvardsson, U. Friberg, and T. Nilsson. "Sexual Conflict Promotes Speciation in Insects." *Proceedings of the National Academy of Sciences* 97 (2000): 10460–10464.

Arnqvist, G., and L. Rowe. "Antagonistic Coevolution between the Sexes in a Group of Insects." *Nature* 415 (2002): 787–789.

Bateman, A. J. "Intra-Sexual Selection in *Drosophila*." *Heredity* 2 (1948): 349–368.

Bercovitch, F. B. "Time Budgets and Consortships in Olive Baboons *(Papio anubis)*." *Folia Primatologica* 41 (1983): 180–190.

———. "Female Cooperation, Consortship Maintenance, and Male Mating Success in Savanna Baboons." *Animal Behaviour* 50 (1995): 137–149.

Bisazza, A., G. Vaccari, and A. Pilastro. "Female Mate Choice in a Mating System Dominated by Male Sexual Coercion." *Behavioral Ecology* 12 (2001): 59–64.

Brereton, A. R. "Coercion-Defense Hypothesis: The Evolution of Primate Sociality." *Folia Primatologica* 64 (1995): 207–214.

Chapman, T., G. Arnqvist, J. Bangham, and L. Rowe. "Sexual Conflict." *Trends in Ecology and Evolution* 18 (2003): 41–47.

Chapman, T., L. R. Liddle, J. M. Kalb, M. F. Wolfner, and L. Partridge. "Cost of Mating in *Drosophila melanogaster* Females Is Mediated by Male Accessory Gland Products." *Nature* 373 (1995): 241–244.

Civetta, A., and A. Clark. "Correlated Effects of Sperm Competition and Postmating Female Mortality." *Proceedings of the National Academy of Sciences* 97 (2000): 13162–13165.

Clutton-Brock, T. H., and G. A. Parker. "Potential Reproductive Rates and the Operation of Sexual Selection." *The Quarterly Review of Biology* 67 (1992): 437–456.

———. "Sexual Coercion in Animal Societies." *Animal Behaviour* 49 (1995): 1345–1365.

Clutton-Brock, T. H., and A. C. J. Vincent. "Sexual Selection and the Potential Reproductive Rates of Males and Females." *Nature* 351 (1991): 58–60.

Connor, R. C., A. F. Richards, R. A. Smolker, and J. Mann. "Patterns of Female Attractiveness in Indian Ocean Bottlenose Dolphins." *Behaviour* 133 (1996): 37–69.

Connor, R. C., R. A. Smolker, and A. F. Richards. "Two Levels of Alliance Formation among Male Bottlenose Dolphins (*Tursiops* sp.)." *Proceedings of the National Academy of the Sciences USA* 89 (1992): 987–990.

Cordero, A., and J. Andrés. "Male Coercion and Convenience Polyandry in a Calopterygid Damselfly." *Journal of Insect Science* 2 (2002): 1–7.

Cordero, C., and W. G. Eberhard. "Female Choice of Sexually Antagonistic Male Adaptations: A Critical Review of Some Current Research." *Journal of Evolutionary Biology* 16 (2003): 1–6.

Cox, C. R., and B. J. LeBoeuf. "Female Incitation of Male Competition: A Mechanism in Sexual Selection." *The American Naturalist* 111 (1977): 317–335.

Crean, C. S., and A. S. Gilburn. "Sexual Selection as a Side-Effect of Sexual Conflict in the Seaweed Fly, *Coelopa ursina* (Diptera: Coelopidae)." *Animal Behaviour* 56 (1998): 1405–1410.

Cunningham, E. "Female Mate Preferences and Subsequent Resistance to Copulation in the Mallard." *Behavioral Ecology* 14 (2003): 326–333.

Cunningham, E. J. A., and T. R. Birkhead. "Sex Roles and Sexual Selection." *Animal Behaviour* 56 (1998): 1311–1321.

Dunn, D., C. Crean, and A. Gilburn. "Male Mating Preference for Female Survivorship in the Seaweed Fly *Gluma musgravei* (Diptera: Coelopidae)." *Proceedings of the Royal Society of London, B* 268 (2001): 1255–1258.

Emlen, S., and L. Oring. "Ecology, Sexual Selection, and the Evolution of Mating Systems." *Science* 197 (1977): 215–223.

Fox, E. A. "Female Tactics to Reduce Sexual Harassment in the Sumatran Orangutan *(Pongo pygmaeus abelii)*." *Behavioral Ecology and Sociobiology* 52 (2002): 93–101.

Galimberti, F., L. Boitani, and I. Marzetti. "Female Strategies of Harassment Reduction in Southern Elephant Seals." *Ethology Ecology and Evolution* 12 (2000): 367–388.

Gavrilets, S. "Rapid Evolution of Reproductive Barriers Driven by Sexual Conflict." *Nature* 403 (2000): 886–889.

Gavrilets, S., G. Arnqvist, and U. Friberg. "The Evolution of Female Mate Choice by Sexual Conflict." *Proceedings of the Royal Society of London, B* 268 (2001): 531–539.

Gowaty, P. A., and N. Buschhaus. "Ultimate Causation of Aggressive and Forced Copulation in Birds: Female Resistance, the CODE Hypothesis, and Social Monogamy." *American Zoology* 38 (1998): 207–225.

Hackländer, K., and W. Arnold. "Male-Caused Failure of Female Reproduction and Its Adaptive Value in Alpine Marmots *(Marmota marmota)." Behavioral Ecology* 10 (1999): 592–597.

Haig, D., and A. Wilczek. "Sexual Conflict and the Alternation of Haploid and Diploid Generations." *Philosophical Transactions of the Royal Society B* 361 (2006): 335–343.

Hohmann, G., and B. Fruth. "Intra- and Inter-Sexual Aggression by Bonobos in the Context of Mating." *Behaviour* 140 (2003): 1389–1413.

Holland, B., and W. R. Rice. "Chase Away Sexual Selection: Antagonistic Seduction versus Resistance." *Evolution* 52 (1998): 1–7.

———. "Experimental Removal of Sexual Selection Reverses Intersexual Antagonistic Coevolution and Removes a Reproductive Load." *Proceedings of the National Academy of Sciences* 96 (1999): 5083–5088.

Hosken, D. J., T. W. J. Garner, and P. I. Ward. "Sexual Conflict Selects for Male and Female Reproductive Characters." *Current Biology* 11 (2001): 489–493.

Jaeger, R. G., J. R. Gillette, and R. C. Cooper. "Sexual Coercion in a Territorial Salamander: Males Punish Socially Polyandrous Female Partners." *Animal Behaviour* 63 (2002): 871–877.

Hrdy, S. B., and G. Hausfater. "Comparative and Evolutionary Perspectives on Infanticide: Introduction and Overview." In *Infanticide*, eds. G. Hausfater and S. B. Hrdy. New York: Aldine, 1984.

Jormalainen, V., S. Merilaita, and J. Riihimäki, J. "Costs of Intersexual Conflict in the Isopod *Idotea baltica." Journal of Evolutionary Biology* 14 (2001): 763–772.

Kokko, H. "Treat 'Em Mean, Keep 'Em (Sometimes) Keen: Evolution of Female Preferences for Dominant and Coercive Males." *Evolutionary Ecology* 19 (2005): 123–135.

Kokko, H., R. Brooks, M. D. Jennions, and J. Morley. "The Evolution of Mate Choice and Mating Biases." *Proceedings of the Royal Society of London, B* 270 (2003): 653–664.

Kokko, H., and M. Jennions. "It Takes Two to Tango." *Trends in Ecology and Evolution* 18 (2003): 103–104.

Kokko, H., and P. Monaghan. "Predicting the Direction of Sexual Selection." *Ecology Letters* 4 (2001): 159–165.

Lande, R. "Sexual Dimorphism, Sexual Selection and Adaptation in Polygenic Characters." *Evolution* 34 (1980): 292–305.

LeBoeuf, B. J., and S. Mesnick. "Sexual Behavior of Male Northern Elephant Seals: I. Lethal Injuries to Adult Females." *Behaviour* 116 (1990): 143–162.

Lessells, C. "The Evolutionary Outcome of Sexual Conflict." *Philosophical Transactions of the Royal Society B* 361 (2006): 301–317.

Linklater, W. L., E. Z. Cameron, E. O. Minot, and K. J. Stafford. "Stallion Harassment and the Mating System of Horses." *Animal Behaviour* 58 (1999): 295–306.

Luttbeg, B. "Female Mate Assessment and Choice Behavior Affect the Frequency of Alternative Male Mating Tactics." *Behavioral Ecology* 15 (2004): 239–247.

Manson, J. "Mating Patterns, Mate Choice, and Birth Season Heterosexual Relationships in Free-Ranging Rhesus Macaques." *Primates* 35 (1994): 417–433.

McKinney, F., S. R. Derrickson, and P. Mineau. "Forced Copulation in Waterfowl." *Behaviour* 86 (1983): 250–294.

McLain, D. K., and A. E. Pratt. "The Cost of Sexual Coercion and Heterospecific Sexual Harassment of the Fecundity of a Host-Specific, Seed-Eating Insect *(Neacoryphus bicrucis)*." *Behavioral Ecology and Sociobiology* 46 (1999): 164–170.

Mesnick, S. L., and B. J. Le Boeuf. "Sexual Behavior of Male Northern Elephant Seals: II. Female Response to Potentially Injurious Encounters." *Behaviour* 117 (1991): 262–280.

Michiels, N. K., and L. J. Newman. "Sex and Violence in Hermaphrodites." *Nature* 391 (1998): 647.

Mitani, J. C. "Mating Behaviour of Male Orangutans in the Kuai Game Reserve, Indonesia." *Animal Behaviour* 33 (1985): 392–402.

Møller, A. P., and R. V. Alatalo. "Good-Genes Effects in Sexual Selection." *Proceedings of the Royal Society of London, B* 266 (1999): 85–91.

Olsson, M. "Forced Copulation and Costly Female Resistance Behavior in the Lake Eyre Dragon, *Ctenophorus maculosus*." *Herpetologica* 51 (1995): 19–24.

Palombit, R. A., D. L. Cheney, and R. M. Seyfarth. "Female-Female Competition for Male 'Friends' in Wild Chacma Baboons, *Papio cynocephalus ursinus*." *Animal Behaviour* 61 (2001): 1159–1171.

Parker, G. "Sexual Conflict over Mating and Fertilization: An Overview." *Philosophical Transactions of the Royal Society B* 361 (2006): 235–259.

Partridge, L., and L. Hurst. "Sex and Conflict." *Science* 281 (1998): 2003–2008.

Perry, S. "Male-Female Social Relationships in Wild White-Faced Capuchins *(Cebus capucinus)*." *Behaviour* 134 (1997): 477–510.

Pitnick, S., G. T. Miller, J. Reagan, and B. Holland. "Males' Evolutionary Responses to Experimental Removal of Sexual Selection." *Proceedings of the Royal Society of London, B* 268 (2001): 1071–1080.

Réale, D., P. Boussés, and J. Chapuis. "Female-Biased Mortality Induced by Male Sexual Harassment in a Feral Sheep Population." *Canadian Journal of Zoology* 74 (1996): 1812–1818.

Rice, W. "Sexually Antagonistic Genes—Experimental Evidence." *Science* 256 (1992): 1436–1439.

———. "Male Fitness Increases When Females Are Eliminated from Gene Pool: Implications for the Y Chromosome." *Proceedings of the National Academy of Sciences* 95 (1998): 6217–6221.

Rice, W., and B. Holland. "The Enemies Within: Intergenomic Conflict, Interlocus Contest Evolution (ICE), and the Intraspecific Red Queen." *Behavioral Ecology and Sociobiology* 41 (1997): 1–10.

Roldan, E. R. S., and M. Gomendio. "The Y Chromosome as a Battle Ground for Sexual Selection." *Trends in Ecology and Evolution* 14 (1999): 58–62.

Rowe, L. "The Costs of Mating and Mate Choice in Water Striders." *Animal Behaviour* 48 (1994): 1049–1056.

Rowe, L., G. Arnqvist, A. Sih, and J. J. Krupa. "Sexual Conflict and the Evolutionary Ecology of Mating Patterns: Water Striders as a Model System." *Trends in Ecology and Evolution* 9 (1994): 289–293.

Ryan, M. J. "Sexual Selection, Receiver Biases, and the Evolution of Sex Differences." *Science* 281 (1998): 1999–2003.

Scott, E., J. Mann, J. Watson-Capps, B. Sargeant, and R. Connor. "Aggression in Bottlenose Dolphins: Evidence for Sexual Coercion, Male-Male Competition, and Female Tolerance through Analysis of Tooth-Rake Marks and Behaviour." *Behaviour* 142 (2005): 21–44.

Shine, R., D. O'Connor, and R. T. Mason. "Sexual Conflict in the Snake Den." *Behavioral Ecology and Sociobiology* 48 (2000): 392–401.

Smuts, B. "Male Aggression against Women." *Human Nature* 3 (1992): 1–44.

———. "The Evolutionary Origins of Patriarchy." *Human Nature* 6 (1995): 1–32.

Smuts, B. B., and R. W. Smuts. "Male Aggression and Sexual Coercion of Females in Nonhuman Primates and Other Mammals: Evidence and Theoretical Implications." *Advances in the Study of Behavior* 22 (1993): 1–63.

Snook, R. R. "Sexual Selection: Conflict, Kindness and Chicanery." *Current Biology* 11 (2001): R337–R341.

Sterck, E. H. M., D. P. Watts, C. P. van Schaik. "The Evolution of Female Social Relationships in Nonhuman Primates." *Behavioral Ecology and Sociobiology* 41 (1997): 291–309.

Stone, G. N. "Female Foraging Responses to Sexual Harassment in the Solitary Bee *Anthophora plumipes*." *Animal Behaviour* 50 (1995): 405–412.

Stutt, A. D., and M. T. Siva-Jothy. "Traumatic Insemination and Sexual Conflict in the Bed Bug *Cimex lectularius*." *Proceedings of the National Academy of Sciences* 98 (2001): 5683–5687.

Thornhill, R. (1980) "Rape in *Panorpa* Scorpionflies and a General Rape Hypothesis." *Animal Behaviour* 28:52–9.

Tregenza, T., N. Wedell, and T. Chapman. "Introduction. Sexual Conflict: A New Paradigm?" *Philosophical Transactions of the Royal Society B* 361 (2006): 229–234.

Watson, P. J., G. Arnqvist, and R. R. Stallmann. "Sexual Conflict and the Energetic Costs of Mating and Mate Choice in Water Striders." *The American Naturalist* 151 (1998): 46–58.

Watts, D. P. "Coalitionary Mate Guarding by Male Chimpanzees at Ngogo, Kibale National Park, Uganda." *Behavioral Ecology and Sociobiology* 44 (1998): 43–55.

Welch, A. M., R. D. Semlitsch, and H. C. Gerhardt. "Call Duration as an Indicator of Genetic Quality in Male Gray Tree Frogs." *Science* 280 (1998): 1928–1930.

Wigby, S., and T. Chapman. "Female Resistance to Male Harm Evolves in Response to Manipulation of Sexual Conflict." *Evolution* 58 (2004): 1028–1037.

Wiley, R. H., and J. Poston. "Indirect Mate Choice, Competition for Mates, and Co-evolution of the Sexes." *Evolution* 50 (1996): 1371–1381.

3

Intersexual Conflict in Primates: Infanticide, Paternity Allocation, and the Role of Coercion

Parry Clarke, Gauri Pradhan, and Carel van Schaik

Following Darwin's (1871) lead, scholars have traditionally viewed sexual selection as a positive process that enhances individual fertility and, consequently, population fitness. More recently, however, our understanding of the intersexual dynamic has changed (Arnqvist and Rowe 2005). It is now widely acknowledged that, owing to divergent reproductive optima, the interests of the sexes will invariably be at odds across almost all facets of reproduction. Indeed, it appears that sexual conflict has pervaded everything from sex allocation to life history (for reviews, see Lessels 1999; Arnqvist and Rowe 2005; Wedell et al. 2006). Early theoretical treatments of this conflict (e.g., Borgia 1979; Parker 1979) concluded that males were the likely "winners" because females lacked the evolutionary ammunition (i.e., variable reproductive success) with which to respond (Gowaty 2004). More contemporary models, however, now recognize the female's capacity to reply and consequently view the conflict as an antagonistic coevolutionary cycle, or "arms race," in which each sex is continually evolving strategies to overcome the negative effects inflicted by those of the opposing sex (Rice 1996, 2000; Rice and Holland 1997; Holland and Rice 1999). Because of the costs associated with further escalation, arms races may frequently lead to equilibrium (Härdling et al. 1999; Gavrilets et al. 2001), although the winners and losers may often be difficult to predict (Parker 2006).

To date, studies of intersexual conflict have relied mainly on invertebrates, notably water striders (e.g., Arnqvist and Rowe 2002a, b) and fruit flies (e.g., Rice 1996; Holland and Rice 1999). As a result, little is known of the role of intersexual conflict in shaping mammalian mating systems. It is our aim here to elucidate the role of antagonistic coevolution in the mating system of primates. In particular, we propose that the majority of coercion within the order is inherently linked to a fundamental conflict of interest between the sexes stemming from infanticide. Thus we argue that understanding variation in coercion requires that we identify the causes of variation in infanticide risk.

Infanticide

Infanticide by males is a source of intersexual conflict in many species (Schneider and Lubin 1996; Schneider 1999; Blumstein 2000; Wolff and Macdonald 2004; Bellemain et al. 2006), albeit not all (Kunz and Ebensperger 1999). This is particularly true of primates, for whom slow life history and a propensity to resume cycling when dependent infants are lost provide ideal conditions for reproductive gain to be derived from such behavior (van Schaik 2000a). Indeed, all available evidence from the order suggests that infanticide is a sexually selected strategy (*pace* e.g., Bartlett et al. 1993; Bartlett 2003) that serves to increase the rate at which males encounter receptive females: (i) only unrelated infants are ever killed, (ii) death of an infant results in a premature return to estrus, and (iii) the perpetrator frequently gains subsequent access to the mother (Hrdy 1979; van Schaik 2000b).

Because infanticide imposes high costs on the infant's mother and sire, it is expected to have driven selection for counterstrategies (Hrdy 1979; Hrdy et al. 1995; de Ruiter 1996; van Schaik et al. 1999, 2004). Here, we focus on the counterstrategies of females, in particular those concerning their sexuality. Although infanticide can represent a significant source of infant mortality (e.g., Borries 1997; Palombit et al. 2000; Henzi and Barrett 2003), it is generally uncommon in many of the species in which it is likely to be a viable strategy (see van Schaik 2000a), thus attesting to the effectiveness of the counterstrategies.

Female Counterstrategies

Although it is thought that females may have evolved both social and sexual counterstrategies to the risks of infanticide (see Hrdy 1979; van Schaik and Dunbar 1990; van Schaik and Kappeler 1997; Treves 1998; Palombit 1999, 2000, Chapter 15; van Schaik et al. 1999, 2000, 2004; Sterck and Korstjens 2000; van Schaik 2000a, b), only the sexual counterstrategies are relevant to models of intersexual conflict and thus in turn the expression of coercion. However, it must be noted that social and sexual counterstrategies interact in that association with protector males depends on their paternity assessment.

A male's response to an infant is expected to depend on his perception of paternity (Soltis 2002; van Schaik et al. 2004): only if he has a high enough probability of paternity will selection favor his offering protection, whereas, conversely, a chance of paternity at (or near) zero is a precondition for attacking an infant. Therefore, female sexual counterstrategies have been proposed to focus on (i) manipulating the distribution of paternity throughout a pool of males, and (ii) manipulating the information available to males on which to base their paternity estimates (van Schaik et al. 1999, 2000, 2004; Pradhan et al. 2006).

If an intratroop male can be assured of zero paternity, then he will benefit from infanticide as much as an extratroop male, assuming a high enough probability of siring the female's future offspring. To reduce this risk, females should adopt a promiscuous strategy (the first function) that allows them to "concentrate" paternity in the male best placed to protect their infant, while simultaneously "diluting" it among the rest of the male cohort in an effort to prevent infanticidal attacks should any of these males become dominant. Therefore, the female strategy is longitudinal and, contrary to traditional models of mate choice, partner preference should change with the probability of ovulation, with dominant males being favored during periods of high ovulation probability and subordinates during periods of lower probability.

At the same time, there is selection on females to raise the paternity estimates to higher values than the actual probabilities (the second function: van Schaik et al. 1999, 2004; Pradhan et al. 2006). Obviously, any sexual activity during pregnancy or early lactation must serve this function alone (van Noordwijk and van Schaik 2000; van Schaik et al. 2004). This includes "situation-dependent receptivity" (Hrdy 1979), which is seen in situations where females encounter males who have zero probability of having sired their born or unborn infant. This nonconceptive receptivity can be signaled behaviorally or morphologically (for reviews see Hrdy 1979; van Schaik et al. 1999).

Support for the role of female sexuality in reducing the risk of infanticide comes from the significantly higher incidence of multimale mating seen in species vulnerable to infanticide than in those that are not (van Schaik et al. 1999; van Noordwijk and van Schaik 2000; Wolff and Macdonald 2004), as well as the fact that invariably sires are likely to protect infants during attacks (e.g., Borries et al. 1999; Palombit et al. 2003). Furthermore, in two studies that have examined female mating behavior longitudinally, in terms of their estrous cycle, females were reported to be indiscriminately promiscuous during periods of low ovulation probability, but increasingly selective, in favor of dominant males, as this probability increased (Matsumoto-Oda 1999; Stumpf and Boesch 2005). However, given the limited number of studies, the significance of this finding cannot be confirmed without further work.

Females are thought to be able to manipulate male paternity estimates because males themselves appear unable to directly recognize their infants (Elwood and Kennedy 1994). Consequently, it is assumed that they are forced to use indirect measures based on the quality of their sexual experience, such as copulation frequency and timing (relative to ovulation), relative to those of their rivals (van Schaik et al. 2004). The fact that males appear able to recognize their juvenile offspring (Buchan et al. 2003) does not violate this assumption (*pace* Sherman and Neff 2003), but simply indicates that selection on masking paternity may exist just for infants (Pagel 1997; see also Mateo 2006).

Polyandry and Intersexual Conflict

In the face of infanticide risk, then, a female's reproductive success is expected to be in part dependent on the number of males she mates with, as well as the frequency and order, relative to ovulation, in which she mates with them. Of course, this places her at odds with males, who will be selected to secure exclusive mating access. Consequently, intersexual conflict over mating exclusivity or consort duration ensues (Nunn 1999; van Noordwijk and van Schaik 2000; van Schaik et al. 1999, 2000, 2004; see also Box 3.1). Because males are assumed to be competing in a "priority-of-access" system (*sensu* Altmann 1962), and thus subordinate males are often largely excluded from mating, this conflict will usually be most pronounced between breeding females and dominant males (van Schaik et al. 2004). Thus, any conflict between females and subordinate males is expected to be expressed only weakly, and so classic "three-way tugs-of-war" (*sensu* Rice 1998; Gavrilets and Hayashi 2006) will be largely absent. However, fission-fusion systems, discussed below, may represent a notable exception.

With regard to males, this conflict is predicted to drive counterselection for male strategies that improve paternity certainty. The most salient of these strategies is likely to be sexual coercion either in the form of persistent attempts to mate (harassment: *sensu* Clutton-Brock and Parker 1995b), attacks against females if they attempt to leave (i.e., coercive mate guarding), "punishment" (*sensu* Clutton-Brock and Parker 1995a) of female promiscuity, and/or forced matings (Smuts and Smuts 1993). Since dominant males are most likely to monopolize access to females and consequently most likely to lose out under the female strategy, they are also the most likely to show such sexual coercion (see van Schaik et al. 2004 for a review). Because female preference will change with the underlying probability of ovulation (see above), the form of coercion seen will depend on the timing of a given mating event, as well as the identity of the female's partner in this event. For example, we may expect dominant males to "punish" females whenever they mate with subordinates, but to do so proportional to the probability of ovulation.

In response to the constraints imposed by (dominant) male coercion, selection is expected to have favored a series of female counterstrategies involving behavior and reproductive physiology that serve to break the dominant male's monopoly, enabling their promiscuous objectives to be secured passively (indirect mate choice: Wiley and Poston 1996). These counterstrategies function by (i) attracting other males by signaling attractiveness and (ii) reducing monop-

---

**Box 3.1: Modeling Primate Sexual Conflict
(van Schaik et al. 2004)**

Models are formulated for a multimale, multifemale group in which males compete in a priority-of-access system, so that there is one dominant and one or more subordinates. For simplicity's sake, it is assumed that only one female is in estrus during the period of interest and it is only the strongest subordinate that poses an infanticidal threat. In addition, it is assumed that infant survival is an additive function of maternal care (milk and general protection), the protective attentions of the likely sire, and the risk of being killed by males with low or zero probability. Modeling focuses on the male contribution and assumes that infant survival is an increasing function of paternity (see Figure B.3.1) and thus considers male behavior to range from protective to indifferent to attacking, depending on his probability of paternity.

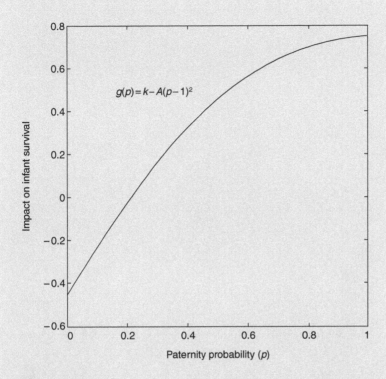

Figure B.3.1 The relationship between paternity certainty (of a single male) and infant survival as described by the function $g(p) = k - A(-1)^m(p-1)^m$ (see van Schaik et al. 2004), where $g(p)$ denotes the probability of infant survival, k and A are constants describing the maximum positive impact of likely sires and maximum negative impact of nonsires, respectively, and m is a shape parameter determining the rate of ascendancy and saturation as p increases.

Based on these assumptions, fitness curves (see Figure B.3.2) for dominant males and breeding females are developed in terms of the optimum paternity of dominant males (q). Female fitness is maximized at levels of q that maximize infant survival, while dominant male fitness is an increasing function of fertilization success (i.e., paternity certainty) and thus is maximized at levels of q well above the optimum of females. This indicates that, as long as females are subject to infanticide, mating conflict between females and dominant males will be an ever present feature of multimale troops.

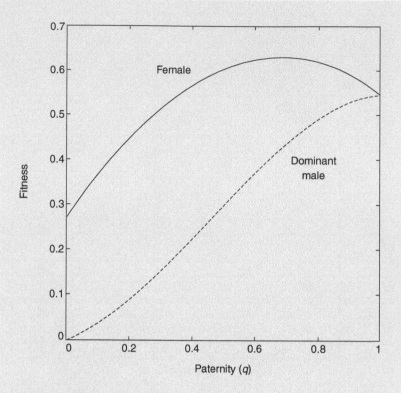

Figure B.3.2 Fitness of females and dominant males as a function of the probability of dominant male paternity (q).

From these curves we can then assess the intensity and direction of any conflict by determining the *potential fitness loss* (PFL) for a given player if forced to operate at the opposing player's optimum q (see Pradhan and van Schaik 2008). When PFL is substantial for both players, we may expect the intensity of the conflict to be high because neither will be expected to acquiesce. If, however, only one player will experience high PFL in a given setting, we may expect conflict intensity to be minimal as the benefits of resistance will be marginal for the opposing player. In addition, we may conclude that selection for resisting the opposition's optimum will be greatest on the player with the greatest PFL. In Figure B.3.2 it can be seen that if females were forced to mate at the dominant male optimum, they would experience a PFL of around 14%, while for males mating at the female optimum would result in a PFL of 22%. Therefore, in the situation depicted, we may conclude that neither player will acquiesce, but that selection to resist movement away from their optimum will be strongest on males.

olization by the dominant male. First, exaggerated sexual swellings, as well as possibly copulation calls (calls emitted during or immediately following copulation), provide a graded signal of ovulation probability, by which males can determine when it is best to guard females (Nunn 1999). Because males are expected to compete in a priority-of-access system, these swellings ensure that females encounter males in ascending order of their rank relative to ovulation, so that paternity becomes concentrated in dominant males without entirely excluding subordinate males. Second, females may reduce the dominant male's mate-guarding potential by extending the mating period and having multiple mating periods per conception (Zinner and Deschner 2000), either before conception or both before and after it. These features mean that the loss of energy and time owing to mate guarding (e.g., Alberts et al. 1996) becomes prohibitive relative to the benefits. Together, these counterstrategies ensure female access to multiple males in the required proportion and relative timing (van Schaik et al. 1999, 2004; van Noordwijk and van Schaik 2000; Wrangham 2002). Of course, all counterstrategies assume that ovulation is unpredictable at the time scale of the polyandry (van Schaik et al. 2000).

Support for this model comes from the observation that the mating periods of species where sexual coercion is likely to be effective are significantly longer than in those where it is not (van Schaik et al. 1999, 2004). Furthermore, the duration of the follicular phase in these species also appears to be longer, and the point at which ovulation occurs more unpredictable (Heistermann et al. 1996, 2001; Whiten and Russell 1996; Reichert et al. 2002; Deschner et al. 2003; Engelhardt et al. 2005; Möhle et al. 2005; Aranda et al. 2006). Finally, a growing body of work suggests that sexual swellings do facilitate the consorting of males in ascending order of rank relative to the likelihood of ovulation (e.g., Berenstain and Wade 1983; Noë and Sluyter 1990; Setchell and Wickings 2004; Deschner et al. 2004).

Intersexual Conflict and the Expression of Coercion

We have suggested that sexual coercion is a behavioral expression of an intersexual conflict triggered by infanticide. Its occurrence will therefore be a function of the intensity of this conflict. However, it will also depend on the male's potential to coerce a female, as well as the social context. We will examine these issues in turn.

To estimate the intensity of intersexual conflict, we develop fitness curves that allow us to calculate the *potential fitness loss* each player will experience if the opposing player achieves its own optimum paternity allocation (see Box 3.1). When one player's potential fitness loss is minimal, while the other's is substantial, we would predict equilibrium closer to the latter's optimum and consequently minimal conflict. If, however, potential fitness loss is significant for both, then ostensible mating conflict may be expected. Thus, mutually high potential fitness loss is a prerequisite for the behavioral expression of mating conflict.

Perhaps surprisingly, actual sexual coercion may often be modest and not substantially different across mating and social systems because the intensity of mating conflict is generally matched by the extent of female countermeasures, leaving only limited room for males to improve their mating success through coercive behaviors. We will discuss this coevolution in the next section. However, because these counteradaptations are costly, there will always be some residual "conflict load" (*sensu* Lessels 2006). Hence, some coercion should be observed, and although variation in actual rates may be low, it should be possible to account for it. The extent to which a male has the potential to reach his own optimum paternity through coercion depends on various factors. A dominant male's reproductive monopoly may be the result of factors either extrinsic or intrinsic to the conflict. Examples of extrinsic factors are ecology (e.g., reproductive seasonality), demography (e.g., female group size), and intrasexual competition (e.g., intermale coalitions), while intrinsic factors represent the effects of female physiology (e.g., length of conceptive window) and behavior (e.g., mate choice). Because a male can control neither the extrinsic factors through behavior (apart form male competition tactics, but they do not affect the female) nor most aspects of female reproductive physiology, his potential for sexual coercion depends largely on the degree to which females manipulate his paternity through their behavior. The larger the role of behavior, the more likely the male can coerce.

Given that the potential for coercion exists, what affects its actual expression? Smuts and Smuts (1993) suggest that the most important influence is the extent of the physical asymmetry between male and female in size or weapons. However, more mechanical constraints may also operate. Thus, substrate use plays a role because males in fully terrestrial species may find it easier to subdue females than those in arboreal ones (Pradhan and van Schaik in prep. b). Activity period is also expected to be important, because females in nocturnal species may find it easier to hide from coercive males (cf. Plath et al. 2003). Therefore, even if counterstrategies to coercion are found, we still expect residual levels of sexual harassment in those species that are highly sexually dimorphic, terrestrial, and diurnal. Accordingly, all reports of effective harassment in primates come from the Catarrhini (van Schaik et al. 2004), an infra-order dominated by species exhibiting these traits.

Perhaps, however, the most important factor mediating the expression of coercion is the social context. So far, we have assumed that the mating conflict is largely between the dominant male and the female. However, when females are unable to rely on intermale competition to facilitate the attainment of their polyandrous objectives, we expect coercion from males of all ranks. This applies to fission-fusion systems, in which females spend at least some time alone, or if associated with males, not necessarily with the dominant male or at the preferred time relative to ovulation. In this situation, females may often be forced to resist matings/consortships by males and because the efficacy of a strategy operating through "indirect mate choice" relies on all members of the male cohort being present, they will be unable to mitigate the risks of these males adopting a coercive strategy in response. This, then, explains why both subordinate and dominant males are often seen to use coercion in the chimpanzee mating system (Tutin 1979; Goodall 1986; Muller 2002; Newton-Fisher 2006) and why chimpanzee females, unlike macaque females (Nikitopoulos et al. 2005), may possibly exhibit a preference for mating with the dominant male (Stumpf and Boesch 2005). Similarly, it accounts for why these same features are even more pronounced among orangutans, which are even less gregarious. Thus, female orangutans exhibit marked preference for dominant flanged males, and subordinate males often adopt the most extreme form of coercion, forced copulation (Schürmann 1982; Galdikas 1985; Mitani 1985; Schürmann and van Hooff 1986; Fox 2002). This framework also predicts that, within both chimpanzees and orangutans, coercion increases as mean female gregariousness decreases (van Schaik in prep).

We also suggest that the intensity of mating conflict (and thus, in turn, potentially sexual coercion) covaries predictably with (i) the tenure prospects of the dominant male and (ii) the source of the males taking over top dominance in a group. We will, however, discuss these separately in the section after the next.

Graded Countermeasures as Evidence of Antagonistic Coevolution

To demonstrate the occurrence of an intersexual "arms race," one must provide evidence of concurrent evolution between traits. We therefore expect evidence of an arms race in the form of the accumulation of ever more costly countermeasures by females in response to an escalating intensity/severity of coercion by males. Of course, it is impossible to say a priori which strategies are more costly than others, but it seems safe to assume that physiological alterations (e.g., sexual swellings and extended follicular phase) are more expensive than purely behavioral ones (e.g., surreptitious mating and copulation calls) and will be deployed if such behavioral measures prove ineffective. Furthermore, if we assume that the costs of countermeasures are relatively constant across species and the conflict starts at the same point for all species, then we may expect evidence of escalation in the form of nested counterstrategies: cheap countermeasures will be common to all species, but increasingly more expensive ones will be seen in ever smaller subsets of species. Although much of the relevant information is not known, we can still deduce the plausible sequelae of escalation.

Active mate choice may be the cheapest way to achieve an optimum level of multimale mating for females. Accordingly, in all regularly studied species, active promiscuity is generally seen (Hrdy and Whitten 1987; Paul 2002). Of course, given the possible costs associated with the act of mating itself (Nunn et al. 2000; Nunn 2002) even this strategy is not cost-free, and so selective rather than indiscriminate promiscuity appears to be the norm (Manson 2007). Under the nested counterstrategy model, we must expect that in those species where coercion is ineffective this strategy will suffice and thus no further countermeasures are required. The infant-carrying lemurs and the indrids, many of which are known or suspected to exhibit infanticide (Hood 1994; Wright 1995; Erhart and Overdorff 1998; Jolly et al. 2000; Ichino 2005), are generally seasonal breeders. They are arboreal and characterized by either near-monomorphism or even female-biased dimorphism, suggesting that coercion is likely to impose little limitation on female behavior. Accordingly, they exhibit an actively promiscuous strategy that is clearly aimed at manipulating paternity estimates (e.g., Periera and Weiss 1991; Sauther 1991; Brockman and Whitten 1996; Brockman 1999), but possess none of the traits implicated in "indirect mate choice" (van Schaik et al. 1999, 2004). Therefore, it appears that they can achieve their anti-infanticidal objectives via the least costly means.

The next most costly strategy available may be the use of surreptitious or sneaky mating. For the nested model to hold, such behavior must be seen in all species where males are, to some degree, able to manipulate female behavior, but not in those where they are not. Although data are incomplete, sneaky copulations have been reported for neither lemurs nor indrids, despite being reported in a wide range of other species (*Cebus apella:* Lynch Alfaro 2005; *Cercocebus torquatus:* Range et al. 2005; *Macaca fascicularis:* Gygax 1995; *Macaca fuscata:* Huffman 1992; *Macaca mulatta:* Berard *et al.* 1994; *Macaca arctoides:* Nieuwenhuijsen et al. 1986; *Macaca sylvanus:* Paul 1989; *Pan troglodytes:* Tutin 1979; Goodall 1986; *Papio cynocephalus:* Alberts et al. 2006; *Papio ursinus:* Clarke personal observation; *Presbytis entellus:* Hrdy 1977). Of course, absence of evidence is not *evidence* of absence, but given that female dominance over males is generally the norm in these families (Kappeler 1990; Pereira et al. 1990; Radespiel and Zimmerman 2001; Pochron et al. 2003), the need for sneaky copulations seems minimal. When focusing on those species that do engage in sneaky copulations, as predicted, we find some, such as the stump-tailed macaques *(Macaca arctoides),* that do not exhibit many of the potentially more costly traits. Importantly, these species exhibit substantially more sexual dimorphism, in favor of males, than the prosimians, but less than many of those with costlier traits. This finding suggests that the differences in vulnerability to coercion have led to the differential accumulation of countermeasures.

In species where the constraints of coercion are absent or minimal, females appear able to achieve their objectives through either open or surreptitious solicitation. In contrast, in species where coercion is effective, they will be forced to adopt more passive strategies (i.e., through indirect mate choice: Wiley and Poston 1996). Of those traits facilitating indirect mate choice, copulation calls are likely to be the least costly, whereas extended mating periods, increased number of cycles to conception, and sexual swellings must all impose increasingly greater burdens. Therefore, we also expect evidence of nesting within these traits, with the extent of this nesting correlating with the efficacy of coercion. Accordingly, we find that although all species possessing sexual swellings exhibit copulation calls, at least four species emit calls and do not possess swellings (van Schaik et al. 1999). Furthermore, we know that, though relatively dimorphic and in many cases terrestrial, all these species are characterized by highly pronounced mating

Table 3.1   Duration of mating period relative to number of cycles to conception.

Copulation Calls	Sexual Swellings	Cycle Number	Mating Period (days)
Y	N	1–2	8[3]
Y	Y	1–2	9.67[3]
Y	Y	3–4	10[4]
Y	Y	5–6	11[4]

Data taken from Table 3.3. Figures in parentheses indicate number of species for which relevant information is known.

seasonality, indicating greater female behavioral freedom owing to high levels of intermale competition.

Given that increased cycle number will involve a greater temporal investment than an increased duration of receptivity per cycle and that sexual swellings will be ineffective without a prolonged period of mating, it seems parsimonious to conclude that an extension of the mating period represents the least costly physiological alteration a female may evolve. Therefore, if the nesting model is to hold, we must expect to find that across the set of species exhibiting copulation calls, first, the length of mating period is positively correlated with cycle number and, second, species exhibiting sexual swellings have longer mating periods than those that do not. Although small sample size precludes the opportunity for any formal analysis, the trends confirm this expectation (see Tables 3.1 and 3.2). If sexual swellings are to be "nested" within the countermeasure of "increased cycles to conception," then we would expect that species possessing sexual swellings would have more cycles to conception than those that do not, which is also what we see (see Table 3.2).

Overall, though preliminary, this analysis suggests that across the order we can see a positive correlation between male coercion efficacy and the cost of female countermeasures, in approximately the following order: active mate selection,

Table 3.2   Duration of mating period and number of cycles to conception relative to the possession of sexual swellings.

Copulation Calls	Sexual Swellings	Mating Period (days)	Number of Cycles
Y	N	8[3]	1.65[2]
Y	Y	10.27[12]	3.50[12]

sneaky copulations, copulation calls, longer mating periods, sexual swellings, and increased number of ovarian cycles. This pattern supports the notion that (the avoidance of) sexual coercion has been a key component of the intersexual conflict over female polyandry (i.e., number of mates). Of course, there is clearly a strong taxonomic component to the pattern observed, and so it may be argued that the correlation seen simply reflects underlying "grade shifts" within the order. However, the fact that variance in many of the countermeasures is also seen within infra-orders and superfamilies clearly suggests that phylogeny cannot explain all of the correlation (see also van Schaik et al. 2004).

The Variable Nature of Intersexual Conflict

We have suggested that coercion within primate mating systems is primarily an expression of conflict over female mate number, which in turn is a function of the degree of infanticide risk. Thus, if we are to develop a greater understanding of the variance in the occurrence of coercion, we must determine the manner in which infanticide risk varies across primate mating systems.

For infanticide to be reproductively advantageous, the perpetrator must have some chance to gain subsequent mating access; otherwise the potential costs of the behavior cannot be offset (Broom et al. 2004). Therefore, given that males are not all competitively equal (e.g., Cowlishaw and Dunbar 1991; Kutsukake and Nunn 2006), only a subset of males are likely to pose a threat. This leads us to expect that female strategies will vary in relation to a number of key factors; consequently, so too will the degree of intersexual conflict and the resulting expression of coercion. Recently, the effects of some of these factors have been modeled (Pradhan and van Schaik 2008; see also Box 3.2), and it is the findings and implications of this work that will be the focus of the remainder of the chapter.

Takeover Frequency: The Importance of Alpha Male Tenure

The extent to which infants will be exposed to attack will be correlated with the rate of dominance takeovers (Janson and van Schaik 2000). Consequently, the female mating strategy may be expected to fluctuate in relation to the length of alpha tenure. Assuming females favor a low rate of attack (Janson and van Schaik 2000), they should then prefer dominant males to have extended tenures. Because reproductive output is likely to be positively correlated with tenure, dominant males should also favor this situation. Thus, across species, conflict is

Box 3.2: Modeling Variance in Sexual Conflict
(Pradhan and van Schaik 2008)

This new round of modeling focused on the effects of two new parameters: (1) the relative strength of the dominant male and (2) the probability of dominance takeovers occur from inside the troop. As before (van Schaik et al. 2004), the model was developed for a multimale, multifemale troop, where only one female is in estrus. In addition, for simplicity's sake it was assumed that only the strongest subordinate male is capable of usurping the existing dominant male in an inside takeover situation. The starting point for the model was the general fitness equations developed by van Schaik et al. (2004; see also Box 3.1). Although the fitness equations for males remained untouched during this new round of modeling, a number of additional terms were introduced into the female's equations.

First, a term assessing the contribution made by variations in the tenure of the existing dominant male was included. This term comprised a parameter denoting the relative strength of the dominant male (a), which was scaled to the likelihood of being replaced by a challenger to give probability of a dominance takeover occurring. This new term combined with his probability of paternity (q) described the dominant male's contribution to infant survival. The second term to be added contained the parameter measuring the probability of a dominance takeover occurring from within the troop (ξ; 1 minus which gives the probability of dominance takeover from outside the troop), weighted against the probability that the strongest subordinate successfully protects the infant. The protective abilities of the strongest subordinate will be an approximately linear function of his paternity probability, weighted for his ability to fend off the challenging male, which will, of course, depend on the latter's relative strength.

expected to decline with tenure length, although within a given tenure it is likely to increase over time as the probability of takeover increases.

Figure 3.1 shows the effects of dominant male tenure (a) on the optimum assignment of paternity (q) from the perspective of both the breeding female and the dominant male. It can be seen that while the optimum for the dominant male is invariably 1, regardless of his underlying strength, for females optimum q is relatively low at low-tenure length, but rapidly increases as it increases (although it asymptotes below the male optimum, indicating that some conflict will always remain). For females, this suggests that in situations of prolonged dominant male tenure, the threat of infanticide is reduced and the need for multimale mating is lessened, so that their optimum assignment of paternity is less at odds with that of males. This leads us to two predictions, one interspecific and one intraspecific. First, females should be less promiscuous in species where alpha tenure is generally long (in terms of multiples of interbirth intervals) than

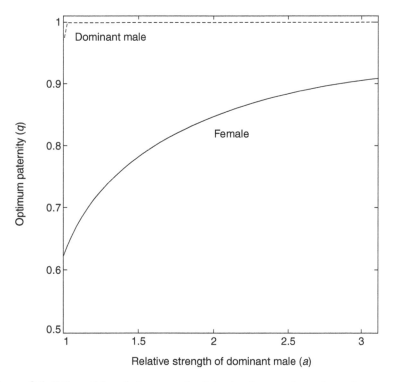

Figure 3.1 Effect of the relative strength of the dominant male on the optimum paternity to dominant males from the perspective of breeding females and dominant males, respectively. Taken from Pradhan and van Schaik (2008).

in species where alpha tenure can be shorter. Second, within species, female promiscuity should increase as the dominant male's alpha tenure progresses and should peak immediately prior to takeover.

The implication from Figure 3.1 is that the disparity in the optimum assignment of paternity for each player will decline with increasing a, which is confirmed by an examination of the potential fitness loss associated with the trait (Figure 3.2). This indicates that the intensity of conflict will decrease with the increasing strength of the dominant male and that across the entire range of the parameter, selection will be strongest on males to resist moving to the opposing player's optimum. Therefore, when alpha tenures are long (high a), the female may be expected to acquiesce, and thus the optimum paternity value for dominant males may be observed. Thus, we expect low polyandry and little conflict between the breeding female and the dominant male. The implications for the occurrence of coercion are obvious. Across species, we expect a negative correla-

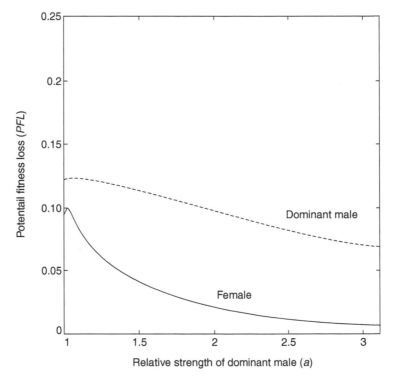

Figure 3.2 The effects of the relative strength of the dominant male on the potential fitness loss (PFL) for dominant males and females. Taken from Pradhan and van Schaik (2008).

tion between the frequency of coercion by dominant males and their tenure length. Within species, females may favor a more promiscuous strategy toward the end of a male's tenure, while he may respond with a more coercive strategy.

Enormous variability exists in dominance structure/stability, not just among species but also within them. In chimpanzees, for example, alpha tenures as long as 15 years have been reported (Uehara et al. 1994), yet some are as short as 20 months (Goodall 1986). Similarly, within the Amboseli baboon population, alpha tenures range from around 8 to 159 months (Alberts et al. 2003). In light of this variability, females may be expected to estimate a dominant male's relative strength directly (as opposed to estimating on an evolutionary scale), using traits such as his propensity to engage in agonistic interactions with potential challengers or his ability to defeat these challengers. These estimates will be indicative of the probability of a dominance takeover and in turn the risks of infanticide. Indeed, in some species this is exactly what we find. In Thomas langurs, for example, where one-male units have a clearly defined life cycle whose ontogeny is negatively correlated with the ability of group males to prevent intrusions by outsider males (Steenbeek et al. 2000), females exhibit a greater tendency to either actively avoid or seek out new males with which to start a new group, depending on whether they are anestrous or estrous, respectively (Steenbeek 1999).

Takeover Source: Familiarity with the Threat

Just as we may expect female strategies to be sensitive to the frequency of dominance takeovers, we may also expect them to exhibit sensitivity to their source. Two sources of dominance acquisition by males can be recognized (van Noordwijk and van Schaik 1985, 2001). The first is characterized by recently immigrated males, invariably in prime condition, acquiring the alpha position (henceforth referred to as an "outside takeover" situation or species)—a system common to the savannah baboons (Packer 1979; Hamilton and Bulger 1990). In contrast, the second source involves immigrating males entering troops as subordinates and then attaining dominance only after several years of residency (henceforth referred to as "inside takeover" situation or species. This source is often seen in a number of macaques (van Noordwijk and van Schaik 1985, 1988, 2001; Sprague 1992, 1998) and, by necessity, the mating system of male philopatric species, such as the chimpanzees, bonobos, and spider monkeys.

Given that only males capable of vying for dominance are likely to attempt infanticide (Hrdy 1979), we may expect a female's optimum distribution of paternity, and therefore also the intensity of sexual conflict, to differ across these two sources. In outside takeover species, where the threat of infanticide is faced blind, the female strategy should be primarily sensitive to the need to concentrate paternity in the existing dominant male so that he is predisposed to offer protection. In contrast, in inside takeover species, where the threat can be seen approaching, females should still ensure that the dominant male will protect her infant, but she must also dispel any infanticidal tendencies subordinate males may harbor following their attainment of dominance. If these expectations hold, then the disparity between the reproductive interests of females and dominant males, in terms of the distribution of paternity, is expected to vary. Indeed, modeling confirms this expectation (see Figure 3.3): as the probability of takeovers occurring from the inside increases, so a female's optimum assignment of paternity to dominant males decreases. In practice, the contrast in female strategies need not always be absolute: first, both sources may be possible

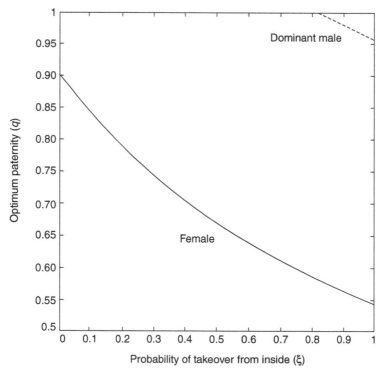

Figure 3.3 Optimum q for dominant males and females relative to the probability of dominance takeovers from the inside (ξ). Taken from Pradhan and van Schaik (2008).

(Sprague et al. 1996; van Noordwijk and van Schaik 2001), and, second, alpha males may die in outside takeover species, leading to a "succession" from within.

Figure 3.4 illustrates the impact of differences in the source of dominance acquisition (the probability of takeover from the inside: ξ) on potential fitness loss for each player. When all takeovers are from the outside (i.e., low ξ), females have little to lose from moving to the dominant male's optimum paternity certainty (q), thus reflecting the females' shift in emphasis toward paternity concentration. As the likelihood of takeovers from the inside increases, it becomes increasingly costly for females to mate at the optimum q for dominant males, to the point that their potential fitness loss rises above that of dominant males at high values of ξ. For dominant males, mating at the optimum q for females is costly across the entire range of ξ, but is increasingly so as its value rises. In short, the intensity of the conflict rises with increasing ξ, and thus so too may the occurrence of coercion.

Although a lot remains unknown, we can still provide an initial assessment of

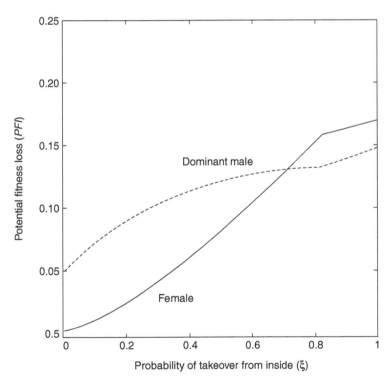

Figure 3.4 PFL for each sex in relation to the probability of takeover from the inside (ξ). Taken from Pradhan and van Schaik (2008).

the expectation that the degree of conflict will vary across systems with differing sources of dominance acquisition. Figure 3.5 shows the extent of dominant male reproductive monopoly (as measured by molecular analysis) in inside and outside takeover species. As predicted, the degree of skew is significantly reduced in inside takeover situations compared to those characterized by outside takeovers (log-ANOVA: $F_{1,13} = 6.661$, $p = 0.023$), indicating that the degree of conflict between these two situations is also significantly different and thus so too may be the expression of coercion.

When considering multimale species characterized by outside takeovers, we may predict that dominant males will rarely display coercion because female partner number will be minimal. Indeed, Clarke (2006) recently found that sexual aggression either in the form of the direct coercing of consort partners or the punishing of female promiscuity was virtually absent in wild chacma baboons. In contrast, because of the greater need for multimale mating, we must expect coercion to be regularly seen in inside takeover species. Confirming this expectation, a study of aggression in the Cayo Santiago population of rhesus

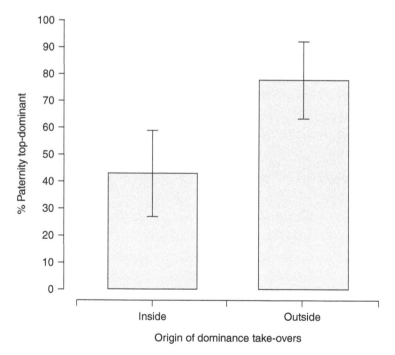

Figure 3.5 Percentage of maximum potential reproductive success, as determined by molecular analysis, secured by dominant males in species typically characterized by either inside or outside takeovers. For data see Table 3.3.

macaques, where takeovers are generally from the inside (Berard 1999), found that dominant males often "punished" female attempts to mate with subordinate males (Manson 1994). Similarly, numerous reports of severe sexual aggression in Japanese macaques are available (Tokuda 1961; Enomoto 1981), which is to be expected given the common occurrence of inside takeovers (Sprague et al. 1996; Sprague 1998), although, the actual function of this aggression may be a point for debate (Soltis et al. 1997).

Although much of the specific data on rates of coercion are lacking, we should still find evidence of differential rates of expression through an examination of variance in female physiology. Specifically, we expect to find that, -because of their greater need for multiple mates, females in inside takeover species should exhibit more of the traits associated with indirect mate choice and in a more exaggerated form than those in outside takeover species. Among species that can be reliably categorized as inside takeover species (see Table 3.3), 84.62% (11/13) exhibit sexual swellings compared to only 62.5% (5/8) of outside takeover species. Furthermore, the mean length of the mating periods in inside takeovers is 10.1 days, while in outside takeover species it is 7.1 days (see Figure 3.6; log-ANOVA: $F_{1,24} = 1.006$, $p = 0.326$). These differences are not

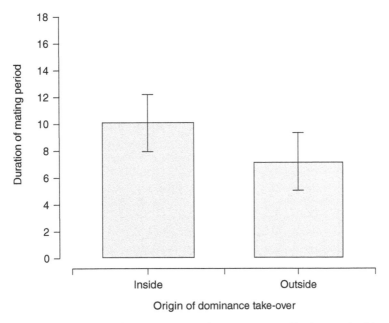

Figure 3.6 Duration (days) of mating period in species typically characterized by either inside or outside takeovers. For data see Table 3.3.

Table 3.3

Superfamily	Genus	Species	Takeover Source	Sexual Swellings	Copulation Calls	Median Mating Period	Cycles to Conception	Reproductive Monopoly of Alpha Male
Lemuroidea	Eulemur	*fulvus*	O	N	N	2	?	67
	Lemur	*catta*	I	N	N	1	?	?
	Propithecus	*verreauxi*	O	N	N	2	?	61
Ceboidea	Alouatta	*caraya*	O	N	N	3	?	?
	Alouatta	*palliata*	O	N	N	3	?	?
	Alouatta	*seniculus*	O	N	N	3	?	95
	Brachyteles	*arachnoides*	I	N	N	2	?	?
	Lagothrix	*poeppigii*	I	N	N	8	?	?
	Cebus	*apella*	O	N	N	5	?	?
	Cebus	*capicinus*	O	N	N	?	?	80
	Saimiri	*oerstedii*	I	N	N	2	?	?
Cercopithecoidea	Cercocebus	*albigena*	I	Y	Y	4	3	42.35
	Cercocebus	*torquatus*	I	Y	Y	?	?	62
	Cercopithecus	*aethiops*	I	N	N	33	2	?
	Cercopithecus	*campbelli*	O	N	N	?	?	?

Colobus	badius	I	Y	Y	5	2	?
Macaca	arctoides	O	N	N	30	?	96
Macaca	cyclopis	O	Y	Y	?	?	?
Macaca	fascicularis	I	Y	Y	15	2	70.75
Macaca	fuscata	I	N	Y	11	1.3	40
Macaca	mulatta	I	Y	Y	9	2	28.57
Macaca	nemestrina	I	Y	Y	13	3.3	?
Macaca	radiata	I	N	Y	5	2	?
Macaca	sylvanus	I	Y	Y	14	2.2	7
Mandrillus	sphinx	O	Y	Y	?	?	87
Papio	anubis	O	Y	Y	6	5.5	?
Papio	cynocephalus	O	Y	Y	9	3	?
Papio	ursinus	O	Y	Y	9	4.25	?
Presbytis	entellus	O	N	N	6	5.4	57
Procolobus	verus	I	Y	Y	?	?	?
Hominoidea							
Pan	paniscus	I	Y	Y	15	5.6	50
Pan	troglodytes	I	Y	Y	14	5.6	42.5

Takeover Source: O=Challenger male originates from outside troop; I=Challenger male originates from inside troop.
Median Mating Period: Measured in days.
Sexual Swellings and Copulation Calls: Y=seen; N=not seen
Reproductive Monopoly of Alpha Male: Percentage of total available paternities, as measured by molecular analysis, secured by dominant male.
All references for data in table available from the authors on request.

significant, perhaps because they do not control for other factors, in particular a female's intrinsic behavioral freedom. Nevertheless, the trends are in the expected direction.

Overall, then, the data suggest that conflict over the allocation of paternity to dominant males is generally stronger in inside takeover species and, consequently, coercion is more salient in these species as well. Thus, as expected, differences in the risks of infanticide and the counterstrategies it favors in females do lead to predictable changes in the expression of coercion.

Conclusions

In this chapter, we propose that infanticide in primates has precipitated a series of antagonistic coevolutionary events, comparable to those seen in the classic invertebrate models (e.g., Rice 1996; Holland and Rice 1999; Arnqvist and Rowe 2002a, b). First, holding intensity of mating conflict constant, we predict that the efficacy of coercion predicts the costliness of female counterstrategies. Empirically, we find some evidence in favor of stepwise increase across species in the cost of counterstrategies, depending on the coercion potential of males, from active mate choice, through surreptitious mating and copulation calls to sexual swellings and longer mating periods, and finally more cycles per conception. Second, *ceteris paribus*, we also predict that male coercion potential and thus male sexual coercion increases with sexual dimorphism, terrestriality and diurnality, as well as in fission-fusion sociality, especially toward the solitary end of the spectrum. This prediction, especially with respect to fission-fusion, is strongly supported empirically. Third, *ceteris paribus*, we predict that mating conflict is more intense where male dominance takeovers are from inside the group and where alpha tenures are short or toward the end of the dominant's tenure. Tests of these predictions are still preliminary, but so far the fit between prediction and evidence is good. Future tests should attempt to take multiple factors into account simultaneously, rather than rely on bivariate tests, and should also consider phylogenetic nonindependence.

As always, we do not suggest that infanticide represents the sole selection pressure operating in primate mating systems, nor we do consider female mate number to be the only source of conflict promoting the expression of coercion. Indeed, female partner preference may vary relative to other parameters, such as genetic compatibility (Zeh and Zeh 2001), and therefore so too may the occurrence of coercion. In addition, the use of sexual aggression may sometimes be unrelated to female behavior per se, but instead may represent an alternative mating strategy (cf. Hogg 1984). Nevertheless, we feel that the infanticide-based model provides the most coherent predictive framework that helps us account for the occurrence of coercion in primates. In addition, by couching our analysis in the emerging coevolutionary models of intersexual conflict theory, we allow equilibria to be identified and thus can not only predict when coercion should be seen, but, perhaps more importantly, also account for why it is often not seen. Finally, of course, we do not suggest that coercion is the only counterstrategy to female promiscuity a male can adopt. Indeed, the ability to assess the probability of ovulation may also represent a trait evolving from this conflict (Weingrill et al. 2003; Alberts et al. 2006; Ostner et al. 2006).

Throughout our treatment of antagonistic coevolution, we have assumed that a male's coercive potential is determined by factors extraneous to sexual conflict. In other words, if coercion is particularly effective in a given mating system, it is simply a fortuitous by-product of that species' ecological niche (substrate, activity period, group size, and seasonality) and the intensity of intrasexual selection (dimorphism) that has historically operated. Thus in a particular species, females are essentially expected to evolve countermeasures to a static male phenotype, which once attained minimize the threat of coercion. Given the well-documented role of intersexual conflict in influencing body size and shape in other species (e.g., Arnqvist and Rowe 2002a, b; Maklakov et al. 2006), this may well be an oversimplification. Indeed, although intrasexual selection is almost universally invoked to account for sexual dimorphism, in mammals in particular, it is often noted that much variance is left unexplained (e.g., Plavcan 2004). Therefore, given the clear evidence for intersexual conflict within primate mating systems, it may be conjectured that some of this variance in dimorphism is the result of antagonistic coevolution favoring larger, more coercive males. Thus, dimorphism is the result of both intra- and intersexual selection (Pradhan and van Schaik 2009). If so, this would further highlight how both intra- and interlocus conflict have played a role in the evolution of primate, and indeed mammalian, mating systems (see also Lindenfors 2002).

References

Alberts, S. C., J. Altmann, and M. L. Wilson. "Mate Guarding Constrains Foraging Activity of Male Baboons." *Animal Behaviour* 51 (1996): 1269–1277.

Alberts, S. C., H. E. Watts, and J. Altmann. "Queuing and Queue-Jumping: Long-Term Patterns of Reproductive Skew in Male Savannah Baboons, *Papio cycnocephalus.*" *Animal Behaviour* 65 (2003): 821–840.

Alberts, S. C., J. C. Buchan, and J. Altmann. "Sexual Selection in Wild Baboons: From Mating Opportunities to Paternity Success." *Animal Behaviour* 72 (2006): 1177–1196.

Altmann, S. A. "A Field Study of the Socio-Biology of the Rhesus Monkey, *Macaca mulatta.*" *Annals of the New York Academy of Sciences* 102 (1962): 338–435.

Anderson, M. J., J. K. Hessel, and A. F. Dixson. "Primate Mating Systems and the Evolution of Immune Response." *Journal of Reproductive Immunology* 61 (2004): 31–38.

Aranda, J. D., E. Brindle, K. D. Carey, K. Rice, M. Tatar, and K. A. O'Connor. "Sexual Swelling Relative to Occurrence and Timing of Ovulation in *Papio sp.*" *American Journal of Physical Anthropology* 42 (Supplement) (2006): S57.

Arnqvist, G., and L. Rowe. "Antagonistic Coevolution between the Sexes in a Group of Insects." *Nature* 415 (2002a): 787–789.

———. "Correlated Evolution of Male and Female Morphologies in Water Striders." *Evolution* 56 (2002b): 936–947.

———. *Sexual Conflict.* Princeton, N.J.: Princeton University Press, 2005.

Bartlett, T. Q. "Sexually Selected Infanticide: Still Room for Doubt?" *American Journal of Primatology* 59 (2003): 93–96.

Bartlett, T. Q., R. W. Sussman, and J. M. Cheverud. "Infant Killing in Primates—A Review of Observed Cases with Specific Reference to the Sexual Selection Hypothesis." *American Anthropologist* 95 (1993): 958–990.

Bellemain, E., J. E. Swenson, and P. Taberlet. "Mating Strategies in Relation to Sexually Selected Infanticide in a Non-Social Carnivore: The Brown Bear." *Ethology* 112 (2006): 238–246.

Berenstain L., and T. D. Wade. "Intrasexual Selection and Male Mating Strategies in Baboons and Macaques." *International Journal of Primatology* 4 (1983): 201–235.

Berard, J. D. "A Four-Year Study of the Association between Male Dominance Rank, Residency Status, and Reproductive Activity in Rhesus Macaques *(Macaca mulatta).*" *Primates* 40 (1999): 159–175.

Berard, J. D., P. Nürnberg, J. T. Epplen, and J. Schmidtke. "Alternative Reproductive Tactics and Reproductive Success in Male Rhesus Macaques." *Behaviour* 127 (1994): 177–201.

Blumstein, D. T. "The Evolution of Infanticide in Rodents: A Comparative Analysis. In *Infanticide by Males and its Implications,* eds. C. P. van Schaik, and C. H. Janson, pp. 178–197. Cambridge: Cambridge University Press, 2000.

Borgia, G. Sexual Selection and the Evolution of Mating Systems. In *Sexual Selection and Reproductive Competition in Insects,* eds. M. S. Blum, and N. A. Blum, pp. 19–80. New York: Academic Press, 1979.

Borries, C. "Infanticide in Seasonally Breeding Multimale Groups of Hanuman Langurs *(Presbytis entellus)* in Ramnagar (South Nepal)." *Behavioural Ecology and Sociobiology* 41 (1997): 139–150.

Borries, C., K. Launhardt, C. Epplen, J. T. Epplen, and P. Winkler. "Males as Infant Protectors in Hanuman Langurs *(Presbytis Entellus)* Living in Multimale Groups: Defence, Paternity and Sexual Behaviour." *Behavioural Ecology and Sociobiology* 46 (1999): 350–356.

Brockman, D. K. "Reproductive Behavior of Female *Propithecus verreauxi* at Beza Mahafaly, Madagascar." *International Journal of Primatology* 20 (1999): 375–398.

Brockman, D. K., and P. L. Whitten. "Reproduction in Free-Ranging *Propithecus verreauxi:* Estrus and the Relationship between Multiple Matings and Fertilization." *American Journal of Physical Anthropology* 100 (1996): 57–69.

Broom, M., C. Borries, and A. Koenig. "Infanticide and Infant Defence by Males—Modelling the Conditions in Primate Multi-Male Groups." *Journal of Theoretical Biology* 231 (2004): 261–270.

Clarke, P. M. R. "Inter-Sexual Mating Conflict in the Chacma Baboon *(Papio hamadryas ursinus)*." Ph.D. Thesis, University of Bolton, 2006.

Clutton-Brock, T. H., and G. A. Parker. "Punishment in Animal Societies." *Nature* 373 (1995a): 209–216.

———. "Sexual Coercion in Animal Societies." *Animal Behaviour* 49 (1995b): 1345–1365.

Cowlishaw, G., and R. I. M. Dunbar. "Dominance Rank and Mating Success in Male Primates." *Animal Behaviour* 41 (1991): 1045–1056.

Darwin, C. *The Descent of Man, and Selection in Relation to Sex.* London: John Murray, 1871.

de Ruiter, J. R. "Infanticide Counter-Strategies." *Evolutionary Anthropology* 5 (1996): 5.

Deschner, T., M. Heistermann, K. Hodges, and C. Boesch. "Timing and Probability of Ovulation in Relation to Sex Skin Swelling in Wild West African Chimpanzees, *Pan troglodytes verus.*" *Animal Behaviour* 66 (2003): 551–560.

———. "Female Sexual Swelling Size, Timing of Ovulation, and Male Behaviour in Wild West African Chimpanzees." *Hormones and Behavior* 46 (2004): 204–215.

Elwood, R. W., and H. F. Kennedy. "Selective Allocation of Parental and Infanticidal Response in Rodents: A Review of Mechanisms." In *Infanticide and Paternal Care,* eds. S. Parmigiani and F. S. vom Saal, pp. 397–425. Chur, Switzerland: Harwood Academic, 1994.

Engelhardt, A., J. K. Hodges, C. Niemitz, and M. Heistermann. "Female Sexual Behaviour, but not Sex Skin Swelling, Reliably Indicates the Timing of the Fertile Phase in Wild Long-Tailed Macaques *(Macaca fascicularis)*." *Hormones and Behavior* 47 (2005): 195–204.

Enomoto, T. "Male Aggression and the Sexual Behavior of Japanese Monkeys." *Primates* 22 (1981): 15–23.

Erhart, E. M., and D. J. Overdorff. "Infanticide in *Propithecus diademia edwardsi:* An Evaluation of the Sexual Selection Hypothesis." *International Journal of Primatology* 19 (1998): 73–81.

Fox, E. A. "Female Tactics to Reduce Sexual Harassment in the Sumatran Orangutan *(Pongo pygmaeus abelii)." Behavioral Ecology and Sociobiology* 52 (2002): 93–101.

Galdikas, B. M. F. "Adult Male Sociality and Reproductive Tactics among Orangutans at Tanjung Puting." *Folia Primatologica* 45 (1985): 9–24.

Gavrilets S., G. Arnqvist, and U. Friberg. "The Evolution of Female Mate Choice by Sexual Conflict." *Proceedings of the Royal Society of London Series B—Biological Sciences* 268 (2001): 531–539.

Gavrilets, S., and T. I. Hayashi. "The Dynamics of Two- and Three-Way Sexual Conflict over Mating." *Philosophical Transactions of the Royal Society: Biological Sciences, Series B* 361 (2006): 345–354.

Goodall, J. *The Chimpanzees of Gombe: Patterns of Behavior.* Cambridge, MA: Belknap Press, 1986.

Gowaty, P. A. "Sex Roles, Contests for the Control of Reproduction, and Sexual Selection." In *Sexual Selection in Primates: New and Comparative Perspectives,* eds. P. Kappeler, and C. van Schaik, pp. 37–54. Cambridge: Cambridge University Press, 2004.

Gygax, L. "Hiding Behaviour of Long-Tailed Macaques, *Macaca fasicularis.* I. Theoretical Background and Data on Mating." *Ethology,* 101 (1995): 10–24.

Hamilton, W. J. III, and J. B. Bulger. "Natal Male Baboon Rank Rises and Successful Challenges to Resident Alpha Males." *Behavioral Ecology and Sociobiology* 26 (1990): 357–362.

Härdling, R., V. Jormalainen, and J. Tuomi. "Fighting Costs Stabilize Aggressive Behavior in Intersexual Conflicts." *Evolutionary Ecology* 13 (1999): 245–265.

Henzi, S. P., and L. Barrett. "Evolutionary Ecology, Sexual Conflict and Behavioral Differentiation among Baboon Populations." *Evolutionary Anthropology* 12 (2003): 217–230.

Heistermann, M., U. Möhle, H. Vervaecke, L. van Elsacker, and J. K. Hodges. "Application of Urinary and Fecal Steroid Measurements for Monitoring Ovarian Function and Pregnancy in the Bonobo *(Pan paniscus)* and Evaluation of Perineal Swelling Patterns in Relation to Endocrine Events." *Biology of Reproduction* 55 (1996): 844–853.

Heistermann, M., T. Ziegler, C. P. van Schaik, K. Launhardt, P. Winkler, and J. K. Hodges. "Loss of Oestrus, Concealed Ovulation and Paternity Confusion in Free-Ranging Hanuman Langurs." *Proceedings of the Royal Society of London Series B—Biological Sciences* 268 (2001): 2445–2451.

Hogg, J. T. "Mating in Bighorn Sheep: Multiple Creative Strategies." *Science* 225 (1984): 526–528.

Holland, B., and W. R. Rice. "Experimental Evolution of Sexual Selection Reverses Intersexual Antagonistic Coevolution and Removes Reproductive Load." *Proceedings of the National Academy of Sciences USA* 96 (1999): 5083–5088.

Hood, L. C. "Infanticide among Ringtailed Lemurs *(Lemur catta)* at Berenty Reserve, Madagascar." *American Journal of Primatology* 33 (1994): 65–69.

Hrdy, S. B. *The Langurs of Abu.* Cambridge, Mass.: Harvard University Press, 1977.

———. "Infanticide among Animals: A Review, Classification, and the Implications for the Reproductive Strategies of Females." *Ethology and Sociobiology* 1 (1979): 13–40.

Hrdy, S. B., and P. L. Whitten. "Patterning of Sexual Activity." In *Primate Societies,* eds. B. B. Smuts, D. L. Cheney, R. M. Seyfarth, R. W. Wrangham, and T. T. Struhsaker, pp. 370–384. Chicago: University of Chicago Press, 1987.

Hrdy, S. B., C. H. Janson, and C. P. van Schaik. "Infanticide: Let's Not Throw out the Baby with the Bath Water." *Evolutionary Anthropology* 3 (1995): 151–154.

Huffman, M. A. "Influences of Female Partner Preferences on Potential Reproductive Outcomes in Japanese Macaques." *Folia Primatologica* 59 (1992): 77–78.

Ichino, S. "Attacks on Wild Infant Ring-Tailed Lemurs *(Lemur catta)* by Immigrant Males at Berenty, Madagascar: Interpreting Infanticide by Males." *American Journal of Primatology* 67 (2005): 267–272.

Janson, C. H., and C. P. van Schaik. "The Behavioural Ecology of Infanticide by Males." In *Infanticide by Males and its Implications,* eds. C. P. van Schaik, and C. H. Janson, pp. 469–494. Cambridge: Cambridge University Press, 2000.

Jolly, A., S. Caless, A. Cavigetti, L. Gould, M. E. Pereira, R. E. Pride, H. D. Rabenandrassana, J. D. Walker, and T. Zafison. "Infant Killing, Wounding and Predation in *Eulemur* and *Lemur.*" *International Journal of Primatology* 21 (2000): 21–40.

Kappeler, P. M. "Female Dominance in *Lemur catta:* More than Just Feeding Priority?" *Folia Primatologica* 55 (1990): 92–95.

Kunz, T. H., and L. A. Ebensperger. "Why Does Non-Parental Infanticide Seem So Rare in Bats?" *Acta Chiropterologica* 1 (1999): 17–29.

Kutsukake, N., and C. L. Nunn. "Comparative Tests of Reproductive Skew in Male Primates: The Roles of Demographic Factors and Incomplete Control." *Behavioral Ecology and Sociobiology* 60 (2006): 695–706.

Lessels, C. M. "Sexual Conflict in Animals." In *Levels of Selection in Evolution,* ed. L. Keller, pp. 75–99. Princeton, N.J.: Princeton University Press, 1999.

———. "The Evolutionary Outcome of Sexual Conflict." *Philosophical Transactions of the Royal Society: Biological Sciences, Series B* 361 (2006): 301–317.

Lindenfors, P. "Sexually Antagonistic Selection on Primate Size." *Journal of Evolutionary Biology* 15 (2002): 595–607.

Lynch Alfaro, J. W. "Male Mating Strategies and Reproductive Constraints in a Group of Wild Tufted Capuchin Monkeys *(Cebus apella nigritus).*" *American Journal of Primatology* 67 (2005): 313–328.

Maklakov, A. A., T. Bilde, and Y. Lubin. "Inter-Sexual Combat and Resource Allocation into Body Parts in the Spider, *Stegodyphus lineatus.*" *Ecological Entomology* 31 (2006): 564–567.

Manson, J. H. "Male Aggression: A Cost of Female Mate Choice in Cayo Santiago Rhesus Macaques." *Animal Behaviour* 48 (1994): 473–475.

———. "Mate Choice." In *Primates in Perspective,* eds. C. J. Campbell, K. C. MacKinnon, M. Panger, A. Fuentes, and S. Bearder, pp. 447–463. Oxford: Oxford University Press, 2007.

Mateo, J. M. "Development of Individually Distinct Recognition Cues." *Developmental Psychobiology* 48 (2006): 508–519.

Matsumoto-Oda, A. "Female Choice in the Opportunistic Mating of Wild Chimpanzees *(Pan troglodytes schweinfurthii)* at Mahale." *Behavioural Ecology and Sociobiology* 46 (1999): 258–266.

Mitani, J. C. "Mating Behaviour of Male Orangutans in Kutai Game Reserve, Indonesia". *Animal Behaviour* 32 (1985): 392–402.

Möhle, U., M. Heistermann, J. Dittami, V. Reinberg, and J. K. Hodges. "Patterns of Anogenital Swelling Size and Their Endocrine Correlates during Ovulatory Cycles and Early Pregnancy in Free-Ranging Barbary Macaques *(Macaca sylvanus)* of Gibraltar." *American Journal of Primatology* 66 (2005): 351–368.

Muller, M. N. "Agonistic Relations among Kanyawara Chimpanzees." In *Behavioural Diversity in Chimpanzees and Bonobos,* eds. C. Boesch, G. Hohman, and F. L. Marchant, pp. 112–124. Cambridge: Cambridge University Press, 2002.

Newton-Fisher, N. E. "Female Coalitions against Male Aggression in Wild Chimpanzees of the Budongo Forest." *International Journal of Primatology* 27 (2006): 1589–1599.

Nieuwenhuijsen, K., K. J. de Neef, and A. K. Slob. "Sexual Behaviour during Ovarian Cycles, Pregnancy and Lactation in Group-Living Stump-Tail Macaques, *Macaca arctoides.*" *Human Reproduction* 1 (1986): 159–169.

Nikitopoulos, E., M. Heistermann, H. de Vries, J. A. R. A. M. van Hooff, and E. H. M. Sterck. "A Pair Choice Test to Identify Female Mating Pattern Relative to Ovulation in Longtailed Macaques, *Macaca fascicularis.*" *Animal Behaviour* 70 (2005): 1283–1296.

Noë, R., and A. A. Sluijter. "Reproductive Tactics of Male Savanna Baboons. *Behaviour* 113 (1990): 117–170.

Nunn, C. L. "The Evolution of Exaggerated Sexual Swellings in Primates and the Graded-Signal Hypothesis." *Animal Behaviour* 58 (1999): 229–246.

———. "Spleen Size, Disease Risk and Sexual Selection: A Comparative Study in Primates." *Evolutionary Ecology Research* 4 (2002): 91–107.

Nunn, C. L., J. L. Gittleman, and J. Antonovics. "Promiscuity and the Primate Immune System." *Science* 290 (2000): 1168–1170.

Ostner, J., M. K. Chalise, A. Koenig, K. Launhardt, J. Nikolei, D. Podzuweit, and C. Borries. "What Hanuman Langur Males Know about Female Reproductive Status." *American Journal of Primatology* 68 (2006): 701–712.

Packer, C. "Inter-Troop Transfer and Inbreeding Avoidance in *Papio anubis.*" *Animal Behaviour* 27 (1979): 1–36.

Pagel, M. "Desperately Concealing Father: A Theory of Parent-Infant Resemblance." *Animal Behaviour* 53 (1997): 973–981.

Palombit, R. A. "Infanticide and the Evolution of Pair Bonds in Nonhuman Primates." *Evolutionary Anthropology* 7 (1999): 117–129.

———. "Infanticide and the Evolution of Male-Female Bonds in Animals." In *Infanticide by Males and Its Implications,* eds. C. P. van Schaik, and C. H. Janson, pp. 239–268. Cambridge: Cambridge University Press, 2000.

———. "Male Infanticide in Savanna Baboons: Adaptive Significance and Intraspecific Variation." In *Sexual Selection and Reproductive Competition in Primates: New Perspectives and Directions,* ed. C. B. Jones, pp. 367–412. Norman, OK: American Society of Primatologists, 2003.

Palombit, R. A., D. L. Cheney, J. Fuscher, S. Johnson, D. Rendall, R. M. Seyfarth, and J. B. Silk. "Male Infanticide and the Defense of Infants in Chacma Baboons." In *Infanticide by Males and its Implications,* eds. C. P. van Schaik, and C. H. Janson, pp. 123–152. Cambridge: Cambridge University Press, 2000.

Pandit, S. A., and C. P. van Schaik. "A Model for Leveling Coalitions among Male Primates: Towards a Theory of Egalitarianism." *Behavioral Ecology and Scoiobiology* 55 (2003): 161–168.

Parker, G. A. "Sexual Selection and Sexual Conflict." In *Sexual Selection and Reproductive Competition in Insects,* eds. M. S. Blum, and N. A. Blum, pp. 123–166. New York: Academic Press, 1979.

———. "Sexual Conflict over Mating and Fertilization: An Overview." *Philosophical Transactions of the Royal Society: Biological Sciences, Series B* 361 (2006): 235–259.

Paul, A. "Determinants of Male Mating Success in a Large Group of Barbary Macaques, *Macaca sylvanus,* at Affenberg Salem." *Primates* 30 (1989): 461–476.

———. "Sexual Selection and Mate Choice." *International Journal of Primatology* 23 (2002): 877–904.

Pereira, M. E., R. Kaufman, P. M. Kappeler, and D. Overdorff. "Female Dominance Does Not Characterize All of the Lemuridae." *Folia Primatologica* 55 (1990): 96–103.

Pereira, M. E., and M. L. Weiss. "Female Mate Choice, Male Migration, and the Threat of Infanticide in Ringtailed Lemurs." *Behavioral and Sociobiology* 28 (1991): 141–152.

Plath, M., J. Parzefall, and I. Schlupp. "The Role of Sexual Harassment in Cave and Surface Dwelling Populations of the Atlantic Molly, *Poecilia mexicana* (Poeciliidae, Teleostei)." *Behavioral Ecology and Sociobiology* 54 (2003): 303–309.

Plavcan, J. M. "Sexual Selection, Measures of Sexual Selection, and Sexual Dimorphism in Primates." In *Sexual Selection in Primates: New and Comparative Perspectives,* eds. P. M. Kappeler, and C. P. van Schaik, pp. 230–252. Cambridge: Cambridge University Press, 2004.

Pochron, S. T., J. Fitzgerald, C. C. Gilbert, D. Lawrence, M. Grgas, G. Rakotonirina, R. Ratsimbazafy, R. Rakotosoa, P. C. Wright. "Patterns of Female Dominance in

Propithecus diadema edwardsi of Ranomafana National Park, Madagascar." *American Journal of Primatology* 61 (2003): 173–185.

Pradhan, G. R., A. Engelhardt, C. P. van Schaik, and D. Maestripieri. "The Evolution of Female Copulation Calls in Primates: A Review and a New Model." *Behavioral Ecology and Sociobiology* 59 (2006): 333–343.

Pradhan, G. R., and C. P. van Shaik. "Infanticide—Driven Intersexual Conflict over Matings in Primates and Its Effects on Social Organization." *Behaviour* 145 (2008): 251–275.

Pradhan, G. R., and C. P. van Schaik. "Why Are There Ornaments at All? The Coercion-Avoidance Hypothesis." *Biological Journal of the Linnean Society* 96 (2009): 372–382.

Range, F. "Female Sooty Mangabeys *(Cercocebus torquatus atys)* Respond Differently to Males Depending on the Male's Residence Status: Preliminary Data." *American Journal of Primatology* 65 (2005): 327–33.

Reichert, K. E., M. Heistermann, J. K. Hodges, C. Boesch, and G. Hohmann. "What Females Tell Males about Their Reproductive Status: Are Morphological and Behavioural Cues Reliable Signals of Ovulation in Bonobos *(Pan paniscus)*?" *Ethology* 108 (2002): 583–600.

Rice, W. R. "Sexually Antagonistic Male Adaptation Triggered by Experimental Arrest of Female Evolution." *Nature* 381 (1996): 232–234.

———. "Intergenomic Conflict, Interlocus Antagonistic Coevolution and the Evolution of Reproductive Isolation." In *Endless Forms: Species and Speciation,* eds. D. J. Howard, and S. H. Berlocher, pp. 261–270. New York: Oxford University Press, 1998.

———. "Dangerous Liaisons." *Proceedings of the National Academy of Science USA* 97 (2000): 12953–12955.

Rice, W. R., and B. Holland. "The Enemies Within: Intergenomic Conflict, Interlocus Contest Evolution, and the Intraspecific Red Queen." *Behavioral Ecology and Sociobiology* 41 (1997): 1–10.

Sauther, M. L. "Reproductive Behavior of Free-Ranging *Lemur catta* at Beza Mahafaly Special Reserve, Madagascar." *American Journal of Physical Anthropology* 84 (1991): 463–477.

Schneider, J. M. "Delayed Oviposition: A Female Strategy to Counter Infanticide by Males?" *Behavioral Ecology* 10 (1999): 567–571.

Schneider, J. M., and Y. Lubin. "Infanticidal Male Eresid Spiders." *Nature* 381 (1996): 655–656.

Schürmann, C. "Mating Behaviour of Wild Orang-utans." In *The Orang-Utan: Its Biology and Conservation,* ed. L. E. M. de Boer, pp. 269–284. The Hague: Junk, 1982.

Schürmann, C. L., and J. A. R. A. M. van Hooff. "Reproductive Strategies of the Orang-utan: New Data and a Reconsideration of Existing Sociosexual Models." *International Journal of Primatology* 7 (1986): 265–287.

Setchell, J. M., and E. J. Wickings. "Sexual Swellings in Mandrills *(Mandrillus sphinx):* A Test of the Reliable Indicator Hypothesis." *Behavioral Ecology* 15 (2004): 438–445.

Sherman, P. W., and B. D. Neff. "Father Knows Best." *Nature* 425 (2003): 136–137.

Smuts, B. B., and R. W. Smuts. "Male Aggression and Sexual Coercion of Females in Nonhuman Primates and Other Mammals: Evidence and Theoretical Implications." *Advances in the Study of Behavior* 22 (1993): 1–63.

Soltis, J. "Do Primate Females Gain Nonprocreative Benefits by Mating with Multiple Males? Theoretical and Empirical Considerations." *Evolutionary Anthropology* 11 (2002): 187–197.

Soltis, J., F. Mitsunaga, K. Shimizu, and M. Nozaki. "Sexual Selection in Japanese Macaques I: Female Mate Choice and Male Sexual Coercion." *Animal Behaviour* 54 (1997): 737–746.

Sprague, D. S. "Life History and Male Intertroop Mobility among Japanese Macaques *(Macaca fuscata)*." *International Journal of Primatology* 13 (1992): 437–453.

———. "Age, Dominance Rank, Natal Status, and Tenure among Male Macaques." *American Journal of Physical Anthropology* 105 (1998): 511–521.

Sprague, D, S. Suzuki, and T. Tsukahara. "Variation in the Social Mechanism by which Males Attained Alpha Rank among Japanese Macaques." In *Evolution and Ecology of Macaques Societies*, eds. J. E. Fa, and D. C. Lindburg, pp. 444–458. Cambridge: Cambridge University Press, 1996.

Steenbeek, R. "Female Choice and Male Coercion in Wild Thomas's Langurs *(Presbytis thomasi)*. Ph.D. Thesis, University of Utrecht, 1999.

Steenbeek, R., E. H. M. Sterck, H. De Vries, and J. A. R. A. M. van Hooff. "Cost and Benefits of the One-Male, Age-Graded, and All-Male Phases in Wild Thomas's Langur Groups." In *Primate Males: Causes and Consequences of Variation in Group Composition*, ed. P. M. Kappeler, pp. 130–145. Cambridge: Cambridge University Press, 2000.

Sterck, E. H. M., and A. H. Korstjens. "Female Dispersal and Infanticide Avoidance in Primates." In *Infanticide by Males and Its Implications*, eds. C. P. van Schaik, and C. H. Janson, pp. 293–321. Cambridge: Cambridge University Press, 2000.

Stumpf, R. M., and C. Boesch. "Does Promiscuous Mating Preclude Female Choice? Female Sexual Strategies in Chimpanzees *(Pan troglodytes verus)* of the Taï National Park, Côte d'Ivoire." *Behavioral Ecology and Sociobiology* 57 (2005): 511–524.

Tokuda, K. "A Study of the Sexual Behavior in the Japanese Monkey Troop." *Primates* 3 (1961): 1–40.

Treves, A. "Primate Social Systems: Conspecific Threat and Coercion-Defense Hypothesis." *Folia Primatologica* 69 (1998): 81–88.

Tutin, C. E. G. "Mating Patterns and Reproductive Strategies in a Community of Wild Chimpanzees *(Pan troglodytes schweinfurthii)*." *Behavioral Ecology and Sociobiology* 6 (1979): 39–48.

Uehara, S., M. Hiraiwa-Hasegawa, K. Hosaka, and M. Hamai. "The Fate of Defeated Alpha Male Chimpanzees in Relation to their Social Networks." *Primates* 35 (1994): 49–55.

van Noordwijk, M. A., and C. P. van Schaik. "Male Migration and Rank Acquisition in Wild Long-Tailed Macaques, *Macaca fasicularis.*" *Animal Behaviour* 33 (1985): 849–861.

———. "Male Careers in Sumatran Long-Tailed Macaques." *Behaviour* 107 (1988): 24–43.

———. "Reproductive Patterns in Eutherian Mammals: Adaptations against Infanticide?" In *Infanticide by Males and Its Implications,* eds. C. P. van Schaik, and C. H. Janson, pp. 322–360. Cambridge: Cambridge University Press, 2000.

———. "Career Moves: Transfer and Rank Challenge Decisions by Male Long-Tailed Macaques." *Behaviour* 138 (2001): 359–395.

———. "Sexual selection and the Careers of Primate Males: Paternity Concentration, Dominance-Acquisition Tactics and Transfer Decisions." In *Sexual Selection in Primates: New and Comparative Perspectives,* eds. P. M. Kappeler, and C. P. van Schaik, pp. 208–229. Cambridge: Cambridge University Press, 2004.

van Schaik, C. P. "Social Counterstrategies against Male Infanticide in Primates and Other Mammals." In *Primate Males: Causes and Consequences of Variation in Group Composition,* ed. P. M. Kappeler, pp. 34–52. Cambridge: Cambridge University Press, 2000a.

———. "Infanticide by Male Primates: The Sexual Selection Hypothesis Revisited." In *Infanticide by Males and Its Implications,* eds. C. P. van Schaik, and C. H. Janson, pp. 27–71. Cambridge: Cambridge University Press, 2000b.

van Schaik, C. P., and R. I. M. Dunbar. "The Evolution of Monogamy in Large Primates: A New Hypothesis and Some Crucial Tests." *Behaviour* 115 (1990): 30–62.

van Schaik, C. P., and P. M. Kappeler. "Infanticide Risk and the Evolution of Male-Female Association in Primates." *Proceedings of the Royal Society of London Series B—Biological Sciences* 264 (1997): 1687–1694.

van Schaik, C. P., M. A. van Noordwijk, and C. L. Nunn. "Sex and Social Evolution in Primates." In *Comparative Primate Socioecology,* ed. P. C. Lee, pp. 204–240. Cambridge: Cambridge University Press, 1999.

van Schaik, C. P., J. K. Hodges, and C. L. Nunn. "Paternity Confusion and the Ovarian Cycles of Female Primates." In *Infanticide by Males and Its Implications,* eds. C. P. van Schaik, and C. H. Janson, pp. 361–387. Cambridge: Cambridge University Press, 2000.

van Schaik, C. P., G. R. Pradhan, and M. A. van Noordwijk. "Mating Conflict in Primates: Infanticide, Sexual Harassment and Female Sexuality." In *Sexual Selection in Primates: New and Comparative Perspectives,* eds. P. M. Kappeler, and C. P. van Schaik, pp. 131–150. Cambridge: Cambridge University Press, 2004.

van Schaik, C. P., S. A. Pandit, and E. R. Vogel. "Toward a General Model for Male-Male Coalitions in Primate Groups." In *Cooperation in Primates and Humans,* eds. P. M. Kappeler, and C. P. van Schaik, pp. 151–171. New York: Springer Publications, 2005.

Wedell, N, C. Kvarnemo, C. M. Lessels, and T. Tregenza. "Sexual Conflict and Life Histories." *Animal Behaviour* 71 (2006): 999–1011.

Weingrill, T., J. E. Lycett, L. Barrett, R. A. Hill, and S. P. Henzi. "Male Consortships in Chacma Baboons: The Role of Demographic Factors and Female Conceptive Probabilities." *Behaviour* 140 (2003): 405–427.

Whitten, P. L., and E. Russell. "Information Content of Sexual Swellings and Fecal Steroids in Sooty Mangabeys *(Cercocebus torquatus atys)." American Journal of Primatology* 40 (1996): 67–82.

Wiley, R. H., and J. Poston. "Indirect Mate Choice, Competition for Mates, and Co-evolution of the Sexes." *Evolution* 50 (1996): 1371–1381.

Wolff, J. O., and D. W. Macdonald. "Promiscuous Females Protect Their Offspring." *Trends in Ecology and Evolution* 19 (2004): 127–134.

Wrangham., R. W. "Costs and Benefits of Female Sexual Attractiveness in Chimpanzees and Bonobos: A Scramble Competition Hypothesis." In *Behavioural Diversity in Chimpanzees and Bonobos,* eds. C. Boesch, G. Hohmann, and L. F. Marchant, pp. 204–215. Cambridge: Cambridge University Press, 2002.

Wright, P. L. "Demography and Life-History of Free-Ranging *Propithecus diadema edwardsi* in Ranomafana National Park, Madagascar." *International Journal of Primatology* 16 (1995): 835–854.

Zeh, J. A., and D. W. Zeh. "Reproductive Mode and the Genetic Benefits of Polyandry." *Animal Behaviour* 61 (2001): 1051–1063.

Zinner, D. P., and T. Deschner. "Sexual Swellings in Female Hamadryas Baboons after Male Take-overs: "Deceptive" Swellings as a Possible Counter-Strategy against Infanticide." *American Journal of Primatology* 52 (2000): 157–168.

II

SEXUAL COERCION AND MATE GUARDING IN NONHUMAN PRIMATES

4

Orangutans: Sexual Coercion without Sexual Violence

Cheryl D. Knott

It is perhaps not surprising, given the prevalence of male aggression against women in humans, that the role of sexual coercion as a male mating strategy in animals emerged into the mainstream academic consciousness after the women's movement of the 1970s. With the publication of Smuts and Smuts's seminal paper on the subject in 1993, the importance of sexual coercion as a form of sexual selection was proposed, and sexual coercion has received increasing, though still somewhat limited, attention from scientists. Clutton-Brock and Parker (1995) extended the argument of sexual coercion as sexual selection by proposing a theoretical framework consisting of three male coercive strategies: forced copulation, harassment, and intimidation. These two reviews, as well as subsequent research, have highlighted the conditions under which sexual coercion is manifested in primates. However, one question that has received little or no attention is what causes differences in the *degree of force* used during sexual coercion. That subject is explored here through an examination of forced copulations in orangutans.

Orangutans are unusual among the primates, and indeed among mammals, for their use of force during copulation. At some orangutan sites a majority of copulations are forced. Therefore, orangutans apparently represent one of the most extreme cases of sexual coercion in the animal kingdom. Particularly because of presumed parallels between orangutan forced copulations and human rape, this behavior has been viewed as one of the most violent expressions of coercive male behavior. Smuts and Smuts (1993), for example, detail the prevalence of sexual coercion in the primate literature, concluding that "the most dramatic examples of apparent sexual coercion come from wild orangutans" (p. 6). However, perhaps counterintuitively, despite the high rates of forced copulation in orangutans and the often prolonged struggle of females, severe wounding of females has never been reported (MacKinnon 1974; Rijksen 1978; Schurmann 1981, 1982; Galdikas 1981, 1985a, b; Mitani 1985a; Schurmann and van Hooff 1986; Fox 1998; Utami 2000). In fact, Galdikas (1981:289) reports that "as soon as raping males stopped thrusting, they invariably released the female. Nor did any female sustain injuries or wounds as a result of rape." Thus, I argue here that orangutan males use force as a way to accomplish copulation but do not intentionally wound females. Orangutan coercive sexual behavior is direct coercion and is not used as an indirect means to influence or control future female sexual behavior, as seen in species such as chimpanzees and humans.

In this chapter, I explore orangutan sexual coercion through four objectives. First, I review the evidence of sexual coercion in orangutans and highlight the actual diversity in expression of this behavior exhibited between populations. I argue that copulations can rarely be classified as either forced or consensual because most mating involves elements of both proceptivity and resistance. I also demonstrate that factors other than male morphotype (flanged versus unflanged), such as female hormonal status, contribute to the frequency of forced copulation, and I present a number of hypotheses to be tested. Second, I argue that orangutan sexual coercion may be driven by females resisting males in order to avoid the energetic and disease costs of multiple matings, as well as the cost of insemination by a genetically inferior male. Females may benefit from resisting because it decreases the total length of the mating; sometimes it may also prevent ejaculation and may garner support from the dominant male. Third, I highlight the finding that, although copulations are often forced in orangutans, the degree of physical *wounding* is extremely low. In particular, I evaluate the degree of force associated with orangutan sexual coercion in the light of comparative data from chimpanzees and humans. Finally, I discuss how measures of the *degree of physical force* used in coercion may help us distinguish between direct and indirect coercion and the motivations behind that coercion, and I present further hypotheses for investigation.

Forced Copulations

Defining Forced Copulation in Orangutans

Forced copulations have been described by all orangutan researchers, starting with the earliest pioneers (MacKinnon 1974; Galdikas 1978; Rijksen 1978; Galdikas

1979, 1981; Schurmann 1981, 1982; Mitani 1985a; Schurmann and van Hooff 1986). Galdikas (1981:288) provides the first operational definition: "Rape occurred when a male attempted to copulate or copulated with a female who resisted his efforts to position her for intromission. A female's struggles ranged in intensity and duration all the way from brief tussles with squalling and some pushing and slapping at the male's hand to protracted violent fights in which the female struggled through the length of the copulation, emitted loud rape-grunts and bit the male whenever she could."

Despite the ubiquitous observation of forced copulations, most orangutan researchers report that some copulations are ambiguous and cannot be easily designated as either forced or cooperative (Rijksen 1978; Galdikas 1981; Fox 2002; Knott et al. in review). Often a copulation can have elements of both male force and female proceptivity and can switch from being cooperative to resistant or resistant to cooperative. For example, at Gunung Palung we have seen females that run away, screaming from a male, are chased and then grabbed by him but then become very proactive during the copulation, sitting on top of the male and facilitating intromission and copulatory thrusting. Fox (1998), working at Suaq in Sumatra, reports a number of matings that began cooperatively but eventually turned coercive, as females struggled to escape. For example, she described the mating of the adult female "Ani" with an unflanged male. The mating began with the female showing no resistance and continuing to feed on fruit that was within her reach. After three minutes the male grabbed her feeding hand in an attempt to reposition her. This was followed by 12 minutes of struggle, which included the male biting and hitting the female until he achieved ejaculation.

Despite this frequent mix of coercive and cooperative elements, orangutan copulations are usually characterized as either forced or consensual based on an overall impression of male actions and female responses. This subjective dichotomization masks some of the interesting variation in mating behavior that may help us better understand its origin. In the next sections, I discuss the factors, based on earlier studies as well as a reexamination of the literature and our more recent detailed analysis (Knott et al. in review), that determine whether a copulation is categorized as forced, consensual, or ambiguous. Following Clutton-Brock and Parker (1995), I first consider forced copulations, then harassment, and finally intimidation and nonmating aggression.

Flanged vs. Unflanged Males

Orangutan males are commonly divided into two distinct morphological forms based on the presence or absence of protruding cheek flanges. Flanged males who sport these prominent, fibrous protrusions are very large, typically weighing over 80 kg in the wild (Markham and Groves 1990, Figure 4.1a). They give loud bellowing long calls audible for up to 800 m in the forest (Mitani 1985b) that are used to advertise their presence to other flanged males and perhaps to females (Knott and Kahlenberg 2007). In contrast, unflanged males are smaller, do not give long calls, do not have cheek flanges (Knott and Kahlenberg 2007, Figure 4.1b), and are subordinate to flanged males. Flanged males are almost entirely solitary, except when consorting with females, and show absolute intolerance of other flanged males. Unflanged males commonly travel in small bands and normally are not aggressive toward other males (Knott and Kahlenberg 2007). Physiologically, both male morphs are sexually mature, as evidenced by their ability to father offspring and their similar levels of follicle-stimulating hormone (FSH)—responsible for viable sperm production (Maggioncalda et al. 1999). Testosterone levels are lower in unflanged males in both the wild (Knott in prep) and captivity (Maggioncalda et al. 1999, 2000), accounting for the lack of secondary sexual characteristics in the unflanged males. This condition of male bimaturism is extremely rare in mammals, with perhaps the possible other exception of mandrills (Atmoko and van Hooff 2004).

The developmental status of the male is the key variable that has been investigated in relation to forced copulation. Forced copulation is often depicted as the typical mating strategy of unflanged males, whereas matings with flanged males are said to be consensual (Rijksen 1978; Galdikas 1979, 1981; Schurmann 1982; Fox 2002). However, the data show much more variability than is commonly appreciated (Mitani 1985a). Figure 4.2 distinguishes forced versus consensual copulations according to male "type." At some sites, such as Suaq, all of the copulations with flanged males were consensual, whereas at Kutai and Gunung Palung, approximately 50% of flanged copulations were forced. The data for unflanged males are similar. At Gunung Palung, Kutai, and Suaq a significant proportion of copulations with unflanged males were consensual. Thus, flanged status of the male alone is not sufficient to explain the majority of orangutan mating data.

Figure 4.1 Photos of a (a) flanged and (b) unflanged male from Gunung Palung National Park. Photos by Tim Laman.

A Flanged Males

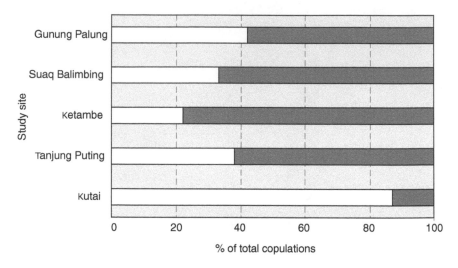

B. Unflanged Males

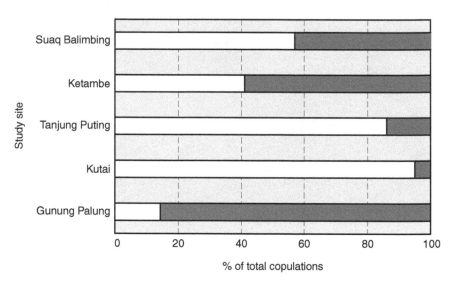

Figure 4.2 Forced (white bars) and cooperative (black bars) copulations involving (A) flanged and (B) unflanged orangutan males at five study sites. Sample sizes are for copulations of known outcome and are as follows (unflanged/flanged): Suaq Balimbing (N=90/66): Fox, 2002; Ketambe (N=38/50, 94/70; averaged between two studies): Schürmann and van Hooff 1986, Utami Atmoko 2000; Tanjung Puting (N=22/30): Galdikas 1985a, b; Kutai (N=151/28): Mitani 1985a; Gunung Palung (N=24/19): Knott et al. (in review).

Male Status

In addition to being classified as flanged or unflanged, within these categories males can be further broken down by other aspects of their status or rank. As described in Knott (in prep), flanged males may be prime or past prime. Past-prime males display greatly diminished cheek flanges, they give significantly fewer long calls, have significantly lower testosterone levels and energy expenditure, mate infrequently, are less aggressive toward other males, and are subordinate to prime males (Knott, in review). This new distinction may explain some of the variation in flanged male matings. For example, at Gunung Palung, although matings with past-prime males were rare, females showed high levels of resistance to them (Knott et al. in review). This was particularly true during the periovulatory period. Females were also significantly less proceptive to past-prime males than to prime males (Knott et al. in prep). At Kutai, where the highest percentage of forced copulations by flanged males has been reported, Mitani (1985a) found that 62% of these flanged forced copulations were by one male that had the least developed cheek flanges. He was never seen to mate nonforcibly. It is possible that rather than being young, this male had small cheek flanges because he was a past-prime male.

The unflanged category also encompasses several types of male. Some of these males are true adolescents whose cheek flanges have not yet developed, others are adults who have remained in this status for many years, and still others are in the process of developing cheek flanges (Crofoot and Knott in press). In wild study populations, many of these unflanged males are unhabituated, transients who briefly visit study areas. Their ages and the "type" of unflanged male they represent are thus unknown. To females, however, these three types of unflanged male likely elicit very different responses to mating attempts. In her study, Fox (1998) classified unflanged males into three size categories (small, medium, and large) and found that mating behavior differed significantly between the categories. Females may be able to recognize males who are on the verge of becoming flanged. At Ketambe, the consensual matings with an unflanged male observed by Utami Atmoko (2000) were with a male who developed cheek flanges and displaced the dominant male six months after these matings were observed. In fact, this male achieved a majority of the copulations attributed to unflanged males, and 50% of paternity at this site was attributed to him (Utami 2002).

Residential Status

Orangutan males range over large areas, and residential status may heavily influence female preference. Mitani (1985a) found that 90% of matings by the dominant flanged male were consensual, whereas only 34% of the matings by the other six nonresident flanged males were. Similarly, Fox (1998, 2002) describes a strong female preference for the resident flanged male, and Utami Atmoko (2000) describes how females that normally preferred the dominant flanged male copulated cooperatively with lower ranked flanged males and unflanged males during periods of male rank instability. At Gunung Palung, among flanged males, the dominant prime flanged male received the majority of the copulations over nonresident flanged males. Thus, residential status and male rank, within a given male "type," is an important determinant of female choice.

Female Reproductive and Ovarian Status

In orangutans, the periovulatory period (POP) when conception occurs cannot be distinguished visually because of the lack of a sexual swelling. Thus, hormonal analysis must be used to determine a female's ovulatory status. At Gunung Palung we (Knott et al. in review) have shown that female ovulatory status interacts with male type to influence whether the copulation is forced. Copulations near or during the POP were almost exclusively with flanged prime males (Figure 4.3) and all copulations with flanged males during the POP were cooperative (Figure 4.4). As a female's fecundity decreased, unflanged males were more likely to mate successfully. Adult females during the POP were significantly more attractive to males than were non-POP or pregnant females. Interestingly, this is not just a reaction to female proceptivity, as the most proceptive females were those in the early stages of pregnancy. All copulations with females during early pregnancy were cooperative.

Copulatory Sequence

Mitani (1985a) suggested that the first encounter between a male and female often leads to a forced copulation, but subsequent copulations are more likely to be consensual. He found this pattern in three of five associations. At Gunung Palung, we only found this pattern in one out of ten such associations (Knott et al. in review). Consecutive copulations on the same day included every possible pattern: cooperative:cooperative, cooperative:forced, forced:forced, and

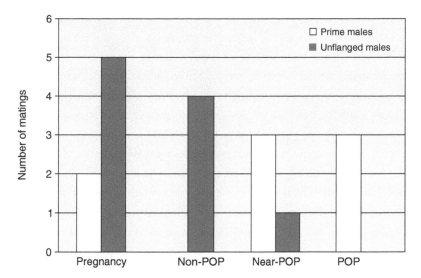

Figure 4.3 Male type during matings of known female ovulatory status.

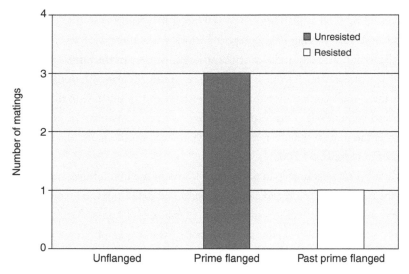

Figure 4.4 Male type and resistance of matings during the periovulatory period (POP).

forced:cooperative. In one such instance of sequential matings at Gunung Palung, a prime flanged male forced a female that was traveling with an unflanged male to copulate despite her very active resistance throughout the mating. Soon after, they mated again, and the female offered no resistance whatsoever. In another case, an unflanged male mated with another adult female, and the first copulation involved genital inspection by the male and no resistance by the female. After three hours they mated again. This time the female ran away, and the male had to force her legs apart to mate with her. Likewise, during the course of these long matings, copulations can switch from forced to consensual and vice versa (average eight minutes of copulatory thrusting for matings at Gunung Palung). Whether the female in these situations eventually accepts the male, gives up, was just "testing" him, or simply has changed her mind is unclear. It may be that during the mating, or between consecutive matings, females are reassessing the relative costs of resisting vs. acquiescing, and behaving accordingly. Thus, in Fox's (1998) description of a female who resisted after initially cooperating, the cost of the mating changed when she was prevented from continuing to eat.

Male:Female Ratio and Number of Copulations

In orangutans, the adult sex ratio in a study area is expected to vary with the availability of fecund females, which itself is related to female energetic status influenced by food availability (Knott 1998a, 2001; Knott et al. 2009). Indeed, both Kutai and Gunung Palung report periods when there were several fecund females simultaneously and a concurrent large increase in the number of males using the study site. As the operational sex ratio increases, I would predict more male-male competition for fecund females and an increase in the number of forced copulations. This relationship has been demonstrated in insects and birds; as the male to female ratio increases, males become more likely to force females to mate (Low 2005).

To test for a relationship in orangutans between the ratio of males to females and the number of forced copulations, I examined data from the literature from the five study sites where this information is available. I found that the ratio of males to females was a significant predictor of the degree of female resistance ($p < 0.01$, Pearson correlation $= 0.957$) for both flanged and unflanged males. Indeed, there was a significant correlation between forced matings by flanged males vs. forced matings by unflanged males ($p < 0.005$, Pearson correlation $= 0.982$). There was no relationship between the rate of consensual copulations and the male:female ratio ($p = 0.624$, Pearson correlation $= -0.300$). I would predict, based on the endocrine data cited above, that these forced matings are occurring at different times in the females' cycle depending on male type. Thus, whereas normally a flanged male may be able to "sit and wait" for ovulating females to come to him, with increased male competition he may become more likely to try and mate with a non-POP female who, because of her ovulatory status, resists him, resulting in a forced copulation. Similarly, with a high operational sex ratio it may be increasingly difficult for flanged males to prevent unflanged males from mating with females during the POP. Thus, there may be more successful matings of unflanged males during the POP, which I would predict should be noncooperative because ovulating females show a clear preference for flanged males.

Female Strategies

Intriguingly, the relationship between increased male density and increased female resistance in orangutans is opposite of that found in some insects and birds, where the term *convenience polyandry* has been used to describe why a female may resist some matings and acquiesce for others. Convenience polyandry occurs when a female mates because the cost of acquiescing is less than the benefits she would gain by resisting or escaping that mating. Factors involved in the assessment of this cost-benefit analysis include risk of mortality, predation, disease transmission, loss of feeding opportunities, and the impact of fertilization. For both damselflies (Rivera and Andres 2002) and New Zealand stitchbirds (Low 2005), the operational sex ratio as it relates to the number of mating attempts is an important factor in determining female mating decisions. In these species, however, females resist matings below a certain threshold, but, as the rate of male harassment increases beyond that threshold, females acquiesce and mate with nonpreferred males in order to reduce the cumulative costs of mating. For females below the threshold, resisting the occasional mating is less costly than mating. When harassment by nonpreferred males is infrequent, females resist the matings, but as mating attempts increase, the costs of resisting becomes too great and females acquiesce. The opposite is true for orangutans because as the number of mating attempts increases, the level of female resistance increases. Although the risks of mortality and predation are not applicable in the case of the orangutan, disease transmission, loss of feeding opportunities, and the impact of fertilization by a nonpreferred male are all potential risks a female assumes during a mating. Thus, increased female resistance in response to an increased number of mating attempts is not explained by convenience polyandry, but may be observed because the benefits of escaping or trying to escape still outweigh the costs of mating for orangutans.

Another argument for female resistance is provided by van Schaik (2004), who argues that females resist matings with nonpreferred males in order to attract the attention of the dominant flanged male and to let him know that she is mating unwillingly. If the dominant male is nearby, he may be able to intervene in the mating. However, because of the dispersed ranging behavior of orangutans, the female is likely to be too far away from the dominant male for him to either hear or assist her. Van Schaik (2004) argues that a responding male may initiate a consortship with the female or assess that she was not ovulating. However, she may indeed be ovulating, and it would seem a better strategy to *not* attract the attention of the dominant male so as not to decrease his paternity certainty. The ability of a dominant male to be aware of matings by females with other males, and to intervene in these matings, likely varies depending on orangutan density.

Why do females mate with nonpreferred males at all? The endocrinological data from Gunung Palung are intriguing because they suggest that the least fecundable females, those during the non-POP, are the most likely to mate with unflanged males. Pregnant females exhibited the most proceptive behavior toward all male types. One interpretation of this behavior is that females are using a paternity confusion strategy to avoid the possibility of infanticide from these males (Hrdy 1981; van Schaik et al. 1999; Knott et al. in review). It may benefit a female to occasionally mate cooperatively with nonpreferred males in the population in order to confuse paternity and avoid potential infanticide. This interpretation is problematic; although orangutans are predicted to be vulnerable to infanticide (van Schaik and Kappeler 1997), thus far it has never been observed (Delgado and van Schaik 2000). However, other evidence is suggestive of this strategy as well. Utami Atmoko's (2000) observation at Ketambe that when the dominant male was being challenged females were much more likely to copulate with otherwise nonpreferred males (subdominant flanged males, unflanged males, and stranger males) and that there was a highly significant increase in matings with pregnant females during this period is also consistent with an anti-infanticide interpretation. Furthermore, Delgado (2003) in playback experiments showed that females ignored the long calls of resident flanged males, but that females and their infants became very upset when long calls of stranger males were played and they often fled in the opposite direction.

*Temporal Variation in Fruit Availability and Its Effects
on Female Fecundity*

The copulatory rate at Gunung Palung is associated with increased food availability, increased caloric intake, and consequently positive female energetic status (Knott 1999; Knott et al. 2009). In turn, positive female energy balance is significantly correlated with higher estrogen production and thus higher fecund-ability. Thus, because of the long interbirth interval in orangutans of seven to nine years (Galdikas and Wood 1990; Knott 2001; Wich et al. 2004) and the frequent clumping of conception years (Singleton and van Schaik 2002), the number of fecund females using a given study area is likely to vary considerably. Thus copulation rates and rates of sexual coercion may fluctuate substantially between study periods. In Borneo, where fruit availability varies dramatically, the rate of copulations may be explicitly tied to varying fruit production, and thus study periods may show pronounced variation in the number of copulations.

Therefore, another interesting distinction between study sites is the number of copulations witnessed per observation hour (Figure 4.5). Gunung Palung stands out as having a very low copulatory rate compared to Kutai with a very high rate. Mitani (1985a) points to the availability of three fecund females during his study at Kutai where he witnessed 179 copulations in 3600 observation hours. By contrast, Rodman (1977), also working in Kutai, witnessed only two copulations in 1644 observation hours. At Gunung Palung 28% of our total matings were observed during a three-month mast period of high fruit availability, representing just 4% of our total sampling time. In most months no matings were observed,

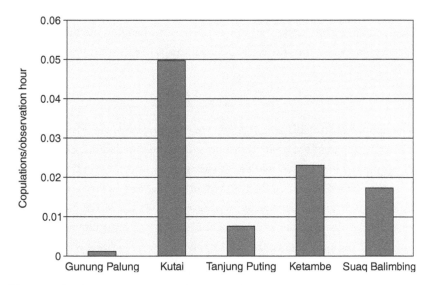

Figure 4.5 Number of copulations observed per hour of observation at five study sites. Samples sizes: Gunung Palung (N = 43): Knott et al., in prep; Kutai (N = 179): Mitani 1985a; Tanjung Puting (N = 52): Galdikas 1985a, b; Ketambe (N = 252): Schürmann and van Hooff 1986, Utami Atmoko 2000; Suaq Balimbing (N = 156): Fox, 2002.

despite intensive search effort and focal observation time. Thus, at some sites orangutan matings are highly clustered. In contrast, Wich et al. (2006) suggest that at Ketambe (Sumatra) diet composition, copulations, and conception rates show little temporal variation.

Furthermore, the overall ecological productivity of a site is predicted to determine the total density of orangutans, and the number of fecund females may influence the relative proportion of flanged vs. unflanged males, which may in turn influence the total number of copulatory attempts and how these break down between forced and consensual categories. Thus, relative food availability may have an important effect on sexual coercion in orangutans, and analyses of mating behavior should control for food availability. Indeed, at Suaq in Sumatra, Fox (2002) observed almost double the rate of forced copulations by unflanged males during periods of high fruit abundance, although this difference was not significant. These forced copulations were due to an increase in the time these females spent in parties with multiple unflanged males (Fox 2002). Females were more likely to consort with flanged males during months of high fruit availability, or within a week before or after a high fruit abundance month (Fox 2002). Based on inference from evidence at Gunung Palung (Knott 1999), Fox (2002) predicts that these were periods of high conception risk, although this was not explicitly measured. Thus, as food availability changes, female reproductive status may follow suit, changing the dynamics of social encounters and thus the rates of forced versus consensual copulations.

Sampling

Finally, sampling biases inherent in orangutan research should also be considered in evaluating apparent study site differences in forced copulations. Orangutans commonly travel alone or in mother–offspring units, and thus, individual animals must be targeted for focal follows. Mitani (1985a) was specifically targeting males in his study, and this strategy may have partially led to the high rate of copulations he witnessed. In support of this finding, at my study at Gunung Palung, where the apparent copulation rate is low, 74% of focal follows have been on females. Furthermore, differences in the way that forced and consensual copulations are categorized may be important. I would recommend that future studies quantify the components of male and female behavior during mating as described in Knott et al. (in review) in order to use comparable measures between study sites.

Summary of Variables

The above factors, in combination, help explain a large part of the variation in the degree of resistance and proceptivity shown by female orangutans, as well as the level of male aggression. Matings of ovulating females with resident, prime, flanged males during periods of high food availability involve high levels of female proceptive behavior and often no resistance and no aggression. Females may resist these males, however, if they mate with them during the non-POP period. In contrast, POP matings between unflanged or past-prime males that are low-ranking and nonresident involve no proceptive behavior by the female and are often severely resisted. Females may mate cooperatively with these males, possibly as a paternity confusion strategy, particularly during the non-POP period. Females during the first trimester of pregnancy show extreme proceptivity to males and usually mate cooperatively, again as a possible paternity confusion strategy. As the ratio of males to females increases, likely owing to the availability of fecund females, higher rates of forced copulations are seen by both flanged and unflanged males. These complex mating decisions in orangutans can be further illuminated by increased study of endocrine physiology and mating data across sites, controlling for the male:female ratio and food availability at the time of mating.

Male Harassment

Male harassment, the second form of sexual coercion described by Clutton-Brock and Parker (1995), occurs in orangutans within the context of unwanted mating attempts, and sometimes forced consortships, by nonpreferred males. Fox (2002) provides considerable evidence that females seek spatial association with flanged males as a way to distance themselves from unflanged males as a mode of protection. Orangutan females who maintained a spatial association with flanged males received lower rates of harassment (Fox 2002). Fox's (2002) data show that unflanged males were successful during 71% of copulation attempts during days when the female was consorting with the resident flanged male compared to a 95% successful copulation rate for unflanged males when she was not accompanied by the resident flanged male.

Consortships between females and flanged males occur under two contexts: mating and nonmating. During mating consortships Fox (2002) found that unflanged males attempted more matings and had more attempts that were successful compared to nonmating consortships. This was presumably because these consortships occurred when females had a higher probability of conception, and thus the unflanged males made increased efforts to mate. Indeed, Fox (2002) found that parties of unflanged males of three or more always occurred in the context of these flanged male-adult female consortships. Failed mating attempts by unflanged males were due to the female fleeing to within 10 m of the resident flanged male (56%), or the resident male chasing the unflanged male (44%). Interestingly, when females were consorting with nonresident flanged males, the unflanged males spent significantly more time within 50 m of the consorting pair compared to consortships with the resident flanged male. The success rate of copulations with unflanged males was similar, at 60% (Fox 2002). These data indicate that both resident and nonresident flanged males offer a degree of protection for females.

Protection-seeking behavior is seen in other female animals as well. Female African elephants are known to consort with large, dominant males during estrous periods to avoid harassment from subordinate males (Moss 1983). It seems that some female orangutans behave similarly by consorting with flanged males to receive protection from unflanged males.

The second type of consortship between flanged males and females does not involve mating, and Fox (2002) speculates that females initiate these associations purely to reduce harassment by unflanged males. In Fox's (2002) study, these associations often occurred after an unflanged male had forcibly copulated with a female. These nonmating associations differed from mating consortships with flanged males in several important ways. They formed and disbanded after the arrival and departure of the unflanged male, and the flanged male did not chase away the unflanged male. Females clearly initiated these consortships with flanged males, whereas the initiator of mating consortships between females and flanged males was hard to determine. Unfortunately, we do not know the ovulatory status of these consortships, although Fox (2002) speculates that this female was not ovulating.

Intimidation, Indirect Coercion, and Noncoercive Male Aggression

There is no evidence of male aggression toward females outside of the mating context in orangutans. Adult male and female orangutans have very low rates of interaction. When they do interact, it is almost entirely within a mating or consortship context. There is no evidence that males use aggression as a way to control females or to influence future mating behavior of females. They also do not show other forms of noncoercive aggression. Wild male orangutans have never been seen, for example, to "police" female-female aggressive interactions (which themselves are very rare), something that is commonly seen in chimpanzees (Boehm 1994; Kahlenberg unpublished data). Intriguingly, however, this has been reported in captive orangutans (Zucker 1987).

The Cost of Sexual Coercion

Smuts and Smuts's definition of sexual coercion includes the requirement that such aggression occurs as *"at some cost to the female"* (1993:3; my emphasis). Although females frequently try to escape from forced copulations, these attempts are rarely successful (Galdikas 1985a, b; Fox 1998; Fox 2002), which begs the question—why resist? Thus, we can ask what cost do female orangutans suffer as a result of sexual coercion? Possible costs to investigate include (1) physical injury, (2) energetic costs, (3) disease risk, and (4) insemination by a nonpreferred male.

Physical Injury

In many mammals, male sexual aggression against females results in the physical wounding of females. Male hyenas sometimes inflict serious wounds on females during "baiting behavior," which occurs when females are sexually proceptive and is the only circumstance in which males are aggressive toward females (Szykman et al. 2003). Bottlenose dolphin males are responsible for the majority of attacks on adult females that result in tooth rakes and scarring. Moreover, female bottlenose dolphins are most likely to be attacked when cycling as opposed to other reproductive states (Scott et al. 2005). In sea lions, sexual aggression during breeding can result in female injury or even death (Chilvers et al. 2005). In primates, male chimpanzees commonly inflict wounds on females (Goodall 1986; Kahlenberg 2006). Such aggression can be measured in the elevated cortisol levels seen in individual female chimpanzees who were the most common victims of male aggression (Muller et al. 2007).

Therefore, it is surprising that a close reading of published data and descriptions of sexual coercion in orangutans include no indication of wounding of the female by the male. Although forced copulations may include physically aggressive behaviors from males, such as hitting and biting (Galdikas 1981; Mitani 1985), no investigators report that these bites actually broke the skin, that visible wounds resulted, or that females appeared injured after these encounters. This suggests that these bites may actually be "threat" bites not intended to actually inflict harm.

It is certainly possible that occasional injuries could be missed. However, it is unlikely that any serious wounds would not be seen in habituated animals that are followed after the copulation occurs. Flanged males regularly experience wounds from fights with other flanged males (Utami Atmoko 2000; Galdikas 1985b); thus the intentional biting of a restrained victim would certainly result in serious wounding. Males have been seen to die eventually from these wounds (Knott 1998b). In my study at Gunung Palung (Knott 1996), chemstrip analysis of urine showed that wounded males had significantly higher leukocyte levels, compared to females who showed no wounding and no leukocyte excretion in their urine.

Besides open wounds, females could be receiving soft tissue injuries. Such bruises would be very hard to see in the wild and have not been reported. Other obvious injuries from wounding such as limping and favoring a limb have also not been reported. In contrast, such injuries are regularly reported for flanged males as a result of male-male competition. Thus, although the force used by males may hurt females, males seem to rarely, if ever, inflict lasting injuries that compromise health or mobility. The observed level of force may be enough to force females to copulate, and wounding females may be both unnecessary and deleterious to the female's health, and thus not in the male's interest if she were to conceive. It has seemed puzzling that females, particularly nonovulating females, resist male aggression given that they apparently expose themselves to injury. However, these data suggest that females suffer little physical cost from resisting male copulatory attempts.

Energetic Costs

Orangutan females may face a cost in lost feeding time during matings. The entire mating "episode" may consist of a prolonged pre-mating period of resistance, foreplay, or the female waiting as the male slowly approaches, followed by 2 to 36 minutes of actual copulatory thrusting. Is this a significant energetic cost? At Gunung Palung matings occurred during periods of high food availability when, it could be argued, females could more readily afford to lose a few of these excess calories. However, these periods are also very important for storing excess calories as fat, and although lost feeding time during the high fruit period would not compromise a female's daily caloric requirements, it would certainly lessen the number of excess calories she could store against future caloric deficit.

Females may suffer caloric costs during consortships as well. During consortship, males usually follow the female and may even feed in adjacent trees, presumably to avoid foraging conflict. However, Fox (1998) tested the energetic hypothesis and did not find decreased foraging efficiency on days when females had forced consortships. She points to two factors that may have obscured a possible relationship: (1) she did not measure caloric intake and (2) the gregariousness of females at this site meant that she had to control for group size effects. Van Schaik (2004), working at the same site, argues that unwanted consortships may still be energetically costly. In addition, the cost of fleeing from an undesirable male and seeking out and traveling toward a desirable male should also be factored in. These energetic deficits would be compounded with multiple matings and could potentially represent a significant cost. The cumulative cost of multiple matings may be one of the reasons an increase in the number of males compared to females leads to increased female resistance. Clearly, more studies quantifying the energetic costs associated with resistance vs. nonresistance are needed to assess this hypothesis.

Disease Risk

Increased disease risk, from both micro and macroparasites, including sexually transmitted diseases and viruses, may represent a significant cost of copulation in orangutans. Individuals in most orangutan populations are largely solitary or semisolitary, and thus would be expected to have a low parasite load. Nunn et al. (2000) report that species that have a greater copulatory frequency have significantly greater white blood cell counts, which seem to be driven by increased risk of acquiring sexually transmitted diseases. Furthermore, ectoparasites and infectious diseases increase in species with more sociality (Altizer et al. 2003). Thus, consider the orangutan female who primarily interacts with her offspring or closely related female kin. Every six to eight years she becomes fecund and goes through a several month period of dramatically increased contact with new individuals, that is, males with whom she mates. This would be a period of significantly increased disease risk for a female orangutan. Diseases of all types are more readily passed on with increased and repeated contact with the same individual, and with multiple individual contacts.

If disease avoidance is an important factor in the cost of mating for orangutans, I would predict that females would (1) try to limit the number of partners with whom they copulate, (2) resist repeated matings with the same male, (3) particularly avoid mating with transitory males who may bring in new pathogens, and (4) limit the length of the matings. These predictions are all consistent with the observed orangutan mating data. With an increased number of males at a site, females increase their level of resistance. Females often resist mating with a male with whom they have just mated earlier, even a preferred male. Females show clear preference for resident vs. nonresident flanged males. Finally, although Mitani (1985a) found that the overall mating "episode" (including noncontact periods) was longer for resisted matings (Mitani 1985a), Knott et al. (in review) found that the length of copulatory thrusting was significantly shorter in resisted vs. unresisted matings. Thus, female resistance, *regardless of whether it is successful,* may serve the important function of reducing the risk of disease exposure for the female.

Are orangutans more vulnerable to disease risk than are other species? The higher white blood cell count in *Pan* compared to *Pongo* (Nunn et al. 2000) suggests an increased disease exposure in *Pan.* Indeed, at the site with the highest density of males, Ngogo, estrous female chimpanzees mated on average 42 times/day for a 12-hour day (Watts in press) compared to a rate of 1.44 matings/day of association for unflanged males and 1.72 matings/day for flanged males (Mitani 1985a). Thus, the chimpanzee mating system may have led to the evolution of an immune system capable of mounting a significant response to disease exposure, particularly from sexually transmitted diseases (STDs), whereas the normally low rates of interaction in orangutans may have led to an immune system less able to combat the increased disease risk of multiple matings. Thus, when possible, orangutan females should try and resist "unnecessary" mating attempts.

Insemination by a Nonpreferred Male

In orangutans, female resistance to mating attempts seems to be a clear indicator of female choice (Fox 2002). In Knott et al. (in review) we show that male aggression during mating is highly correlated with female resistance. Thus, females show their preference for particular males through their level of resistance and through their proceptive behavior. Males, if nonpreferred, try to overcome this choice through the use of force when necessary. As reviewed earlier, females use a number of criteria for distinguishing between males. All orangutan studies indicate that females normally prefer to mate with prime, flanged males. The endocrinological data at Gunung Palung combined with a continuous measure of proceptivity and resistance show that this preference is most extreme during the POP, indicating a clear preference for these males as fathers of their offspring (Table 4.1). Although specific data on who initiates these flanged male-POP female consortships is lacking, these data clearly suggest that female choice is an important factor. Females may thus be using concealed ovulation as a way of exercising their choice of male partner.

Do males "know" that a female is ovulating? The strongest evidence against such knowledge comes from Nadler's (1982) classic study of mating behavior in captive orangutans. In this experiment, a cycling female and prime flanged male were placed in adjacent cages with a barred door between them that could be slid up or down and left at varying heights. In the first set of experiments the door was left open, allowing the male to access the female whenever he chose; he entered the female's cage almost daily, with frequent forced copulations. In the second experiment the door was left just wide enough that only the much smaller female could go between cages. Under this condition the male showed extreme frustration, first reacting violently by shaking the cage and then presenting his erect penis to the female in an apparent attempt to induce her to enter his cage and mate. However, the female only entered the male's cage during midcycle as evidenced by her hormonal levels. Thus, a female orangutan's proceptive behavior is clearly influenced by her ovarian status, and her actions are likely sending a signal to males that she is receptive. Furthermore, the fecundability of each cycle may be variable. Fox (1998) reports

Table 4.1 Relationship between female ovarian status and male type for orangutan matings with matched endocrinological samples at Gunung Palung.

	Unflanged	Prime Flanged	Past-Prime Flanged
Pregnant	nonresisted	nonresisted	no matings
Non-POP	resisted and nonresisted	resisted and nonresisted	resisted and nonresisted
Near-POP	resisted and nonresisted	resisted	no matings
POP	no matings	nonresisted	resisted

from Sumatra that during one POP period a female repeatedly copulated with a flanged male, but during a subsequent apparently POP period she was not proceptive and the male did not attempt to mate with her. This may have been a reflection of lower levels of ovarian hormones, and thus lower fecundability, of the second cycle. Finally, because proceptivity is very pronounced in females during early pregnancy, when their reproductive state is not yet visible, it is not always an honest signal of fecundability.

Males may look for other cues as well. Females at Gunung Palung attracted the most genital inspection when they were pregnant (Knott et al. in review). Pregnant females did not measure high on other measures of attractiveness even though they showed the highest levels of proceptivity. Males tend to avoid mating with females with small offspring, who are presumably not ovulating (Fox 2002; Knott 1997). Recent data from chimpanzees (Emery Thompson 2005) indicate that high-status males initiate more copulations than expected during the days of a female chimpanzee's estrous swelling when she is most likely to conceive. The mechanisms by which such a signal could be conveyed have yet to be determined, although the possibility of olfactory or pheromonal cues have been suggested. Male orangutans (and chimpanzees as well) may sometimes choose to ignore these signals and gamble that mating could lead to conception. Thus, nonpreferred males may attempt to mate with females whenever they can, and preferred males may attempt non-POP mating when there is increased male-male competition. In the Nadler (1982) experiment, the male who was restricted from accessing the female may have mated with her whenever given the opportunity, even during non-POP, because he had no expectation that he would have access to her during her POP. When the female controlled access, the captive male behaved as do wild males, adopting a "sit and wait" strategy in which the female approaches when she is ovulating.

These findings of clear female preference raise the question of why flanged, prime males are most often the preferred fathers and whether there is really a cost to mating with a nonpreferred male. I believe that there are significant reasons to suspect that resident, flanged, prime males may indeed be higher quality sires. As I argue in Knott (in review), the prime male phase is significantly more energetically expensive than the unflanged male stage. In order to attain flanged status males need to be in excellent energetic condition, which may explain, at least in part, the highly variable timing of the initiation of male flange development (Knott in review). Thus, females may choose males who are flanged as an honest indicator of genetic quality, as they were able to attain flanged status and successfully warded off other flanged males (van Schaik 2004). Furthermore, mating frequently with dominant flanged males may increase the male's paternity confidence and thus the likelihood that he will protect a female's infant from infanticide (van Schaik 2004). In addition, Fox (2002) has shown that flanged males can provide females with protection from mating harassment by nonpreferred males.

Summary of Costs

The varied responses of female orangutans to male mating attempts make sense if the cost of mating cooperatively is normally higher than the cost of resisting. These mating costs include energetic loss, disease risk, and loss of protection from a dominant male. Thus, even with preferred males, females may resist mating when they are not ovulating. In some cases, they may even resist repeated mating attempts with a preferred male during POP if they have already mated that day. During POP, mating carries the potentially important cost of insemination by a nonpreferred male, and mating attempts by such males are fiercely resisted. However, females do sometimes mate with nonpreferred males. They may do so because the male dominance hierarchy is unstable, as in Utami's (2000) study at Ketambe, or mating may be infrequent and thus the cost of an individual mating may be low, such as at Gunung Palung, or they may occasionally mate cooperatively as a paternity confusion/anti-infanticide strategy. This may be particularly important when females become pregnant—a likely explanation for the high rate of proceptivity observed in newly pregnant females (Stumpf et al. 2008).

Orangutan Sexual Coercion in Comparative Perspective

The preceding data show that despite having forced copulations, matings in orangutans do not generally involve wounding. Thus, although the term *forced copulation* conjures up the ultimate violence against a female, the actual physical wounding experienced by female orangutans during these copulations is very low. Rather, the degree of force is a direct response to female resistance (Fox 1998; Knott et al. in prep). This observation is in stark contrast to other species that use much more physical force, particularly in indirect coercion, which can be much more physically injurious than direct coercion. Two closely related species in which indirect sexual coercion is very common and often very injurious are chimpanzees and humans.

Comparison to Chimpanzees

Chimpanzee sexual coercion is dramatically different from orangutan sexual co-ercion in two major ways: the context of the coercion and the level of violence involved. Chimpanzees very rarely have forced copulations. Instead, coercion is normally in the context of aggression toward estrous females in nonmating con-texts. This type of sexual coercion is best described as indirect, compared to the direct coercion seen in orangutans. Chimpanzee coercion involves control of female sexuality and an attempt to influence future female behavior. There is no evidence of this behavior in orangutans. Furthermore, wounding as a result of chimpanzee sexual coercion regularly occurs at some sites (Goodall 1986; Kahlenberg 2006). By contrast, such injuries have never been reported in wild orangutans.

Comparison to Humans

Given the features of orangutan sexual coercion we have outlined, the analogy between this and many forms of human rape warrants discussion. Emery Thompson (Chapter 14 in this volume) makes the useful distinction between stranger and acquaintance rape in humans. In humans, stranger rape results in the most serious wounding due to the violent resistance of the victim and the criminal intent to do harm or the pathological mental status of the perpetrator. By contrast, males who rape known victims rarely inflict injuries, and victims report that the use of physical aggression is secondary to verbal coercion (Emery Thompson, Chapter 14).

Because of the low levels of wounding, forced copulation in orangutans is not analogous to human stranger rape. Although the level of aggression used during forced copulations in orangutans may be similar to that seen with known partners in humans, there are considerable differences. Because human males sometimes do use extreme physical aggression during forced copulations, the threat of force is a primary reason that females may acquiesce to unwanted mat-ing efforts by males. Based on analogy with humans, it has been suggested that orangutan females may also acquiesce during matings in order to avoid injury, particularly from the much larger flanged males. However, since wounding of females during forced copulations has not been reported in orangutans, females may not have the same level of fear of injury from males as female humans do. The level of violence utilized by orangutan males may be sufficient to force a copulation without wounding, as it is in many cases of human rape. Thus, one of the major differences is that in humans there can be intentional infliction of injury and severe wounding. Human males can also use these forms of extreme force against females as a means of punishment for real or imagined sexual transgressions, and as a way to control female sexual behavior, both within and outside of a mating context. This type of sexual coercion is not seen in orangutans.

Sexual Coercion and Social Relationships: Why Are Orangutans Different?

What leads to these differences in the context of sexual coercion and the level of aggression involved? The most obvious answer is differences in the social system of these species. The above contrast leads to the hypothesis that frequent contact between a female and male within the context of a multimale mating system, where a female could potentially mate with multiple males, is more likely to lead to indirect and more violent coercion. How well do these predictions fit the data? Chimpanzees and humans are clear examples of species where there is injurious male coercion within the context of a multimale mating or social system. Excessive violence and nonmating aggression in these species are associated mostly with control of female sexuality and behavior beyond achieving copulation. Intriguingly, bonobos are an obvious exception to this generalization.

In contrast, a more dispersed social system, such as in orangutans, leads to direct coercion within the mating context but not the excessive use of force. Paradoxically, more solitary species have increased vulnerability to sexual coercion, as predicted by Smuts and Smuts (1993), because they do not have the protection of other conspecifics. However, I would argue that this coercion is of a different nature than that experienced by females living within multimale communities. In many ways, orangutans are more similar to many nonprimates in the type of sexual coercion observed. Similar accounts of forced copulations have been noted for bighorn sheep, which, interestingly, may also display alternative male morphs (Smuts and Smuts 1993).

Recommendations for Future Studies

The review presented in this chapter suggests that future work on orangutan sexual coercion should (1) use endocrinological data to determine ovulatory status, as this is one of the most critical determinants of female behavior, (2) quantify the components of each mating interaction in order to look at the degree of resistance, aggression, prosexuality, and female attractiveness, (3) quantify the costs of mating and the cost of successful and unsuccessful resistance, and (4) examine more closely the extent of injuries received by female orangutans.

Comparative studies of sexual coercion in animals should be oriented toward further testing the following hypotheses about the context for sexual coercion and the degree of force that is used. *Hypothesis 1:* Indirect sexual coercion leads to more injurious attacks on females than does direct sexual coercion. *Hypothesis 2:* Where the cost of resistance is high and the benefits of escaping mating are low, convenience polyandry results. Where the cost of mating is high, particularly when high male density produces an increased number of matting attempts, female resistance should increase. *Hypothesis 3:* As the degree of force increases such that the potential costs of resisting increase beyond the benefits of escaping, female resistance will decline in the form of convenience polyandry.

Conclusions

Rather than only being a strategy of unflanged males, forced copulations in orangutans should be seen as simply a *male* strategy—one that can be used by either flanged or unflanged males to overcome female resistance. Whether a copulation is forced or consensual depends on the level of resistance exerted by the female, which is in turn dependent on whether or not the male is preferred, the female's ovulatory status, and the costs of mating vs. resisting. The costs of mating in orangutans include energetic costs, disease risk, loss of protection from a dominant male, and the cost of insemination by a nonpreferred male, whereas the risk of injury from resistance appears low.

Concealed ovulation has confounded our ability to understand the context of sexual coercion in orangutans, but recent endocrinological analyses have helped to illuminate male and female strategies. Females prefer to mate with prime, flanged males during their periovulatory period. However, because of the potential costs of mating, females may sometimes choose to resist matings from preferred males as well as nonpreferred males. Females may actually resist mating with a flanged male when the mating attempt occurs outside of her ovulatory period or when multiple matings are sought on the same day. Similarly, whereas females normally resist mating with nonpreferred males, they sometimes do choose to mate cooperatively with these males during periods when they are not ovulating (pregnancy and non-POP periods). Because of this context, it is suggested that this represents an anti-infanticide strategy of confusing male paternity. In addition, an occasional mating with a nonpreferred male during a nonovulatory period may be sufficiently low cost that it is sometimes not avoided.

Orangutan males use aggression to the extent that it is necessary to achieve copulation. However, this level of aggression has not been reported to result in physical injuries to the female—thus the aggression is not "excessive." This finding is surprising given the characterization of orangutan forced copulations as one of the most extreme forms of male coercion in animals. In contrast to chimpanzees and humans, orangutan males do not use coercion to control female sexuality and future sexual behavior. Thus, Estep and Bruce's (1981) concept of *resisted mating* may be a more appropriate term for orangutans—reflecting the notion that the degree of resistance during the mating is primarily a function of female behavior.

Finally, the comparative data between orangutans and chimpanzees is revealing as to the conditions that prompt direct (as in orangutans) vs. indirect (as in chimpanzees) sexual coercion. Indirect coercion is most commonly a strategy of males living in multimale communities as a way to control future female sexual behavior, whereas direct coercion is a strategy of males living in dispersed male communities as a means of achieving an immediate copulation in the event of female resistance, such as is the case with many birds and insects. The level of force in indirect coercion is greater and often involves wounding, whereas in direct coercion it reflects the level of female resistance. Clearly, the nature of orangutan sexual coercion has often been misunderstood and warrants continued reevaluation.

References

Altizer, S., C. L. Nunn, P. H. Thrall, J. L. Gittleman, J. Antonovics, A. A. Cunningham, A. P. Dobson, V. Ezenwa, K. E. Jones, A. B. Pederson, M. Poss, and J. R. C. Pulliam. "Social Organization and Parasite Risk in Mammals: Integrating Theory and Empirical Studies." *Annual Review of Ecology, Evolution and Systematics* 34 (2003): 517–547.

Atmoko, S. U., and J. A. R. A. M. van Hooff. "Alternative Male Reproductive Strategies: Male Bimaturism in Orangutans." In *Sexual Selections in Primates: New and Comparative Perspectives,* eds. P. M. Kappeler, and C. P. van Schaik, pp. 196–207. Cambridge: Cambridge University Press, 2004

Boehm, C. "Pacifying Interventions at Arnehm Zoo and Gombe." In *Chimpanzee Cultures,* eds. R. W. Wrangham, W. C. McGrew, F. B. M. de Waal, and P. G. Heltne, pp. 211–226. Cambridge, Mass.: Harvard University Press, 1994.

Chilvers, B. L., B. C. Robertson, I. S. Wilkinson, P. J. Duignan, and N. J. Gemmell. "Male Harassment of Female New Zealand Sea Lions, *Phocarctos hookeri:* Mortality, Injury, and Harassment Avoidance." *Canadian Journal of Zoology* 83 (2005): 642–648.

Clutton-Brock, T. H., and G. A. Parker. "Sexual Coercion in Animal Societies." *Animal Behaviour* 49 (1995): 1345–1365.

Crofoot, M. C., and C. D. Knott. "What We Do and Do Not Know about Orangutan Male Dimorphism." In *Great and Small Apes of the World Volume II: Orangutans and Gibbons,* eds. B. M. F. Galdikas, N. Briggs, L. K. Sheeran, and G. L. Shapiro. In press.

Delgado, R. A. "The Function of Adult Male Long Calls in Wild Orangutans *(Pongo pygmaeus)*." Ph.D. Dissertation, Duke University, 2003.

Delgado, R., and C. P. van Schaik. "The Behavioral Ecology and Conservation of the Orangutan *(Pongo pygmaeus):* A Tale of Two Islands." *Evolutionary Anthropology* 9 (2000): 201–218.

Emery Thompson, M. "Reproductive Endocrinology of Wild Female Chimpanzees *(Pan troglodytes schweinfurthii):* Methodological Considerations and the Role of Hormones in Sex and Conception." *American Journal of Primatology* 67 (2005): 137–158.

Estep, D. Q., and K. Bruce. "The Concept of Rape in Non-Humans—A Critique." *Animal Behaviour* 29 (1981): 1272–1273.

Fox, E. A. "The Function of Female Mate Choice in the Sumatran Orangutan *(Pongo pygameus abelii)*." Ph.D. Dissertation, Duke University, Durham, 1998.

———. "Female Tactics to Reduce Sexual Harassment in the Sumatran Orangutan *(Pongo pygmaeus abelii)*." *Behavioral Ecology and Sociobiology* 52 (2002): 93–101.

Galdikas, B. "Orangutan Adaptation at Tanjung Puting Reserve, Central Borneo." Ph.D. Dissertation, University of California, Los Angeles, 1978.

———. "Orangutan Adaptation at Tanjung Puting Reserve: Mating and Ecology." In *The Great Apes,* eds. D. L. Hamburg, and E. R. McCown, pp. 195–233. Menlo Park, Calif.: Benjamin/Cummings, 1979.

———. "Orangutan Reproduction in the Wild." In *Reproductive Biology of the Great Apes,* ed. C. E. Graham, pp. 281–300. New York: Academic Press, 1981.

———. "Subadult Male Orangutan Sociality and Reproductive Behavior at Tanjung Puting." *American Journal of Primatology* 8 (1985a): 87–99.

———. "Adult Male Sociality and Reproductive Tactics among Orangutans at Tanjung Puting." *Folia Primatologia* 45 (1985b): 9–24.

Galdikas, B. M. F., and J. W. Wood. "Birth Spacing Patterns in Humans and Apes." *American Journal of Physical Anthropology* 83 (1990): 185–191.

Goodall, J. *The Chimpanzees of Gombe: Patterns of Behavior.* Cambridge, Mass.: Harvard University Press, 1986.

Hrdy, S. B. *The Woman that Never Evolved.* Cambridge, Mass.: Harvard University Press, 1981.

Kahlenberg, S. "Female-Female Competition and Male Sexual Coercion in Kanywara Chimpanzees." Ph.D. Dissertation, Harvard University, Cambridge, Mass., 2006.

Knott, C. "Monitoring Health Status of Wild Orangutans through Field Analysis of Urine." *American Journal of Physical Anthropology Supplement* 22 (1996): 139–140.

———. "Interactions between Energy Balance, Hormonal Patterns and Mating Behavior in Wild Bornean Orangutans *(Pongo pygmaeus)*." *American Journal of Primatology* 42 (1997): 124.

———. "Changes in Orangutan Caloric Intake, Energy Balance, and Ketones in Response to Fluctuating Fruit Availability." *International Journal of Primatology* 19 (1998a): 1061–1079.

———. "Orangutans in the Wild." *National Geographic Magazine.* (1998b) August: 30–57.

———. "Reproductive, Physiological and Behavioral Responses of Orangutans in Borneo to Fluctuations in Food Availability." Ph.D. Dissertation, Harvard University, Cambridge, Mass., 1999.

———. "Female Reproductive Ecology of the Apes: Implications for Human Evolution." In *Reproductive Ecology and Human Evolution,* ed. P. Ellison, pp. 429–463. New York: Aldine de Gruyter, 2001.

———. "Testosterone and Behavioral Differences in Bornean Orangutans *(Pongo pygmaeus pygmaeus):* Past Prime Males and the Evolution of Male Bi-maturism." In prep.

Knott, C. D., M. Emery Thompson, R. Stumpf, S. M. Kahlenberg, and M. McIntyre. "Sexual Coercion and Mating Strategies of Wild Bornean Orangutans." In review.

Knott, C. D., M. Emery Thompson, and S. A. Wich. "The Ecology of Reproduction in Wild Orangutans." In *Orangutans Compared: Ecology, Evolution, Behaviour and Conservation,* eds. S. A. Wich, S. S. Utami, T. Mitra Setia, and C. van Schaik, pp. 171–188. Oxford: Oxford University Press, 2009.

Knott, C. D., and S. M. Kahlenberg. "Orangutans in Perspective: Forced Copulations and Female Mating Resistance." In *Primates in Perspective,* eds. C. J. Campbell, A. Fuentes, K. C. MacKinnon, M. Panger, and S. Bearder, pp. 290–305. Oxford: Oxford University Press, 2007.

Low, M. "Female Resistance and Male Force: Context and Patterns of Copulation in the New Zeland Stitchbird *Notiomystis cincta.*" *Journal of Avian Biology* 36 (2005): 436–448.

MacKinnon, J. R. "The Behaviour and Ecology of Wild Orang-utans *(Pongo pygmaeus)*." *Animal Behaviour* 22 (1974): 3–74.

Maggioncalda, A. N., N. M. Czekala, and R. M. Sapolsky. "Growth Hormone and Thyroid Stimulating Hormone Concentrations in Captive Male Orangutans: Implications for Understanding Developmental Arrest." *American Journal of Primatology* 50 (2000): 67–76.

Maggioncalda, A. N., R. M. Sapolsky, and N. M. Czekala. "Reproductive Hormone Profiles in Captive Male Orangutans: Implications for Understanding Developmental Arrest." *American Journal of Physical Anthropology* 109 (1999): 19–32.

Markham, R. J., and C. P. Groves. "Brief Communication: Weights of Wild Orang-utans." *American Journal of Physical Anthropology* 81 (1990): 1–3.

Mitani, J. "Mating Behaviour of Male Orangutans in the Kutai Game Reserve, Indonesia." *Animal Behaviour* 33 (1985a): 391–402.

———. "Sexual Selection and Adult Male Orangutan Long Calls." *Animal Behaviour* 33 (1985b): 272–283.

Moss, C. J. "Oestrous Behavior and Female Choice in the African Elephant." *Behaviour* 86 (1983): 167–195.

Muller, M. N., S. M. Kahlenberg, M. Emery Thompson, and R. W. Wrangham. "Male Coercion and the Costs of Promiscuous Mating from Female Chimpanzees." *Proceedings of the Royal Society B* 274 (2007): 1009–1014.

Nadler, R. D. "Laboratory Research on Sexual Behavior and Reproduction of Gorillas and Orangutans." *American Journal of Primatology Supplement* 1 (1982): 57–66.

Nunn, C. L., J. L. Gittleman, and J. Antonovics. "Promiscuity and the Primate Immune System." *Science* 290 (2000): 1168–1170.

Rijksen, H. D. *"A Field Study on Sumatran Orang-utans (*Pongo pygmaeus abelii, *Lesson 1827): Ecology, Behaviour, and Conservation."* Ph.D. Dissertation, Wageningen University, The Netherlands, 1978.

Rivera, A. C., and J. A. Andrés. "Male Coercion and Convenience Polyandry in a Calopterygid Damselfly." *Journal of Insect Science* 2 (2002): 14.

Rodman, P. S. "Feeding Behavior of Orangutans in the Kutai Reserve, East Kalimantan." In *Primate Ecology,* ed. T. H. Clutton-Brock, pp. 383–413. London: Academic Press, 1977.

Schurmann, C. L. "Mating Behaviour of Wild Orangutans." In *The Orang utan: Its Biology and Conservation,* ed. L. de Boer, pp. 269–284. The Hague: Dr. W. Junk Publishers, 1982.

———. "Courtship and Mating Behavior of Wild Orang-utans in Sumatra." In *Primate Behavior and Sociobiology,* eds. A. B. Chiarelli, and R. S. Corruccini, pp. 130–135. Berlin: Springer-Verlag, 1981.

Schurmann, C. L., and J. A. R. A. M. van Hooff. "Reproductive Strategies of the Orang-utan: New Data and a Reconsideration of Existing Sociosexual Models." *International Journal of Primatology* 7 (1986): 265–287.

Scott, E. M., J. Mann, J. J. Watson-Capps, B. L. Sargeant, and R. C. Connor. "Aggression in Bottlenose Dolphins: Evidence for Sexual Coercion, Male-Male Competition, and Female Tolerance through Analysis of Tooth-Rake Marks and Behavior." *Behaviour* 142 (2005): 21–44.

Singleton, I. S., and C. P. van Schaik. "The Social Organisation of a Population of Sumatran Orang-utans." *Folia Primatologia* 73 (2002): 1–20.

Smuts, B. B., and R. W. Smuts. "Male Aggression and Sexual Coercion of Females in Nonhuman Primates and Other Mammals: Evidence and Theoretical Implications." *Advances in the Study of Behavior* 22 (1993): 1–63.

Stumpf, R. M., M. Emery Thompson, and C. D. Knott. "A Comparison of Female Chimpanzee and Orangutan Mating Strategies." *International Journal of Primatology* 29 (2008): 865–884.

Szykman, M., A. L. Engh, R. C. Van Horn, E. E. Boydston, K. T. Scribner, and K. E. Holekamp. "Rare Male Aggression Directed toward Females in a Female-Dominated Society: Baiting Behavior in the Spotted Hyena." *Aggressive Behavior* 29 (2003): 457–474.

Utami Atmoko, S. S. "Bimaturism in Orang-utan Males: Reproductive and Ecological Strategies." Ph.D. Dissertation, Utrecht University, 2000.

Utami, S. S. G., B. Goossens, M. W. Bruford, J. R. de Ruiter, and J. A. van Hooff. "Male Bimaturism and Reproductive Success in Sumatran Orang-utans." *Behavioural Ecology* 13 (2002): 643–652.

van Schaik, C. P., and P. M. Kappeler. "Infanticide Risk and the Evolution of Male-Female Association in Primates." *Proceedings of the Royal Society of London B* 264 (1997): 1687–1694.

van Schaik, C. P., M. A. van Noordwijk, and C. L. Nunn. "Sex and Social Evolution in Primates." In *Comparative Primate Socioecology,* ed. P. C. Lee, pp. 204–240. Cambridge: Cambridge University Press, 1999.

van Schaik, C. P. *Among Orangutans.* Cambridge, Mass.: Harvard University Press,, 2004.

Watts, D. P. "Effects of Male Group Size, Parity, and Cycle Stage on Female Chimpanzee Copulation Rates at Ngogo, Kibale National Park, Uganda." *Primates* 48 (2007): 222–231.

Wich, S. A., M. L. Geurts, T. Mitra Setia, and S. S. Utami-Atmoko. "Influence of Fruit Availability on Sumatran Orangutan Sociality and Reproduction." In *Feeding Ecology in Apes and Other Primates,* eds. G. Hohmann, M. Robbins, and C. Boesch, pp. 337–358. Cambridge: Cambridge University Press, 2006

Wich, S. A., S. S. Utami-Atmoko, T. M. Setia, H. D. Rijksen, C. Schurmann, J. A. R. A. M. van Hooff, and C. P. van Schaik. "Life History of Wild Sumatran Orangutans *(Pongo abelii)."* *Journal of Human Evolution* 47 (2004): 385–398.

Zucker, E. L. "Control of Intragroup Aggression by a Captive Male Orangutan." *Zoo Biology* 6 (1987): 219–223.

5

Male Aggression against Females in Mountain Gorillas: Courtship or Coercion?

Martha M. Robbins

Adult male gorillas weigh approximately 200 kg and are roughly twice the size of adult females (see Figure 5.1). Biologists frequently ask why such dimorphism evolved and, in such circumstances, how it influences the behavioral patterns that are used by both males and females to pursue their reproductive strategies. This extreme sexual dimorphism in gorillas enables males to easily exert dominance over females, and therefore they are a likely candidate for sexual coercion. The predominantly one-male or harem social system of gorillas is believed to be an outcome of sexually selected factors. Male-male competition for access to mates leads to high variability and skew in access to mates. Because males may commit infanticide, an extreme form of sexual coercion (Chapters 1, 3 in this volume), the threat of infanticide is believed to lead females to seek protection from the putative father of their offspring against other males, resulting in long-term social relationships among males and females (Watts 1989; van Schaik and Janson 2000; Harcourt and Greenburg 2001). As a result of the threat of infanticide, males' ability to protect females is likely to be one of the key factors influencing female choice. Therefore, males can be viewed as both aggressors and protectors of females because the same behaviors that indicate an opposing male's fighting ability that a female may wish to avoid serve as a measure of her own male's protective abilities. Similarly, such behavior serves the dual interests of the males: to attract females while at the same time protecting mates and dissuading them from seeking another male. The goal of this chapter is to investigate the paradox of males being both protectors and aggressors by examining male aggressive behavior toward females within social groups.

Figure 5.1 Adult male gorillas are roughly twice the size of females and physically dominant to them. Photo by Martha Robbins.

Females may transfer between social groups multiple times in their lives, giving females the opportunity to choose among different males and at the same time leading to male strategies to prevent females from leaving them (herding behavior: Sicotte 1993). Female dispersal occurs only during intergroup encounters, so these are crucial times for males to thwart opponents, attract new females, and retain current mates. During intergroup encounters, silverbacks perform displays that consist of standing, strutting, and running with a stiff-legged and rigid body posture that may culminate in chest-beating (Schaller 1963). These displays are presumably important not only among the males as a measure of their relative fighting ability, but also as a way for females to assess male qualities, such as size, strength, and protective ability. In some cases, male-male aggressive displays escalate into physical fights, which may result in mortality, dispersal of females to the male victor, and infanticide of unweaned offspring of the defeated opponent (Watts 1989; Robbins 2003; Cipolletta 2006). However, in most cases female transfer is voluntary (Sicotte 2001; Robbins et al. 2009).

Although nearly all groups of western gorillas *(Gorilla gorilla gorilla)* contain only one adult male (silverback), as many as 50% of mountain gorilla groups *(Gorilla beringei beringei)* contain two or more silverbacks. The occurrence of both one-male groups and multimale groups in mountain gorillas can lead to differences in behavioral strategies for both sexes, which in turn contribute to the maintenance of this variable social system. Males may either queue for dominance position within multimale groups, or disperse upon maturity and attempt to attract females for a new social unit (Robbins 1995; Watts 2000; Robbins and Robbins 2005). Male dispersal decisions are likely to be influenced by the distribution of females in social groups and by the extent of mating opportunities within their natal group. Dominant males sire approximately 85% of offspring in multimale groups (Bradley et al. 2005; Nsubuga et al., 2007), and those groups exhibit an increase in male-male aggression on days when females are in estrus (versus days they are not; Robbins 2003).

Given the intense competition among males for mates, and the fact that mountain gorilla groups may contain more than one silverback, within-group mating strategies may be as important as between-group strategies (Robbins 1999, 2001, 2003). Within social groups, the relationships that exist among males and females are considered to be "stronger" than those among males and among females, based on patterns of spatial proximity and affiliation (Watts 1996). However, such male-female relationships are characterized not only by affiliative behavior such as grooming, but also by aggressive behavior. That aggression often includes the same strutting displays and chest-beating that males use in contests among themselves.

Why do males exhibit such aggressive behavior toward females in their groups? Although some male aggression directed at females may occur in the context of feeding competition and interventions in female conflicts, Watts (1992) estimated that 60 to 78% of such aggression was linked to male mating tactics. Similarly, in a study of Bwindi gorillas, whereas most aggression among females occurred during feeding, 70% of aggression from males to females occurred outside of the feeding context (Robbins 2008). This distinction likely represents further evidence that such aggression reflects male reproductive strategies: both male policing of female-female aggression that would serve to reduce strong female dominance relations and/or female dispersal (Chapter 1 in this volume; Watts et al. 2000) as well as potential sexual coercion.

If such aggression is a mating strategy, however, then it should intensify in reproductive contexts, such as with newly immigrant females, when females are in estrus, or when more than one male is present (Smuts and Smuts 1993; Muller et al. 2007; Chapter 1 in this volume). In both one-male and multimale groups, recent immigrants received higher rates of aggression than long-term residents, suggesting that such aggression may be a combination of courtship and coercion as males strive to develop relationships with these new females (Harcourt 1979; Watts 1992; Sicotte 2000). Harcourt (1979) observed a *lower* frequency of aggression by males in proximity to estrous females than anestrous females. His observations were limited to one-male groups, however, where the silverback may not need to increase aggression toward females when they are in estrus because they have no choice in mates. In observations of two multimale groups, Sicotte (2002) reported that displays did not comprise a higher proportion of the male aggression toward estrous versus anestrous females, but she did not compare the overall rates of aggression. Robbins (2003) found higher overall rates of aggression (including displays) toward estrous females by most males in multimale groups. In cases where a male harassed a mating pair, he often would attack the female, not the male (Robbins 1999). All of those studies focused on mountain gorillas studied at the Karisoke Research Center in the Virunga Volcanoes. Given the mixed results of previous studies, which may partially reflect small sample sizes, there is the need for additional study, particularly focusing on the reproductive state of the female and the number of males in a group. Also needed are studies from other gorilla populations.

The aim of this study is to investigate within-group male aggressive behavior toward females in another population of mountain gorillas, in Bwindi Impenetrable National Park, Uganda. The population of Bwindi gorillas contains approximately 50% multimale groups (McNeilage et al. 2006). During the 66 months of this study, the one research group was initially a multimale group, but the expulsion of the older, dominant male led to it being a one-male group for the last 17 months. I examined whether the rate of aggression varied when the group type changed, and whether it depended on the reproductive state of the female. I also focused on whether male aggressive behavior as a reproductive strategy is more likely to be considered coercion or courtship.

Methods

Study Group

This study was conducted in Bwindi Impenetrable National Park (331 km^2), located in the southwest corner of Uganda, on one group of fully habituated mountain gorillas, the "Kyagurilo Group," on 672 days between June 2000 and December 2005. The group consisted of 14 to 16 individuals during the course of the study.

The age/sex categories used followed Watts and Pusey (1993). At the onset of the study, the group consisted of 14 gorillas: two silverback/adult males, five adult females, one subadult/nulliparous female, four juveniles, and two infants. During the course of the study, several demographic changes occurred. Most important to this study, the younger silverback (RC) emigrated in November 2000 but rejoined the group in March 2001. The initially dominant silverback (ZS) was deposed by RC and abandoned by the group in July 2004. Therefore, the group transitioned through the following conditions in regards to silverbacks (Figure 5.2): ZS as dominant and RC as subordinate, ZS as dominant in a one-male group, ZS as dominant and RC as subordinate, and RC as dominant in a one-male group. Behavioral data were collected during only 17 days of the five months that ZS was dominant in a one-male group (but the group was monitored daily), so that time period is excluded from analysis. All immature gorillas advanced to older age classes, and there were four births: one to the nulliparous female when ZS was dominant (sired by RC; Nsubuga et al. 2008) and three when RC was dominant (one died at 3 months of age). Genetic analysis has shown that none of the adult females and silverbacks was close kin (Nsubuga 2005).

Four of the adult females went through periods of pregnancy, lactation, and cycling during the study (Figure 5.2). The other two adult females mated very

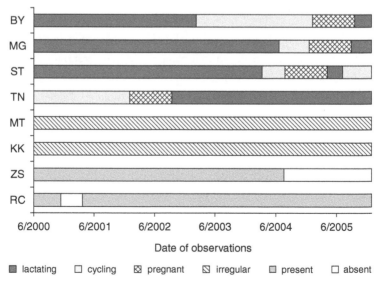

Figure 5.2 Timeline of the reproductive status of each female (BY, MG, ST, TN, MT, and KK), and the presence/absence of both silverbacks (ZS and RC).

infrequently and had no offspring during the course of the study; they are therefore not included in the analysis. Females were classified according to reproductive status as cycling, pregnant, or lactating based on the following information: Gestation length was approximately 255 days (Harcourt et al. 1980) such that the approximate date of conception and time of pregnancy could be calculated by backdating from birthdates. Females then went through a lactational anestrous period for approximately three years while lactating (Stewart 1988). Matings occurred during the one- to two-day estrous periods of the 28-day cycle as well as at irregular intervals during pregnancy (Harcourt et al. 1980; Watts 1991). Mating behavior has been correlated to time of ovulation through hormonal analysis (Czekala and Sicotte 2000). I considered any day when at least one mating was observed to be a "mating day," and I compared the behavioral patterns on these days to those on "nonmating days" during periods when females were cycling and lactating (Sicotte 1994; Robbins 1999, 2003).

Data Collection

The gorillas were observed for an average of four hours per day (range two to nine hours). Data were collected primarily using focal animal sampling (Altmann 1974). Focal observation periods were for a minimum of 15 minutes and a maximum of 60 minutes, because the dense vegetation of Bwindi made it difficult to follow focal animals for longer periods of time. Choice of focal animal was based on giving priority to the individuals with the least amount of focal time, with the aim of collecting equal focal observation on all eight adult individuals. A total of 1468 hours of focal observation time were obtained (mean = 183.4 hours, range 131–196 hours per individual).

Agonistic interactions included displacements (approach-retreat interactions), submissive behavior, and aggression, but only aggression is considered here (see Robbins 2008 for information on dominance and submission behavior). Aggressive interactions were classified into three categories based on increasing intensity of behavior, following Robbins (1996): mild aggression, which included cough grunting and screaming (agonistic signals in Watts 1994); moderate aggression, which included any part of the chest-beating or strut-walking sequence typically given by silverbacks (Schaller 1963) as well as lunges toward the recipient; and high aggression, which included any behavior involving physical contact including hits, bites, kicks, and attacks.

Calculations

I calculated a rate of aggression for each male-female dyad on each day by summing up the number of aggressive acts by the silverback toward the female and dividing by the combined number of focal hours for those two gorillas on that day. For each dyad, I then used the Mann Whitney U test to compare the mating versus nonmating days. Owing to small sample sizes, daily rates for all females were combined in the comparisons for ZS versus RC. Rate-based chi-square tests were used to compare pooled data for lactating versus nonlactating females (Altmann 1977).

Results

The majority of aggression showed by both silverbacks was moderate, indicating that displays were the most commonly used form of aggression. Thirty-two percent of aggression by ZS was mild; 40% was moderate; and 28% was high. Eleven percent of aggression by RC was mild; 66% was moderate; and 23% was high.

Overall, RC exhibited higher rates of aggression toward females that were cycling or pregnant than to lactating females (Figure 5.3, chi-square = 14.54, p = 0.001), whereas for ZS there was only a trend of higher aggression toward cycling and pregnant females (chi-square = 2.87, p = 0.09).

Comparisons of aggression showed by males to females on days they mated vs. nonmating days (while females were cycling or pregnant) were done on a dyadic basis. Rates of aggression given by ZS were higher on mating days for two of the four females (Figure 5.4a; n_1 = nonmating days, n_2 = mating days ZS – BY, n_1 = 156, n_2 = 4, U = 316, p = 0.82; ZS – MG, n_1 = 18, n_2 = 1, U = 10, p = 0.732; ZS – ST, n_1 = 54, n_2 = 7, U = 124, p = 0.012; ZS –TN, n_1 = 136, n_2 = 6, U = 278, p = 0.001). RC did not direct any aggression toward these females on the days that ZS mated with each of them. For RC, rates of aggression were higher on mating days for all four females (Figure 5.4b; RC – BY, n_1 = 319, n_2 = 2, U = 173, p = 0.0001; RC – MG, n_1 = 162, n_2 = 12, U = 561, p = 0.001; RC – ST, n_1 = 238, n_2 = 15, U = 1357, p = 0.001; RC – TN, n_1 = 84, n_2 = 15, U = 286, p = 0.001). RC was subordinate for all days that TN was observed mating, and he was dominant for all the days that he mated with the other three females. ZS was only present in the group on the days RC mated with TN, and only on one of those mating days did he direct any aggression toward her.

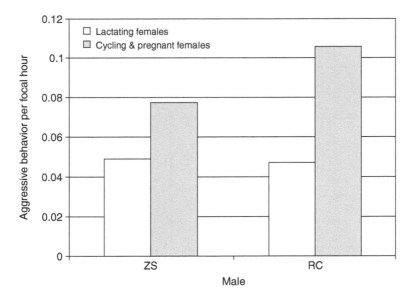

Figure 5.3 Aggression rates of each silverback toward lactating females, versus their rates toward females who were cycling or pregnant.

To determine whether there is a difference in aggression given by alpha males in one-male groups vs. multimale groups, comparisons were made of aggression given by ZS and RC when each was dominant. There was no difference in aggression directed at females who were cycling and pregnant (ZS=0.079 acts per dyad focal hour, RC=0.099 acts per dyad focal hour, n_1=506, n_2=382, U=95022, p=0.274), nor in the rate of aggression given by each male to estrous females on mating days only ((ZS=0.732 acts per focal hour, RC=0.537 acts per focal hour, n_1=17 , n_2=29, U=236 , p=0.779).

The rate of aggression toward pregnant and cycling females was 0.148 acts per dyad-hour during the 507 hours when the group was multimale, which is not significantly higher than the rate of 0.113 during the 415 hours when it was one-male (chi-square=2.0, df=1, p=0.15). However, the multimale group always contained two silverbacks, so its rate per *female*-hour was double the rate per *dyad*-hour, whereas the two values were the same in one-male groups. The rate of 0.296 acts per female-hour in multimale groups is significantly higher than the rate of 0.113 acts per female-hour in one-male groups (chi-square=28.7, df=1, p<0.001). Thus, although males were not significantly more aggressive when the group was multimale, females received more aggression because they were with more silverbacks.

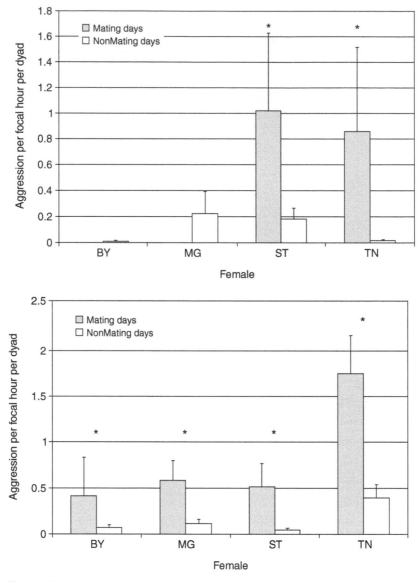

Figure 5.4 Rates of aggression by each silverback toward females on days each male-female dyad mated versus days the dyad did not mate (does not include when they were lactating). (Top) ZS and (bottom) RC.

Discussion

This study shows that silverback males were more aggressive toward females who were cycling and pregnant than toward lactating females (for chacma baboons see also Chapter 6, and for hamadryas baboons, see Chapter 10 in this volume). Furthermore, males were more aggressive toward females on days they mated with them than on days when they did not mate. These results are consistent with expectations that the aggression is related to male reproductive strategies. The majority of the aggression given is moderate, being predominantly components of the chest-beating displays (Schaller 1963) and rarely leading to injuries or wounds. It was expected that aggression by the dominant male would be higher in the group when it was multimale than one-male, but surprisingly, no difference in aggression by the two dominant males was observed. These results lead to two main questions: Why were similar levels of aggression by each dominant male observed when the group was both one-male and multimale? Second, is this aggressive behavior coercive, or is it simply courtship behavior?

Aggression in One-Male vs. Multimale Groups

When the group was multimale, females received more aggression than when it was one-male, but this was only because they were receiving aggression from two males rather than one and not because of an increase in aggression by the dominant male. No difference was found between the aggression rates when ZS was dominant in a multimale group versus when RC was dominant in a one-male group. Those results remained consistent even after adjusting for the females' reproductive status. Males in all groups might use aggression toward females for several reasons, but in multimale groups, silverbacks might use additional aggression to discourage them from mating with other males within the group (Chapter 1) or as an advertisement to females and other males in the group of his quality (Kitchen et al., Chapter 6 in this volume).

This comparison of group type should be viewed with some caution, not only because of the small sample size, but also because of potentially confounding factors in the analysis. First, when RC was subordinate, the nulliparous female (TN) was the only female he mated with and the only female he showed any significant levels of aggression toward (and he sired her first offspring). This differs from observations at Karisoke, where three of out four subordinate males in two groups mated with several females and all increased levels of aggression toward these females on mating days (Robbins 2003). There also was no difference in the rate of aggression toward estrous females by the dominant and subordinate males in these two groups (Robbins 2003). Second, the comparison of when Kyagurilo Group was one-male vs. multimale involved different males at very different times in their reproductive life span: ZS was an older "retiring" male, while RC was forming social/dominance relationships with all females as a newly dominant male. If males are more aggressive at the beginning of their dominance tenure (hamadryas baboons: Swedell and Schreier, Chapter 10 in this volume), or if aggression rates vary from male to male throughout their lives, then those factors may obscure any effect of group type in our current study. Further study is needed to isolate the effects of each factor, by examining (for example) whether RC reduces his rate of aggression after his social relationships are more firmly established in a one-male group. Preliminary analysis of behavior after TN resumed cycling in May 2006 showed that there was no difference in the rate of aggression RC directed toward TN on mating days compared to when he was subordinate ($n_1 = 15$, $n_2 = 7$, $U = 55$, $p = 0.859$). In addition, herding behavior was more common in multimale groups than one-male groups in the Virunga population (Sicotte 1993), but not the Bwindi gorillas (Robbins and Sawyer 2007). All of this variability in behavior emphasizes that results may vary depending on the life history stages of both males and females involved, and the social relationships among them, in a particular study. This is unsurprising for research on a long-lived species for which few habituated groups have been observed.

Courtship or Coercion?

Hypothetically, one could propose that the aggressive displays are simply courtship behavior and not sexual coercion. Therefore, it is useful to consider the three components of the classic definition of sexual coercion by Smuts and Smuts (1993): "(1) the use by a male of force, or threat of force, that (2) functions to increase the chances that a female will mate with him at the time she is likely to be fertile, and to decrease the chances that she will mate with other males, (3) at some cost to the female."

First, while gorillas have only very rarely been observed to commit rape as do orangutans (Sicotte 2001; Knott, Chapter 4 in this volume), nor do they beat and injure females as intensively as have been observed in chimpanzees (Muller et al., Chapter 8 in this volume), given the large sexual dimorphism in gorillas, any aggressive behavior by silverbacks toward females can be deemed as force or as an intimidating threat of force. This is true even of behavior that does not involve physical contact. In contrast to gorillas, chimpanzees live in multimale-multifemale communities and males may exhibit strong, damaging aggression toward females. It is possible that because of the predominantly one-male grouping system and the more extreme sexual dimorphism of gorillas, they evolved to exhibit coercive (or courtship) behavior through displays rather than through actual physical aggression exhibited by chimpanzees. Chest-beating displays typically cause nearby gorillas (and researchers) to move out of the way, presumably to avoid getting caught in the cross fire, and females often show submissive behavior to silverbacks immediately after such displays, perhaps as a way to show subordination and to reduce the likelihood of more intense aggression (Sicotte 2002; Robbins personal observation).

Second, in the absence of experiments, it is difficult to determine whether aggression increases the likelihood that a female will mate with an aggressive male rather than other males. As Smuts and Smuts (1993) point out, we have no way to know what would have happened in the absence of aggressive/seemingly coercive behavior. However, in this study, only the male who mated with the female had higher rates of aggression toward her on that particular day, not the male who did not mate. If herding successfully prevents a female from transferring to another group, it will certainly decrease the chances that she will mate with the males in that group (because extragroup copulations are very rare).

Finally—and perhaps the part of the definition that poses the most difficulty—is determining whether females incur some cost as a result of the males' aggression. Wounds obviously impose costs on females, owing to the risk of infection and energy needed to heal, but they were extremely rare in this study. Interestingly, three of the adult females had bite wounds on their heads just prior to the male dominance turnover; one was the most severe bite wound that I have ever seen on an adult female gorilla (BY). Although it was unknown who caused these injuries, the fact that ZS also had severe bit wounds to his head suggests that RC was the aggressor, and his aim was to coerce the females into viewing him as the dominant male. The other two females who were cycling (ST and MG) at the time that RC became dominant also sustained bite wounds on the head (which are uncommon) within two months of the turnover. Few studies have shown costs incurred by females, although Muller et al. (2007) nicely show that females experience increased cortisol levels when sexually receptive, which could be tested in gorillas. Furthermore, the fact that females often show submissive behavior toward silverbacks (Watts 1994; personal observation) suggests that females seek to minimize aggressive behavior, even displays, directed toward them. It is unknown if male aggression toward females in multimale groups imposes a greater cost of limiting females' ability to exert mate choice (preference to mate with males other than the aggressive male). Even if females in multimale groups do not receive more aggression from dominant males than from subordinates, they may experience more aggression overall than those in one-male groups owing to more than one male being aggressive. If aggression poses a cost to the females, we would expect lower female reproductive success in multimale groups than in one-male groups, but there was no indication of this at Karisoke (Robbins et al. 2007). Thus, the typical cost of coercion may be negligible or offset by some other benefit, such as improved group stability and protection afforded by multimale groups.

Smuts and Smuts (1993) also point out that the functional significance of "ritualized" aggression during courtship is not well understood. Displays such as chest-beating in gorillas may actually serve to demonstrate a male's health and vigor and might thereby facilitate female choice rather than be coercive behavior, especially in the absence of proof for the latter two aspects of their definition of coercion (see also Kitchen et al., Chapter 6 in this volume). Although courtship behavior is any behavior that serves to entice mates, there is no explicit caveat as to whether this behavior would reduce the likelihood of females mating with other males or if it would incur a cost for the female. Therefore, although to be considered coercive a behavior should include these latter two components, such a behavior could still be considered as courtship. A major challenge to field studies examining sexual coercion is to show these latter two components of the Smuts and Smuts definition. Further studies of gorillas should also consider other behavior that may serve as courtship behavior, such as following and neighing behavior (Sicotte 1994), and whether mate guarding occurs.

In sum, we are left with the question of whether coercion is an integral part of courtship in gorillas. Silverbacks have evolved to use their size and strength to outcompete other males, yet males also use their size and strength to coerce/court females. Therefore, females may actually seek to choose the strongest, most protective males. Because males can cause serious aggressive damage to females as well as commit infanticide, females therefore may use/view males as protectors essentially as a result of coercive strategies. Displays of aggressive potential to females indicates not only that silverbacks can protect them and their offspring against competing males, but may also serve to indicate that aggression may escalate into more damaging behavior toward them. This same displaying behavior is used in male-male competition, which sometimes does result in physical aggression. Along these lines, Borgia and Coleman (2000) have shown that female bowerbirds show preferences for displays that are used in male-male competition but were latterly co-opted for courtship. Testing whether male courtship behavior evolved from male-male aggressive displays would require knowledge of the behavioral pattern of species ancestral to gorillas.

Acknowledgments

I thank the Uganda Wildlife Authority and the Uganda National Council of Science and Technology for their permission and long-term support of this research. The Institute of Tropical Forest Conservation (ITFC) and Alastair McNeilage have provided invaluable logistical support. Angela Higginson, Nick Park, and Sarah Sawyer worked tirelessly as research assistants and contributed to the behavioral observations. Thanks go to all the field assistants of ITFC for their dedicated work and data collection with the gorillas, especially Tibenda Emmanuel, Twinomujuni Gaad, Ngamganeza Caleb, Kyamuhangi Narsis, Byaruhanga Gervasio, Tumwesigye Philimon, and Murembe Erinerico. The manuscript benefited greatly from discussions with Christophe Boesch and especially Andrew Robbins. This project was funded by the Max Planck Society.

References

Altmann, J. "Observational Study of Behaviour: Sampling Methods." *Behaviour* 49 (1974): 227–265.

Altmann, S. A., and J. Altmann. "Analysis of Rates of Behaviour." *Animal Behaviour* 25 (1977): 364–372.

Borgia, G., and S. W. Coleman. "Co-option of Male Courtship Signals from Aggressive Display in Bowerbirds." *Proceedings of the Royal Society, B* 267 (2000): 1735–1740.

Bradley, B. J., M. M. Robbins, E. A. Williamson, H. D. Steklis, N. Gerald Steklis, N. Eckhardt, C. Boesch, and L. Vigilant. "Mountain Gorilla Tug-of-War: Silverbacks Have Limited Control over Reproduction in Multi-Male Groups." *Proceedings of the National Academy of Sciences, USA* 102 (2005): 9418–9423.

Czekala, N., and P. Sicotte "Reproductive Monitoring of Free-Ranging Female Mountain Gorillas by Urinary Hormone Analysis." *American Journal of Primatology* 51 (2000): 209–215.

Harcourt, A. H. "Social Relationships between Adult Male and Female Mountain Gorillas in the Wild." *Animal Behaviour* 27 (1979): 325–342.

Harcourt, A. H., D. Fossey, K. G. Stewart, and D. Watts. "Reproduction in Wild Gorillas and Some Comparisons with Chimpanzees." *Journal of Reproduction and Fertility* 28 (1980): 59–70.

Harcourt, A. H., and J. Greenberg. "Do Gorilla Females Join Males to Avoid Infanticide? A Quantitative Model." *Animal Behaviour* 62 (2001): 905–915.

McNeilage, A., M. M. Robbins, M. Gray, W. Olupot, D. Babaasa, R. Bitariho, A. Kasangaki, H. Rainer, S. Asuma, G. Mugiri, and J. Baker. "Census of the Mountain Gorilla Population in Bwindi Impenetrable National Park, Uganda." *Oryx* 40 (2006): 419–427.

Muller, M. N., S. M. Kahlenberg, M. Emery Thompson, and R. W. Wrangham. "Male Coercion and the Costs of Promiscuous Mating for Female Chimpanzees." *Proceedings of the Royal Society, B* 274 (2007): 1009–1014.

Nsubuga, A. M. "Genetic Analysis of the Social Structure in Wild Mountain Gorillas *(Gorilla beringei beringei)* of Bwindi Impenetrable National Park, Uganda." Ph.D. Dissertation, University of Leipzig, Germany, 2005.

Nsubuga, A. M., Robbins, M. M., Boesch, C., and Vigilant, L. "Patterns of Paternity and Group Fission in Wild Multimale Mountain Gorilla Groups." *American Journal of Physical Anthropology* 135 (2008): 263–274.

Robbins, A. M., and M. M. Robbins. "Fitness Consequences of Dispersal Decisions for Male Mountain Gorillas *(Gorilla beingei beringei)*." *Behavioral Ecology and Sociobiology* 58 (2005): 295–309.

Robbins, A. M., Stoinski, T., Fawcett, K., and Robbins, M. M. "Socioecological Influences upon the Dispersal Patterns of Female Mountain Gorillas—Evidence of a Second Folivore Paradox." Behavioural Ecology and Sociobiology 63 (2009): 477–489.

Robbins, M. M. "A Demographic Analysis of Male Life History and Social Structure of Mountain Gorillas." *Behaviour* 132 (1995): 21–47.

———. "Male-Male Interactions in Heterosexual and All-Male Wild Mountain Gorilla Groups." *Ethology* 102 (1996): 942–965.

———. "Male Mating Patterns in Wild Multimale Mountain Gorilla Groups." *Animal Behaviour* 57 (1999): 1013–1020.

———. "Variation in the Social System of Mountain Gorillas: The Male Perspective." In *Mountain Gorillas: Three Decades of Research at Karisoke,* eds. M. M. Robbins, P. Sicotte, and K. Stewart, pp. 29–58. Cambridge: Cambridge University Press, 2001.

———. "Behavioral Aspects of Sexual Selection in Mountain Gorillas." In *Sexual Selection and Reproductive Competition in Primates: New Perspectives and Directions,* ed. C. B. Jones, pp. 477–501. Norman, Okla: American Society of Primatologists, 2003.

———. "Feeding Competition and Female Social Relationships in Mountain Gorillas of Bwindi Impenetrable National Park, Uganda." *International Journal of Primatology* 29 (2008): 999–1018.

Robbins, M. M., A. M. Robbins, N. Gerald-Steklis, and H. D. Steklis. "Female Reproductive Success of the Virunga Mountain Gorillas: Comparisons with the Socioecological Model." *Behavioral Ecology and Sociobiology* 61 (2007): 919–931.

Robbins, M. M., and S. C. Sawyer. "Intergroup Encounters in Mountain Gorillas of Bwindi Impenetrable National Park, Uganda." *Behaviour* 144 (2007): 1497–1519.

Schaller, G. B. *The Mountain Gorilla: Ecology and Behavior.* Chicago: University of Chicago Press, 1963.

Sicotte, P. "Intergroup Encounters and Female Transfer in Mountain Gorillas—Influence of Group Composition on Male Behavior." *American Journal of Primatology* 30 (1993): 21–36.

———. "Effect of Male Competition on Male-Female Relationships in Bi-Male Groups of Mountain Gorillas." *Ethology* 97 (1994): 47–64.

———. "Female Mate Choice in Mountain Gorillas." In *Mountain Gorillas: Three Decades of Research at Karisoke,* eds. M. M. Robbins, P. Sicotte, and K. Stewart, pp. 59–88. Cambridge: Cambridge University Press, 2001.

———. "The Function of Male Aggressive Displays towards Females in Mountain Gorillas." *Primates* 43 (2002): 277–289.

Smuts B. B., and R. W. Smuts. "Male Aggression and Sexual Coercion of Females in Nonhuman Primates and Other Mammals: Evidence and Theoretical Implications." *Advances in the Study of Behavior* 22 (1993): 1–63.

Stewart, K. J. "Suckling and Lacatational Anoestrus in Wild Gorillas *(Gorilla gorilla).*" *Journal of Reproduction and Fertility* 83 (1988): 627–634.

van Schaik, C. P., and C. H. Janson, eds. *Infanticide by Males and Its Implications.* Cambridge: Cambridge University Press, 2000.

Watts, D. P. "Infanticide in Mountain Gorillas—New Cases and a Reconsideration of the Evidence." *Ethology* 81 (1989): 1–18.

———. "Mountain Gorilla Reproduction and Sexual Behavior." *American Journal of Primatology* 24 (1991): 211–225.

———. "Social Relationships of Immigrant and Resident Female Mountain Gorillas 1: Male-Female Relationships." *American Journal of Primatology* 28 (1992): 159–181.

———. "Agonistic Relationships between Female Mountain Gorillas *(Gorilla gorilla beringei).*" *Behavioral Ecology and Sociobiology* 34 (1994): 347–358.

———. "Comparative Socioecology of Gorillas." In *Great Ape Societies,* eds. W. C. McGrew, L. F. Marchant, and T. Nishida, pp. 16–28. Cambridge: Cambridge University Press, 1996.

———. "Causes and Consequences of Variation in Male Mountain Gorilla Life Histories and Group Membership." In *Primate Males,* ed. P. M. Kappeler, pp. 169–180. Cambridge: Cambridge University Press, 2000.

6

The Causes and Consequences of Male Aggression Directed at Female Chacma Baboons

Dawn M. Kitchen, Jacinta C. Beehner, Thore J. Bergman,
Dorothy L. Cheney, Catherine Crockford, Anne L. Engh,
Julia Fischer, Robert M. Seyfarth, and Roman M. Wittig

At least once in every 10 hours of observation, a female chacma baboon *(Papio hamadryas griseipes)* can be heard screaming as she runs from a hostile adult male. Why do fairly protracted attacks like this occur so frequently outside of a feeding context? To the casual observer, it may not seem remarkable that male baboons assault females—given their huge sexual size dimorphism, it would appear they do so because they can. To determine whether there are any functional explanations for the phenomenon, in this chapter we search for underlying patterns to male aggression directed at females. Are some victims targeted while others ignored? Are all attacks as violent as they could be, or do males show restraint? Do all males use intimidation tactics equally?

Most savannah baboons of sub-Saharan Africa live in large multimale, multifemale groups. Females are philopatric and maintain close bonds with matrilineal kin, forming matrilineally based, stable dominance hierarchies (Silk et al. 1999; Cheney and Seyfarth 2007). Conversely, most male baboons emigrate to neighboring groups at 9 to 11 years of age, and rank is largely determined by condition and fighting ability (Kitchen et al. 2003b, 2005). Although males form linear dominance hierarchies that are stable over the short term, rank reversals are common.

By the time they are 5 or 6, juvenile male baboons typically outrank adult females. As adults, only males possess large, sharp canines, and they outweigh females by approximately 70% (Figure 6.1). Thus, these formidable males are capable of capriciously abusing females within otherwise female-bonded social groups.

Figure 6.1 Baboons throughout Africa are sexually dimorphic in size and other features. Here a male olive-hamadryas hybrid baboon threatens an adult female who is fear-grimacing, a submissive expression. Photo by Jacinta Beehner.

Predation and infanticide are the primary factors affecting reproductive success among female chacma baboons at our study site in the Okavango Delta of northwestern Botswana (Cheney et al. 2004). Although females can reduce the probability of a predator attack by remaining near conspecifics, close proximity to individuals (especially adult males) also increases the likelihood of intragroup aggression. For example, adult males easily supplant or chase females from food resources, although such bouts of malice are usually mild and brief.

Other aggressive encounters with males, seemingly unrelated to feeding competition, can be even more disruptive to a female's daily life. For example, male chacma baboons frequently engage in aggressive loud call ("wahoo") displays that are thought to function in intra- and intergroup male-male competition (Kitchen et al. 2003b). During most wahoo displays, males chase females. When a female is targeted, she often runs up a tree and hangs from a thin branch (Figure 6.2) or cowers out of reach, deep inside a thorny bush. Meanwhile, the displaying male swats at her and shakes the vegetation—eventually

Figure 6.2 An adult male chacma baboon chased this cycling female up a tree. Here he is producing loud "wahoo" vocalizations while swatting at her as she hangs from a branch. He jumped up and down on the branch until she eventually fell from the tree. Photo by Dawn Kitchen.

she may fall from the tree or flee to another bush. Despite sometimes falling from 5 m or more and receiving contact aggression by the males, females rarely exhibit obvious injuries following an assault. Nevertheless, attacks can be protracted (wahoo displays can last for over an hour), and the incessant screaming of the victim, and sometimes members of her family, indicates that it is a disturbing event.

A similar story seems to unfold among East African baboons. In an extensive study of olive baboons *(P. h. anubis)*, Smuts (1985) described the most common contexts in which males attacked females. Although females were frequently assaulted during feeding competition or when a male defended a third-party female, many attacks occurred during male-male competitive contexts (26%) or were seemingly unprovoked (32%). In this chapter, we focus on the latter two categories and ask whether any of three possible functional explanations apply to male-female aggression among chacma baboons. Although closely related, we also highlight relevant differences between eastern and southern savannah baboons in our discussion.

It is possible that male-female hostility is fairly arbitrary. Aggression is commonplace in all age and sex classes, at least in chacma baboons, and frequently observers note a domino effect: an individual attacks a lower-ranking individual, this victim subsequently attacks an even lower-ranking animal, and so on. In many cercopithecine species, this method may be used by each individual to assert and maintain his dominance ("random acts of aggression": Silk 2002). However, this mechanism is unlikely to explain male aggression directed at female baboons because females are not a threat to any adult male's rank. More likely, this pattern could emerge as a by-product of male-male aggression (hereafter, the *redirected aggression hypothesis*). In other words, because all males outrank all females, a male who lost a fight with another male, or who is unwilling to risk injury in an escalated male-male contest, could aggressively target a female either to relieve stress or to focus attention away from himself (e.g., Castles and Whiten 1998). If this hypothesis explains the majority of nonfeeding-related male-female attacks, then we expect that most aggression will be conducted by low-ranking males following male-male aggression, that males will target the nearest adult female regardless of her reproductive state, and that males will also frequently attack juveniles of both sexes.

Male-female attacks may be more than random acts, however, and may instead function to communicate information about the assailant. First, males may be more successful at securing future mating if they harass estrous females with violence or the threat of violence (hereafter, the *sexual coercion hypothesis:* Smuts and Smuts 1993). The sexual coercion hypothesis assumes that females have some ability to control which males they mate with and that there are physical and physiological consequences of attacks that influence their choices. As an extension of the sexual coercion hypothesis, the *mating conflict model* (van Schaik et al. 2004) suggests that high-ranking males in particular may employ aggression to constrain female sexuality. In infanticidal species such as chacma baboons (Palombit et al. 2000), females may benefit from diluting paternity concentration by mating with multiple males. Female reproductive interests conflict most with the dominant male who maximizes his fitness by maintaining exclusive mating access. Thus, high-ranking males should be more motivated than low-ranking males to attack estrous females to prevent them from mating polyandrously. If attacks function to communicate intent to the female, they should also be surreptitious rather than part of an attention-grabbing loud call display, and there is no reason to predict that they should either follow or incite male-male aggression.

Second, male attacks on females may function to communicate information about the male's dominance, stamina, or fighting ability (1) to choosy females (although we currently have no data available to address this hypothesis) or (2) to competitive male rivals (hereafter, the *male-male competition hypothesis*). Elsewhere we report that loud wahoo calls are honest indicators of a male's fighting ability (Kitchen et al. 2003b, 2005; Fischer et al. 2004)—only males in good condition produce frequent, intense, and high-quality wahoo displays. Furthermore, males may use this information to assess rivals; preliminary experimental playback data suggest that males attend to the acoustic quality of their rival's wahoos (Kitchen et al. unpublished), and, following the predictions of game theory, males preferentially target rivals of comparable fighting ability during antiphonal calling contests, chases, and physical altercations. Displays are clearly energetically draining—for example, the acoustic quality of even the highest ranking male's wahoos decline during a protracted calling bout. Probably as a result, males alter their investment in displays depending on the caliber of rival they face. Producing a prolonged bout of good-quality wahoos at a fast rate while also running up and down trees and chasing another individual might provide even more reliable information to rivals.

If chasing females is part of a collection of honest signals that communicate information about a male's prowess to a mixed-sex audience, then it must be energetically costly and thus conducted more often by high-ranking males in the best physical condition. Even so, attacks need not have major fitness consequences to the victims. In addition, attacks should occur as part of an otherwise stereotypical wahoo display and should incite rather than follow male-male aggression. Furthermore, this hypothesis predicts that a male should either be just as likely to choose any lower-ranking animal or, if the victim's screaming bolsters a male's display, he should choose an individual that would generate the most noise. Although adult females and juveniles seldom physically assist each other when a male attacks one of them, they often lend "vocal support" to a victim by grunting, fear-barking, or screaming during attacks (Wittig et al. 2007). Because high-ranking adult females should generate the most vocal support when attacked (Cheney and Seyfarth 2007), males could target these females to maximize the din they generate. Alternatively, given the potential for infanticide in this population, lactating females would be good targets for a male attempting to generate the most commotion.

But if the goal of many such attacks is just to goad other males, why not direct *all* attacks at low-ranking male rivals? Males may sometimes avoid victimizing other adult males in order to reduce the substantial risk of injury through retaliation, even by inferior competitors. For example, the winner of one observed altercation retained his rank but permanently lost vision in his left eye after a subordinate rival punctured it with his canine. When possible, males may avoid the more dangerous, direct confrontations with other males by focusing on females or juveniles. Still, these attacks are not risk free—the plight of the victim may elicit aggression from other males (e.g., Smuts 1985; Palombit et al. 2000; Kitchen et al. 2005) or at least might challenge other males to engage in a prolonged vocal battle.

Table 6.1 summarizes the predictions of these three possible hypotheses. This is not an exhaustive list of hypotheses, nor do all hypotheses have mutually exclusive predictions. In fact, we think it is likely that attacks serve multiple functions, sometimes simultaneously. We discuss this possibility in more depth after we summarize the patterns and consequences of male aggression toward females and review the counterstrategies females use to avoid male aggression. Although we generate our results from nine years of systematically collected data (both published and unpublished), the majority of the data were not collected to specifically address this topic. We hope that our preliminary analysis will direct future studies specifically designed to test the questions raised here.

Study Site, Population, and Methods

The Okavango Delta lies within the Moremi Game Reserve in the northwest corner of Botswana. This area floods from approximately June through October, and only "islands," elevated tree-lined areas (1 to > 100 ha), remain above water. The baboons ford or swim through the water to get from island to island in search of food within their approximately 5 km² home range (described in Cheney and Seyfarth 2007).

The main study group has been observed since 1978 (Bulger and Hamilton 1987; Bulger 1993). The group is fully habituated to observers on foot and has been followed almost daily since 1992 (Cheney and Seyfarth 2007). The age and matrilineal relationships of all natal animals are known. The main study group averages 75 members, including 19 to 26 adult females and 3 to 12 adult males (Cheney et al. 2004).

All researchers at this site follow essentially the same data collection protocols

Table 6.1 Predictions of three possible hypotheses explaining chacma baboon patterns of male aggression directed at females, with symbols indicating which predictions were supported.

	Redirected Aggression	Male–Male Competition	Mating Conflict/Sexual Coercion
Attacker identity	Lower-ranking male	+ Higher-ranking male	+ Higher-ranking male
Victim identity	Any individual, based on proximity	Any, or lactating/high-ranking females	+ Estrous females
Context	Follows male-male aggression	+ Precedes male-male aggression	No relationship predicted
	Unrelated to male vocal behavior	+ Aggression occurs while male vocalizes	Aggression occurs while male silent
Consortship challenges	Lower-ranking males should redirect at female	(+) Males should target each other rather than contested female	(+) Males, particularly higher-ranking, should target contested female rather than each other
Female mate choice	No relationship predicted	No relationship predicted	+ Females should have some control over consort formation
	+ No relationship predicted	+ No relationship predicted	Females should evade higher-ranking consort partners in effort to dilute paternity
Fitness consequences	None necessarily	(+) None to females necessarily; energetic costs to males	(+) Attacks should be costly to females

+ Results support prediction. (+) Results provide partial support for prediction.

(see Cheney and Seyfarth 2007): (1) Handheld computers are used to conduct 10-minute focal samples on target animals, during which all social interactions between the focal animal and others are recorded; (2) a daily census is taken to record any immigrations, emigrations, disappearances, and the number of adult females available in each reproductive category: pregnant, lactating and cycling, with cycling females further subdivided based on whether or not they are in estrus and the size of the perineal swelling (Dixson 1998); (3) all "friendships" (formed between lactating females and some resident males: Palombit et al. 1997), "consortships" (exclusive mate-guarding relationships between males and estrous females: Crockford et al. 2007), and changes in the male or female hierarchy (updated daily using approach-retreat interactions) are recorded; (4) specific information is systematically collected during major events (predation, infanticide, intragroup male competitive displays, and intergroup encounters); (5) digital audio recordings are collected opportunistically of vocalizations of interest for acoustic analysis or playback experiments; and (6) fecal hormones are collected and extracted in the field from targeted individuals (Beehner and Whitten 2004).

Periodically at our site, certain behaviors or different age/sex classes are targeted to address specific research questions. Here we attempt to use the most complete and appropriate data sets to address each of our predictions. Although many of our analyses are conducted on data collected in 1999–2001 and 2006 when male aggression was the focus, we also incorporate relevant data sets from other time periods when possible. In the text, we cite the years of data collection available for each test.

Here we define "attacks" as any intense chase (> 50 m for > 30 s) and/or aggressive physical contact (hitting, biting, grappling). Attacks are often but not always produced as part of a male "wahoo display" (wahoos are two-syllable, loud, repetitive vocalizations). More than one male can participate in a display, and each participant may attack multiple individuals.

Data collected during male competitive displays are considered independent if attacks on females are separated by more than an hour and produced in a new context (i.e., typically involving different participants in a new location). Data from different focal sampling periods are also considered independent because subjects are randomly chosen and no subject is sampled twice until all animals are sampled once. However, as in any study of individuals who live together for long periods of time and who interact repeatedly in various dyadic or larger combinations, the data presented here cannot be perfectly independent. Therefore, other statistical methods employed to avoid pseudoreplication are discussed in this chapter or in the cited references.

Patterns of Male-Female Aggression

To test whether male aggression toward adult females supports the *redirected aggression, male-male competition,* or *sexual coercion hypothesis,* we examine patterns in attacker identity, victim identity, context, and consortships (Table 6.1).

Attacker Identity

At least one male chased at least one female in 138 of 183 independent wahoo displays from 1999 to 2001. To avoid pseudoreplication in our analysis, we only counted males once per display regardless of the number of individuals they attacked, and we used each male's average rank over the entire study period. Even with this conservative approach, we found that high-ranking males were more likely to chase females than low-ranking males ($n = 16$ males, $r_s = -0.72$, $p < 0.050$). To control for individual differences, we examined the chases of females by one male who occupied the most rank positions during the study. The same relationship emerged in this case study, with the subject more likely to chase females when he was high-ranking than low-ranking ($n = 7$ rank positions, $r_s = -0.80$, $p < 0.050$).

    Next, we restricted analysis to the subset of displays involving only two males. This approach eliminated the effect of overall rank (i.e., rivals might both be low- or both be high-ranking) and focused only on the relative rank between two rivals. For analysis, males were counted only once per display regardless of the number of females they chased. Because females were chased by one of the two males in 41 of 61 observed dyadic displays and by both participants in 14 of these displays, we used 56.6% for the expected value (i.e., the likelihood that a male in a dyadic display would chase a female was $= (14 + (41/2))/61$). Contrary to the predictions of the redirected aggression hypothesis, low-ranking males were less likely than expected by chance to chase females ($\chi^2_1 = 5.29$, $p < 0.050$). However, although the higher-ranking of the two males chased females more often (50.8% of 61 displays) than lower-ranking males (39.3%), they did not do so more often than expected by chance ($\chi^2_1 = 0.59$, $p > 0.100$). A likely explanation for these equivocal results is that chasing females serves multiple functions, particularly for males of different rank.

Male rank was previously shown to be unrelated to age or size in this population (Kitchen et al. 2003b). Using data from 1999–2001, we also found no effect of a male's estimated age class (Kruskal-Wallis: $n = 12$ males, $H_2 = 3.73$, $p > 0.100$) or size class (Mann-Whitney: height: $n = 10$, $U_1 = 18$, $p > 0.100$; weight: $n = 8$, $U_1 = 9$, $p > 0.100$) on the likelihood he would attack a female (using the average monthly rate of attacks per male per class).

Victim Identity

1. Attacks on juveniles vs. adults: Despite the fact that the juvenile age class typically constitutes about 60% of a baboon group (e.g., Cheney et al. 2004), males only directed attacks at them during 16% of displays (based on 286 displays in a two-year period: Kitchen et al. 2005, unpublished), whereas they attacked adult females (approximately 30% of the group) during 68% and other adult males (approximately 10% of the group) during 57% of displays. Therefore, although the redirected aggression hypothesis predicts that male baboons should be equally likely to attack anyone ranking below them, they seemed to ignore juveniles of either sex ($\chi_1^2 = 32.27$, $p < 0.001$) and focused their attacks on adult males ($\chi_1^2 = 336.40$, $p < 0.001$) and females ($\chi_1^2 = 24.30$, $p < 0.001$).

Redirected aggression might still explain this pattern of attacks if adults happen to be in closer proximity than juveniles. For 4.5 months in 2006, we collected data to examine this possibility. We used instantaneous scan samples at the beginning and end of each 10-minute focal sample, noting all individuals over one year of age sitting within 5 m of the male subject. We collected 501 samples from 12 adult males. On average, males had 1.9 neighbors within 5 m. Of the 951 neighbors identified, 38% were juveniles, 12% were adult males, and 50% were adult females. Using these data as expected values, adult males were targeted more than expected based on chance ($\chi_1^2 = 261.33$, $p < 0.001$), juveniles were targeted less than expected ($\chi_1^2 = 12.74$, $p < 0.001$), and females were targeted as expected ($\chi_1^2 = 0.98$, ns). Below we explore whether males target specific adult females.

2. Female rank and age: We examined the identity of all female victims in 369 chases by adult males observed in 1999–2001. Each female was only entered once in our analysis based on her average rank or average age throughout the study. Contrary to predictions, males did not target high-ranking females ($n = 28$ females, $r_s = -0.28$, $p > 0.100$). Females were also not targeted based on age ($n = 28$, $r_s = 0.07$, $p > 0.100$).

3. Reproductive state: We used 47,660 min of focal data collected in 2001–2003 to compare the rate of male aggression (lunging, chasing, biting, hitting) toward adult females with the average number of females available in each reproductive category. As predicted by sexual coercion, males attacked females with an estrus swelling more than expected by chance (n = 13 males, observed: 16.8% of 101 acts of aggression; expected: 10%; $\chi_1^2 = 4.62$, p < 0.050), whereas all other cycling (obs: 21.8%; exp: 30%; $\chi_1^2 = 2.24$, p > 0.100), lactating (obs: 33.7%; exp: 40%; $\chi_1^2 = 0.99$, p > 0.100), and pregnant females (obs: 27.7%; exp: 20%; $\chi_1^2 = 2.96$, p > 0.088) were attacked as expected based on their availability. By comparison, we found no evidence that adult females were aggressive to other females based on the victim's reproductive state (n = 33 females; estrous: obs: 12.9% of 186 incidents of aggression; exp: 10%; $\chi_1^2 = 0.84$; nonestrous cycling: obs: 27.4%; exp: 30%; $\chi_1^2 = 0.23$; lactating: obs: 44.1%; exp: 40%; $\chi_1^2 = 0.42$; pregnant: obs: 15.6%; exp: 20%; $\chi_1^2 = 0.97$; all p > 0.100). These data replicate results obtained in 1999–2001, when males targeted estrous females 2.5 to 3.9 times as often (based on availability) as females in any other reproductive category.

We used 36,910 min of nonconsortship focal data collected from 2004 to 2005 to examine aggression toward cycling females in different phases. In further support of the sexual coercion hypothesis, we found that aggressive acts were more likely to be directed at cycling females *with* an estrous swelling (70% of 30 aggressive acts) than at nonestrous cycling females ($\chi_1^2 = 4.80$, p = 0.028). Furthermore, focusing just on females with estrous swellings (for this analysis, it was necessary to combine consorting and nonconsorting data), we found an effect of swelling size and male rank on aggression patterns. Low-ranking males (n = 4) distributed aggression evenly between all estrous females (females in detumescence received 57.1% of 14 aggressive acts, females with an increasing or maximally sized swelling received 42.9%; $\chi_1^2 = 0.29$, p > 0.100). Conversely, the aggressive acts of high-ranking males (n = 5) were more focused on females whose swellings were increasing or maximally sized (83.3% of 12 aggressive acts) than on females in detumescence who were unlikely to conceive (16.7 %; $\chi_1^2 = 5.33$, p = 0.021). This lends support to the mating conflict model, which predicts that high-ranking males should be more likely to use sexual coercion as a reproductive strategy.

4. Proximity to females: Redirected aggression might still explain why males target estrous females if these individuals happen to be in closer proximity to males than females in other reproductive states. Contrary to this hypothesis,

however, we found that females in all reproductive states were equally likely to be within 5 m of an adult male according to their availability within the group; using the same data collected in 2006 described above, 45.6% of a male's female neighbors were lactating (expected: 35.6%; $\chi^2_1=2.81$, p>0.100), 17.9% were pregnant (exp: 26.8%; $\chi^2_1=2.96$, p>0.100), 27.2% were in estrus (exp: 21.0%; $\chi^2_1=1.83$, p>0.100), and 9.3% were cycling but had no sexual swelling (exp: 16.6%; $\chi^2_1=3.21$, p>0.100).

The specific effect of male rank on aggression toward estrous females was also not a result of proximity; of cases where a focal male's nearest neighbor was an estrous female (controlling for number of observation days per male), high-ranking males were no more likely to be near an estrous female (29% of 129 samples) than middle- or low-ranking males (46% and 25%, respectively). Results did not change if samples collected during consortships were examined separately. Thus, we find no evidence that estrous females are targeted simply because they spend more time near high-ranking males than other females.

We did, however, find some evidence that adult males and estrous females were attracted to each other outside of aggressive acts. Using 7410 min of non-consortship focal data on 12 males and 23 females collected in 2006, we found that females with sexual swellings approached or were approached by males twice as often (30.1% of all approaches) as cycling females without a sexual swelling (15.4%). When we examined all reproductive states (estrous, none-strous cycling, pregnant, lactating), we found that only estrous females approached or were approached more than expected based on availability within the group (expected: 16.6%; $\chi^2_1=10.979$, p<0.001). However, we did not record "leave" behavior in this study; thus, we had no information on how long males and females were within 2 m of each other, nor did we know which sex was responsible for maintaining this proximity. In a study that did quantify duration of male-female proximity, it was the lactating females, not cycling or pregnant females, that spent significantly more time within 25 m of males (Palombit et al. 1997). In addition, because it was almost impossible to record data on nearest neighbors when males were displaying, it was unclear whether estrous females in the 2006 study were within 2 m immediately before displays (evidence that might suggest they were targeted by chance). Rather, it was our impression, based on detailed observations of displays in 1999–2001, that males began most displays by first walking through the group and then selecting a female to chase, often from some distance away. To adequately test this proposition, a study needs to be specifically designed to record which females are near a male immediately before a chase begins, as well as victim behavior toward an attacker following a chase (Smuts and Smuts 1993).

5. Intergroup contests: Baboon density is high in this area of Botswana (approximately 24/km^2: Cheney et al. 2004), with the home ranges of at least five groups overlapping our main study group. Perhaps as a result, intergroup interactions are common. These meetings often result in aggressive loud call displays in which males herd the female members of their own group in the opposite direction of the rival group. Adult males often immigrate into new groups following such encounters.

We found that all males were more likely to loud call and chase a female during an intergroup encounter if at least one female in the group was in estrus than if no females were in estrus (Kitchen et al. 2004). As in intragroup contests, estrous females were preferentially targeted for chasing during intergroup contests (17.9% of those available per day, 2 to 12 times more than any other reproductive category), even though most of these estrous female victims were not involved in a consortship during the encounter. We also found that subjects (n=6 males in paired comparisons) were more likely to chase a female during intergroup encounters when they were involved in an exclusive mate-guarding consortship than when they were not. Finally, high-ranking males (whether or not in a consortship) were more likely to attack females than low-ranking males.

Overall, these data suggest that males target estrous females during intergroup encounters and these attacks function to herd females away from other groups (possibly reducing female opportunities for extragroup copulations) and/or to coerce females into mating with the aggressor in the future. Although this is strong support for the sexual coercion hypothesis, the fact that males chase females typically as accompaniment to a loud call display does not negate the male-male competition hypothesis. We address this idea in more detail in the following discussion.

Display Context

Contrary to the sexual coercion hypothesis, we found a contextual relationship for female attacks using 128 independent displays that included chases of both sexes from 1999 to 2001. In support of the male-male competition hypothesis, we found that male-female chases were more likely to precede (64.1% of 128) than to follow male-male chases (35.9%; $\chi_1^2 = 10.13$, p=0.001). Thus, attacks on females seemed more likely to prompt male aggression than vice versa.

Often group members are spread out in tall grass or in wooded areas; as a result, not all baboons can see the details of all the attacks that occur in a group. If a male does not vocalize during an attack, he does not necessarily draw attention to himself. Therefore, males should silently chase females if male attacks function to coerce females into mating with the assailant. On the other hand, males should produce loud wahoo calls while chasing females if male attacks on females are meant to call attention to the aggressor's stamina. Of 294 male-female chases in 1999–2001, 172 (58.5%) were accompanied by loud calls from the attacker ($\chi_1^2 = 8.50$, p = 0.004), lending support to the latter hypothesis. Although a statistically significant result, males still silently attacked females 42.5% of the time; therefore, at least some chases of females might be explained by sexual coercion or redirected aggression.

Patterns in Consortships

In this population, males of all ranks form consortships with estrous females for a period of hours or days. When a low-ranking male is in a consortship with a female nearing ovulation, he is eventually challenged by a higher-ranking male who will likely seize control of the female (a "takeover"). Thus, the alpha male typically monopolizes females when they are maximally swollen and most likely to conceive (Bulger 1993; Cheney and Seyfarth unpublished data). Consortship challenges often escalate to physical fights, and participating males also frequently attack the contested female.

1. Consort formation: Male harassment can only increase the likelihood of future mating if females have some influence over which males they copulate with. There is some published evidence of female choice in baboons (e.g., Seyfarth 1978a, b; Smuts 1985; Beehner and Bergman 2006; Swedell 2006; Palombit, Chapter 15 in this volume). Anecdotal evidence suggests that females in our population allow some consort partners to successfully sequester them. For example, a fourth-ranking male in consort with an estrous female stayed far away from the rest of our study group for several days in 2000, and the pair did not return until she was in detumescence. Thus, the alpha male never had an opportunity to take over the consortship during what turned out to be a conceptive cycle for this female.

To examine the potential effects of females on consort formation and maintenance (as a proxy measure of female choice), we examined the number of times males and females approached each other in 2006. We found that estrous females approached males more often than males approached females when not in a consortship (57.8% of 225 approaches; $\chi_1^2 = 5.44$, p = 0.020) or when cycling but with no sexual swelling (61.7% of 115 approaches; $\chi_1^2 = 6.34$, p = 0.012), whereas males were more than twice as likely to approach females than the reverse when the pair was in consort (68.1% of 342 approaches; $\chi_1^2 = 44.96$, p < 0.001). Overall, these approach data suggest that females influence consort formation, but, once paired, males maintain proximity. As a comparison, approaches were just as likely to be made by either sex when females were pregnant (55.3% of 159 approaches; $\chi_1^2 = 1.82$, p > 0.100), whereas lactating females approached males more often than the reverse (72.6% of 248 approaches; $\chi_1^2 = 50.58$, p < 0.001), most likely in an effort to maintain spatial proximity with a male "friend" as an anti-infanticide strategy (see also Palombit et al. 1997).

2. Female behavior: Because high-ranking males are more likely than low-ranking males to thwart consort takeover attempts and thus sustain exclusive access to a female, the mating conflict model predicts that females should try to evade high-ranking consort partners to increase their ability to mate with multiple males during a given cycle (van Schaik et al. 2004). Using approach data gathered in 2006 to specifically address this question, we found that females tended to approach low-ranking male consorts less than expected (observed: 6.4%; expected: 13.5%; $\chi_1^2 = 3.73$, p = 0.054), but approached mid-ranking (obs: 25.7%; exp: 25.5%; $\chi_1^2 = 0.00$, p > 0.100) and high-ranking males (obs: 67.9%; exp: 61%; $\chi_1^2 = 0.78$, p > 0.100) as often as expected. Therefore, contrary to this variation of the sexual coercion hypothesis, females did not try to avoid high-ranking partners and instead avoided low-ranking partners (Figure 6.3). Outside of consortships, we have no indication that females avoid the advances of low-ranking males in "sneaky" copulations, which can occur when a female temporarily becomes separated from her consort partner (Cheney and Seyfarth 2007).

3. Male behavior: If attacks on females function to coerce them into immediate or future mating, males who attempt to take over a consortship, or males who defend their consort partner from a possible takeover, should target the female rather than the defending or challenging male (van Schaik et al. 2004). Using data collected on 81 consortships from 1999 to 2001 during which aggression was directed at the female and/or at one of the males, we found that 6.2% involved only male-male aggression, 39.5% involved only male-female aggression, and the majority involved both types of aggression (54.3%).

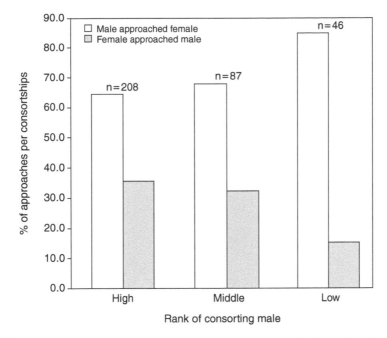

Figure 6.3 The percentage of all approaches to within 2 m by the male or female member of a consorting pair based on the rank of the consorting male.

Thus, results were equivocal; although females were more likely to be targeted overall ($\chi^2_1 = 5.37$, p = 0.021), attacks were as likely to be solely directed at females as they were to be at both sexes ($\chi^2_1 = 1.90$, p > 0.100). In addition, the style of attacks varied with the sex of the victim; aggressive acts toward females were typically threats or short chases, whereas male-male attacks more frequently escalated from chases to grappling and biting. In fact, 27% of the 93 male-male physical fights observed in this study period were in the context of estrous female defense (Kitchen et al. 2005).

Following the mating conflict model, high-ranking males should be more coercive to consort partners than low-ranking males in order to dissuade females from polyandrous mating (van Schaik et al. 2004). We did not find strong support for this prediction using consortship challenge patterns. Although low-ranking consort males were less likely to direct aggression solely at females than at both sexes ($\chi^2_1 = 5.49$, p = 0.019; Figure 6.4), we found that high-ranking consort partners were just as likely to attack both sexes as to focus only on the contested female ($\chi^2_1 = 0.10$, p > 0.100).

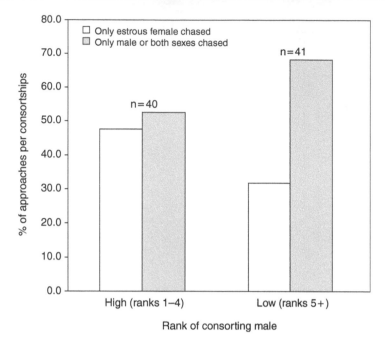

Figure 6.4 The percentage of consortship conflicts resulting in attacks on females alone or on both sexes based on the rank of the consorting male.

Discussion

When feeding, chacma baboons of all age and sex classes frequently redirect aggression—one animal supplants another from a food source, and the victim subsequently threatens or lunges at a third, lower-ranking animal. However, in nonfeeding contexts, the consistent and sometimes violent attacks on females by males seem less random. For example, if these attacks are simply redirected aggression, low-ranking males should be more likely to attack females than high-ranking males, and males should target the closest lower-ranking animal available. In contrast, we found that males virtually ignored some individuals in close proximity and targeted others. Furthermore, high-ranking males were the most common attackers, a result that instead supports both alternative hypotheses for male-female aggression (male-male competition and sexual coercion). By definition, male aggression redirected at females would also have to be *provoked* by male-male aggression, yet this did not explain the majority of cases in which female-male aggression *preceded* rather than followed male-male aggression.

We found strong support for the sexual coercion hypothesis in terms of which female victims were targeted. Contrary to the predictions of the other two hypotheses, attacks were preferentially focused on estrous females during intra- and intergroup displays. We did find some evidence that estrous females and males approached each other more often than other male-female dyads. However, whether this attraction (1) influences male choice of targets and (2) is a result of female mate choice or of successful male harassment will require collection and analysis of proximity data immediately before and after an attack. The pattern of aggression by males of different ranks to females in different phases of their cycles also supports the mating conflict/sexual coercion hypothesis (van Schaik et al. 2004). We found that high-ranking males were more likely than low-ranking males to focus attacks on those estrous females nearing ovulation.

Sexual coercion can only work if females have some choice in their mating partners. Although males were responsible for maintenance of proximity once paired, our approach data did suggest that estrous females were responsible for some portion of consort formation. However, females in a consortship were more likely to approach high-ranking than low-ranking partners, suggesting they did not avoid these males as would be expected under the mating conflict model. This hypothesis also predicts that during a consortship challenge, male aggression should be focused on the contested females. Although we found that females received more attacks overall, we found that attacks on females were relatively mild and that it was common for both male-female and male-male aggression to occur simultaneously during an attempted consort takeover. Thus, our consortship data provide only weak support for the mating conflict/sexual coercion hypothesis.

The strongest support for the male-male competition hypothesis is based on the context of attacks on females. First, male-female aggression typically occurs prior to rather than following male-male aggression, suggesting female attacks might be partially responsible for inciting other males. Second, contrary to the predictions of the sexual coercion hypothesis, the majority of attacks on females are accompanied by male wahoo displays, suggesting males are calling attention to themselves.

Still, enough of the attacks on females follow male-male aggression or occur when males are silent to suggest that attacks could serve more than one function. Furthermore, if males use female screams solely to bolster their displays, they would not need to target estrous females over other females. Moreover, males could just as easily choose from among the numerous juveniles in the group. Yet juveniles are rarely attacked and nonestrous females are only attacked as expected based on proximity and availability. For these and other reasons given above, we suspect that male attacks on females serve a variety of functions. Testing this multifunction hypothesis would require a more detailed analysis than our data presently allow. For example, we predict that (1) *vocal* attacks should *precede* male-male aggression; (2) *silent* attacks should be targeted at *estrous* females; and (3) attacks by *low-ranking* males should *follow* male-male aggression.

Whether attacking a female during a wahoo display incurs sufficient energy cost to be part of a collection of honest signals about a male's stamina requires further investigation. First, a fine-scale analysis of calling bouts could be used to determine whether running and chasing have an effect on wahoo acoustic quality. We would predict that even high-ranking males' wahoos would decline in quality while chasing, but the decline would be less precipitous than among low-ranking males. Second, playback experiments could be used to simulate wahoo bouts with and without female chases (e.g., by pairing male wahoos and threat-grunts with the sound of a female screaming) to see if the inclusion of such attacks increases a rival's interest in the display.

Consequences of Male Coercion

If male attacks on females are coercive in nature, then females should change their mating preference as a direct or an indirect result of an attack. In chacma baboons, aggressive acts directed at estrous females do not immediately result in copulation, nor do they lead to immediate formation of a consortship. Assuming instead that there are "indirect" coercion effects (Muller et al., Chapter 8 in this volume), it is still difficult to test whether targeted harassment of estrous females actually reduces a female's tendency toward polyandrous mating and thereby increases the attacking male's exclusive mating access. Even if we could confirm this using copulation or paternity data, we would still have difficulty distinguishing the mechanism. In other words, are males successful because they frighten uncooperative females with violence, because they use their protracted aggressive displays as an honest signal of quality to impress females, or because they scare off other male contenders with their demonstrations of power?

There is another predicted consequence of sexual coercion that we can begin to address here. That is, female attacks should result in tangible fitness costs to the females involved (Smuts and Smuts 1993). In contrast, we would not anticipate clear fitness costs if attacks were redirected aggression or purely part of male-male competitive displays.

Female-Female Aggression

Using 794.3 h of focal data collected in 2001–2003 during feeding and non-feeding contexts, we found that male-female aggression was no more frequent than female-female aggression. Female-female attacks (0.234 incidents/h) were 1.84 times as common as male-female attacks (0.127/h). Given that there were 2.54 times as many females in the group as males (33 vs. 13), attacks on females were just as likely to be by males or by other females ($\chi_1^2 = 0.19$, p > 0.100).

Physical Injuries

Of the individuals we were able to weigh at this site, males (n = 9) were 1.23 to 2.38 times heavier than females (n = 14; Kitchen et al. 2003b). This size difference makes an attack by a male risky to females in terms of physical injuries. From 1999 to 2001, we noted any visible injuries (cuts, blood) or limping by adult females, adult males, and juveniles. We also described any altercations involving the victim just prior to the injury. Females received more injuries from males than from other females, but overall they were injured less frequently than males. We observed 45 injuries on 12 males, and all were a result of male-male physical fights. We observed 11 injuries on 19 adult females that resulted from male attacks, 1 due to a female attack, 3 presumably from unsuccessful predator attacks, and 18 that were unexplained. Even if all unexplained injures were due to male aggression, this would be a rate of 1.53 injuries per female in 23 months versus 3.75 per male. Injuries to male baboons were also more severe, probably because male victims fight back (see also Smuts 1985). For example, in our population most male injuries were to the face or shoulders (80% of 45 injuries) because males turned to face each other before grappling, whereas females rarely received injuries to the face (24% of 33 injuries). Instead, most female injuries were cuts to the back, probably as the female fled.

Although injuries inflicted by males were more severe and frequent than injuries caused by other females, there have been no known fatal conspecific attacks on females in this population. By comparison, we attributed at least one male death in our group to injuries incurred in a fight with another male (Cheney et al. 2004). Nevertheless, even minor injuries may have impacts on female fitness, such as reduced foraging/traveling efficiency and increased risk of infection.

Infanticide

In addition to the time lost feeding, the risk of physical injury, and the psychological and physiological stress experienced during aggressive attacks by males, females with dependent offspring face a more tangible fitness cost. If one of the displaying males is a high-ranking recent immigrant, he may commit infanticide. Because the average tenure of an alpha male is shorter than the typical interbirth interval, infanticide may be a strategy that potentially increases the number of offspring that a male is able to sire by hastening the mother's return to reproductive condition (Palombit et al. 2000). Infanticide represents the largest source of mortality for infants in this population (Cheney et al. 2004; see also Palombit, Chapter 15 in this volume).

Physiological Effects

One way to examine how stressful an event is to an individual is by measuring changes in glucocorticoid hormones, the so-called stress hormones. Though associated with some costs if prolonged, an acute stress response is adaptive and related to mobilizing glucose for immediate use (mainly, to escape from a predator or another aggressor: reviewed in Sapolsky et al. 2000). Changes in levels of glucocorticoids can provide a direct measure of the physiological (and possibly psychological) responses of females to aggression or the threat of aggression by males (reviewed in Beehner et al. 2005).

A study examining stress and glucocorticoid production was first conducted at our site from 2001 to 2003. Fecal glucocorticoids were collected from 18 females during four consecutive periods, each approximately one month long. Two periods were marked by stability in the upper ranks of the male hierarchy, and there were two "takeover" periods (Beehner et al. 2005). A takeover occurs when the alpha male identity changes, usually accompanied by general male instability and an increase in male-male aggressive displays. One takeover was a natal male that ascended the hierarchy, and one was a recent immigrant male. In a second study, from 2003 to 2004, 21 females were sampled over the course of 16 months (Engh et al. 2006). Several potentially infanticidal males immigrated during this study, and two bouts of actual infanticidal attacks occurred approximately one year apart. In these two studies, the physiological impact of male aggression on females in different reproductive stages was assessed.

Although estrous females are preferentially targeted during male aggressive displays in this group, estrous females exhibited no variation in mean glucocorticoids across time periods—whether stable or unstable. In fact, individual victims targeted during one takeover period in 2002 did not show elevated glucocorticoids when compared with females who were not chased by these immigrant males (Beehner et al. 2005). As a further test, we compared the glucocorticoid levels of cycling females in different phases (n = 209 fecal samples from 17 females not in consortships). We found no difference in the glucocorticoid levels of cycling females with estrous swellings (mean ± SE = 116.95 ± 1.05 ng/g) compared to nonestrous cycling females (109.90 ± 1.06 ng/g; General Linear Mixed Model: $F_{1,195.758} = 0.079$, p > 0.100). Our results remained the same when females were differentiated further based on the phase of their estrous cycle (those with increasing or maximally sized swellings vs. those in detumescence) or on whether consorting females were included. Thus, although being attacked by a male is probably temporarily traumatic for any female, we found no physiological evidence that this aggression causes an enduring stress response in the victims.

Conversely, the glucocorticoids of *lactating* females increased dramatically following the immigration of potentially infanticidal males, and this increase was significantly higher following actual infanticidal attacks than following the mere threat of infanticide. By comparison, the glucocorticoid levels of lactating females did not increase during the takeover by the natal male (in our population, natal males who become the alpha rarely commit infanticide: Palombit et al. 2000).

Finally, lactating females exhibited an increase in glucocorticoids, even though rates of male aggression directed toward them did not increase following male immigration (they even declined in one study). Thus, it seems that physiological changes and perceived danger among females in this population are not due to male attacks or even the threat of violence, but rather to the psychological threat of one particular male behavior: infanticide, the most extreme form of sexual coercion.

Discussion

Many researchers consider infanticide to be the ultimate act of sexual coercion (or at least of "sexual conflict"; see Palombit, Chapter 15 in this volume). As predicted, when immigrant males ascend to the alpha position in our population, the female baboons most at risk of infanticidal attacks—lactating females— exhibit the strongest physiological stress response.

Otherwise, contrary to the predictions of the sexual coercion hypothesis, we found little evidence that male harassment and abuse of females incur a substantial fitness cost. Although attacks are likely an immediately disturbing event (see Silk 2002 for discussion of self-directed behaviors as measures of stress), targeted females do not demonstrate any lasting physiological stress response.

Although males injure females, injuries do not occur at a high rate and male-female aggression is no more common than female-female aggression. Males have the capability of doing great damage with their canines, and their size relative to females means little risk of victim retribution. Yet the tactics of males in their conflicts with females are very different from those they employ with other males. Male attacks on females sometimes appear savage, but looks can be deceiving: males seem to restrain themselves and avoid inflicting injuries that could harm a female's reproductive potential.

It is interesting that, although male chacma baboons are more frequently aggressive to females (once every 10 h including only chases and violent attacks) than male olive baboons (every 17 h including mild threats: Smuts 1985), olive males inflict more severe injuries on females and at higher rates (1/y per female) than chacmas (less than 0.76/y per female). Differences between baboon populations do not end there. Although few systematic comparative studies have been published, accounts between geographic areas differ. For example, male dominance hierarchies below the rank of alpha might be slightly more stable in East African than in chacma baboons (e.g., rank reversals occur at a rate of 1.5/mo in our population vs. approximately 0.75/mo in a yellow baboon population: Alberts et al. 2003). Although dominance rank typically correlates with mating success in baboons from both eastern and southern Africa, it seems that "queue-jumping" is more likely among East African baboons (at least in one long-term study of yellow baboons: Alberts et al. 2003). Alpha male East African baboons might have less exclusive mating privileges based on a potentially stronger influence of female choice than in chacma baboons and on male coalition formation, which does not occur in chacmas (reviewed in Bulger 1993 and Alberts et al. 2003). Furthermore, in contrast to our population, infanticide is rare in East Africa (reviewed in Henzi and Barrett 2003), and male-female friendships there seem to function for more than infant protection (Smuts 1985; Palombit et al. 1997; Lemasson et al. 2008). Therefore, perhaps *direct* female coercion has more impact on mating success in East African baboons, which might explain the severity of male-female attacks in the olive population. Conversely, male mating success in chacma baboons may be more related to placement in the male dominance hierarchy and infanticide strategies (*indirect* coercion). This might explain the common use of repetitive vocal displays as indicators of fighting ability in chacmas—displays that are rarely or never used in most East African baboon populations (e.g., Beehner and Bergman, personal observation). We are not aware of any published reports of vocal contests among East African baboons.

Female Counterstrategies

Assuming that male aggression might have at least some fitness costs, we expect to see female counterstrategies to male aggression. We briefly summarize some here (see also Palombit, Chapter 15 in this volume).

Avoidance

It seems possible that females can monitor when a male display is about to erupt and avoid males during these critical times. Certainly, a human observer can effectively anticipate that a male is about to initiate a display based on several behavioral cues (e.g., a male has a specific stride prior to a display). Our observations suggest that females and juveniles are also sensitive to these visual cues because they usually flee from males just before the onset of a wahoo display. Male loud calls could also alert females to displays, particularly when they cannot see approaching males because of tall grass or trees. Although one set of playback experiments suggests that females respond more strongly to alarm wahoos than to contest wahoos (Kitchen et al. 2003a), these trials were carried out during a period of time when there were no potentially infanticidal males resident in the group. It seems likely that lactating females would respond strongly to the contest wahoos of a high-ranking, recent immigrant.

Female Allies

Female-female alliances against adult male attackers are rare in this population (Cheney and Seyfarth 2007). For example, during a 23-month study on male displays (1999–2001), females only came to the aid of their kin during an attack by an adult male in two instances, one of which was an unsuccessful infanticidal attempt by a new immigrant male (personal observation; see also Palombit et al. 2000). Although females regularly provide vocal support during male-female aggression (Wittig et al. 2007), females' relatively small body size largely prevents them from providing physical support to each other against males.

Male Allies

The support of male friends during attacks considerably reduces the likelihood of infant injuries or deaths (Palombit et al. 1997, 2000). Moreover, fecal gluco-corticoid levels (measured in 2003–2004) increased among lactating females following two separate bouts of infanticidal attacks within the group compared to other time periods, but only among those females *without* male friends (Engh et al. 2006). Thus, it appears that females with male friends perceive themselves to be at reduced risk of infanticide (see also Beehner et al. 2005).

Discussion

Visual and auditory cues exist that females might use to anticipate and avoid the aggression that inherently surrounds male-male competitive displays. However, whether females attend to these cues and successfully avoid displaying males has not been explicitly tested. Although females cannot rely on kin to protect them during male attacks, lactating females with male friends can anticipate support (see also Palombit, Chapter 15 in this volume). Probably as a consequence, the physiological stress levels of these females are lower than those of other lactating females following infanticidal attacks.

Summary and Conclusions

During nonfeeding contexts, the behavioral patterns of chacma baboons suggest that male aggression toward females is not simply explained as redirected aggression. Rather, most acts of aggression toward females are cases of high-ranking males targeting sexually receptive females during loud call displays, and these attacks typically precede and perhaps prompt the involvement of other males. By preferentially pursuing estrous females, males may use female attacks as an indirect method of sexual coercion. However, male aggression toward females may sometimes have less to do with the victim and more to do with the broader audience. Males may incorporate a screaming female into their energetic loud call displays to showcase their stamina and condition to competitive rivals. In other words, can a male continue to produce good quality wahoos at a fast rate while also chasing a female up and down trees? Although other adult males are also targeted in such attacks, including females as potential targets lowers an assailant's immediate risk of injury from victim retaliation, while still serving to impress and challenge other males.

Our results also suggest that this aggression might have more fitness implications for males, and fewer for females, than was previously assumed. Using our preliminary analysis of physical injuries and physiological stress levels of targeted estrous females, we did not find strong evidence that attacks had substantial fitness costs to females. Nevertheless, the baseline stress hormones of lactating females rose significantly under the threat of the most extreme form of sexual coercion: infanticide.

In sum, male aggression toward females in this highly dimorphic species is not as random as expected and probably serves a variety of functions such as impressing rivals and coercing females. Likewise, Rodseth and Novak (Chapter 12 in this volume) suggest that in humans some aggression toward women is "private" violence used to intimidate and dominate, whereas other aggression is meant for the public arena and functions to impress others in the community. Similarly, male baboons probably target females both as a low-cost, noisy addition to a loud call display advertising their endurance and fighting ability *and* as a means of harassing and controlling specific females nearing ovulation. Thus, both males and females in the "audience" (watching or participating) probably use the information conveyed in these displays. This possibility warrants more investigation in other highly competitive multimale species.

Acknowledgments

We thank M. Muller and R. Wrangham for organizing the symposium leading to this volume. We are also grateful for the helpful comments on our chapter by M. Muller, R. Wrangham, and two anonymous reviewers. Our research would not have been possible without the vital assistance in the field from our colleagues K. Hammerschmidt, R. Hoffmeier, M. Metz, A. Mokupi, M. Mokupi, J. Nicholson, C. Seyfarth, L. Seyfarth, and E. Wikberg. We thank the Botswana government and the Department of Wildlife and National Parks for permission to conduct the research, and W. J. Hamilton III and colleagues for initiating research in this area. We are grateful for the support of I. Clark, Game Trackers, the former and current managers and staff at Eagle Island Camp, Mack Air, and our many other dear friends in Botswana over the years.

We thank L. Moscovice for comments on an early version of this manuscript and P. Whitten for assistance with hormonal analysis. Funding was provided at different times by the Ohio State University, the Deutsche Forschungsgemeinschaft, the University of Pennsylvania, the National Institute of Health, and the National Science Foundation.

References

Alberts, S. C., H. E. Watts, and J. Altmann. "Queuing and Queue-Jumping: Long-Term Patterns of Reproductive Skew in Male Savannah Baboons, *Papio cynocephalus.*" *Animal Behaviour* 65 (2003): 821–840.

Beehner, J. C., and T. J. Bergman. "Female Behavioral Strategies of Hybrid Baboons in the Awash National Park, Ethiopia." In *Reproduction and Fitness in Baboons: Behavioral, Ecological, and Life History Perspectives,* eds. L. Swedell and S. R. Leigh, pp. 53–79. New York: Springer, 2006.

Beehner, J. C., T. J. Bergman, D. L. Cheney, R. M. Seyfarth, and P. L. Whitten. "The Effect of New Alpha Males on Female Stress in Free-Ranging Baboons." *Animal Behaviour* 69 (2005): 1211–1221.

Beehner, J. C., and P. L. Whitten. "Modifications of a Field Method for Fecal Steroid Analysis in Baboons." *Physiology and Behavior* 82 (2004): 269–277.

Bulger, J. "Dominance Rank and Access to Estrous Females in Male Savanna Baboons." *Behaviour* 127 (1993): 67–103.

Bulger, J., and W. J. Hamilton III. "Rank and Density Correlates of Inclusive Fitness Measures in a Natural Chacma Baboon *(Papio ursinus)* Troop." *International Journal of Primatology* 6 (1987): 635–650.

Castles, D. L., and A. Whiten. "Post-Conflict Behaviour of Wild Olive Baboons. I. Reconciliation, Redirection and Consolation." *Ethology* 104 (1998): 126–147.

Cheney, D. L., and R. M. Seyfarth. *Baboon Metaphysics.* Cambridge: Cambridge University Press, 2007.

Cheney, D. L., R. M. Seyfarth, J. Fischer, J. Beehner, T. Bergman, S. E. Johnson, D. M. Kitchen, R. A. Palombit, D. Rendall, and J. B. Silk. "Factors Affecting Reproduction and Mortality among Baboons in the Okavango Delta, Botswana." *International Journal of Primatology* 25 (2004): 401–428.

Crockford, C., R. M. Wittig, R. M. Seyfarth, and D. L. Cheney. "Baboons Eavesdrop to Deduce Mating Opportunities." *Animal Behaviour* 73 (2007): 885–890.

Dixson, A. F. *Primate Sexuality: Comparative Studies of the Prosimians, Monkeys, Apes and Human Beings.* Oxford: Oxford University Press, 1998.

Engh, A. L., J. C. Beehner, T. J. Bergman, P. L. Whitten, R. R. Hoffmeier, R. M. Seyfarth, and D. L. Cheney. "Female Hierarchy Instability, Male Immigration and Infanticide Increase Glucocorticoid Levels in Female Chacma Baboons." *Animal Behaviour* 71 (2006): 1227–1237.

Fischer, J., D. M. Kitchen, R. M. Seyfarth, and D. L. Cheney. "Baboon Loud Calls Advertise Male Quality: Acoustic Features and their Relation to Rank, Age, and Exhaustion." *Behavioral Ecology and Sociobiology* 56 (2004): 140–148.

Henzi, S. P., and L. Barrett. "Evolutionary Ecology, Sexual Conflict and Behavioral Differentiation among Baboon Populations." *Evolutionary Anthropology* 12 (2003): 217–230.

———. "The Historical Socioecology of Savanna Baboons *(Papio hamadryas)*." *Journal of Zoology. London* 265 (2005): 215–226.

Kitchen, D. M., D. L. Cheney, and R. M. Seyfarth. "Female Baboons' Responses to Male Loud Calls." *Ethology* 109 (2003a): 401–412.

———. "Factors Mediating Inter-group Encounters in Savannah Baboons *(Papio cynocephalus ursinus)*." *Behaviour* 141 (2004): 197–218.

———. "Contextual Factors Mediating Contests between Male Chacma Baboons in Botswana: Effects of Food, Friends and Females." *International Journal of Primatology* 26 (2005): 105–125.

Kitchen, D. M., R. M. Seyfarth, J. Fischer, and D. L. Cheney. "Loud Calls as Indicators of Dominance in Male Baboons *(Papio cynocephalus ursinus)*." *Behavioral Ecology and Sociobiology* 53 (2003b): 374–384.

Lemasson, A., R. A. Palombit, and R. Jubin. "Friendships between Males and Lactating Females in a Free-Ranging Group of Olive Baboons *(Papio hamadryas anubis)*: Evidence from Playback Experiments." *Behavioral Ecology and Sociobiology* 62 (2008): 1027–1035.

Palombit, R. A., D. L. Cheney, J. Fischer, S. Johnson, D. Rendall, R. M. Seyfarth, and J. B. Silk. "Male Infanticide and Defense of Infants in Chacma Baboons." In *Infanticide by Males and Its Implications,* eds. C. P. van Schaik and C. H. Janson, pp. 123–152. Cambridge: Cambridge University Press, 2000.

Palombit, R. A., R. M. Seyfarth, and D. L. Cheney. "The Adaptive Value of "Friendships" to Female Baboons: Experimental and Observational Evidence." *Animal Behaviour* 54 (1997): 599–614.

Sapolsky, R. M., L. M. Romero, and A. U. Munck. "How Do Glucocorticoids Influence Stress Responses? Integrating Permissive, Suppressive, Stimulatory, and Preparative Actions." *Endocrine Reviews* 21(2000): 55–89.

van Schaik, C. P., G. R. Pradhan, and M. A. van Noordwijk. "Mating Conflict in Primates: Infanticide, Sexual Harassment and Female Sexuality." In *Sexual Selection in Primates: New and Comparative Perspectives,* eds. P. Kappeler and C. P. van Schaik, pp. 131–150. Cambridge: Cambridge University Press, 2004.

Seyfarth, R. M. "Social Relationships among Adult Male and Female Baboons. I. Behaviour during Sexual Consortship." *Behaviour* 64 (1978a): 204–226.

———. "Social Relationships among Adult Male and Female Baboons. II. Behaviour Throughout the Female Reproductive Cycle." *Behaviour* 64 (1978b): 227–247.

Silk, J. B. "Practice Random Acts of Aggression and Senseless Acts of Intimidation: The Logic of Status Contests in Social Groups." *Evolutionary Anthropology* (2002): 221–225.

Silk, J. B., R. M. Seyfarth, and D. L. Cheney. "The Structure of Social Relationships among Female Baboons in the Moremi Reserve, Botswana." *Behaviour* 136 (1999): 679–703.

Smuts, B. B. *Sex and Friendship in Baboons.* Cambridge, Mass.: Harvard University Press, 1985.

Smuts, B. B., and R. W. Smuts. "Male Aggression and Sexual Coercion of Females in Nonhuman Primates and Other Mammals: Evidence and Theoretical Implications." *Advances in the Study of Behavior* 22 (1993): 1–63.

Swedell, L. *Strategies of Sex and Survival in Hamadryas Baboons: Through a Female Lens.* Upper Saddle River, N.J.: Pearson Prentice Hall, 2006.

Wittig, R. M., C. Crockford, R. M. Seyfarth, and D. L. Cheney. "Vocal Alliances in Chacma Baboons *(Papio hamadryas ursinus)*." *Behavioral Ecology and Sociobiology* 61 (2007): 899–909.

7

Female-Directed Aggression and Social Control in Spider Monkeys

Andres Link, Anthony Di Fiore,

and Stephanie N. Spehar

Spider monkeys (genus *Ateles*) have been studied in the wild since the mid-1930s, and, since the late 1970s, long-term studies have provided considerable insight into their ecology, social behavior, and social structure. Spider monkeys live in a fission-fusion social system, characterized by large, stable social groups that divide into smaller foraging or traveling parties that can vary in size and composition (Wrangham 1977; Symington 1987). These flexible grouping patterns are proposed to reflect an optimal balance for each individual between the advantages and disadvantages of group living (Lehmann and Boesch 2004).

One of the more consistent results to emerge from long-term studies of spider monkey sociality is the difference between males and females in many aspects of their social behavior (Fedigan and Baxter 1984; Symington 1987; Shimooka 2005). Spider monkeys are one of the few primate societies in which males remain in their natal groups, while females disperse to other groups as they reach sexual maturity. This dispersal pattern sets up a rare social situation wherein males within a social group are likely to be more closely related to each other, on average, than are females (Morin et al 1994; Di Fiore 2003; Di Fiore in press; Di Fiore et al. in review). In addition, spider monkey males are involved more frequently than females in aggressive behaviors, both toward their own group members and toward spider monkeys in neighboring groups. Males also often range in all-male subgroups, show strong affiliative bonds and cooperative behaviors with one another, and are more actively involved in coordinated territorial defense than are females.

It is widely recognized that a major source of conflict between the sexes among mammals arises from the differential investment that males versus females direct toward offspring care (Trivers 1972), and spider monkeys are no exception. Female spider monkeys not only allocate considerable energy to their offspring during pregnancy, but also nurse and carry them for several years after birth. Females' disproportionate investment in offspring (Link et al. 2006) suggests that access to feeding resources may be a key factor influencing female spider monkey's reproductive success. Male spider monkeys, on the other hand, theoretically should benefit from increasing their chances of mating with any potentially ovulating female.

Nevertheless, the unique features of spider monkey sociality—male philopatry, higher within-group (compared to between-group) male-male relatedness, low home range overlap, and active territorial defense by males—impose a different set of constraints and challenges for males to successfully reproduce than is seen in most primates. Under these circumstances, for example, it is likely that male competition over access to females should be higher between social groups than within them, especially when considering a potential role of kin selection among related males within groups. That is, the reproductive success of spider monkey males may then be determined not only by their direct access to females but also by the successful reproduction of related males. Even if males within a social group are not closely related to one other, strong bonds among males within groups could emerge if the competitive regime for mates is stronger between social groups than within them. In a parallel system to that described by Wrangham (1980) for female-bonded groups (where high between-group competition can lead to tolerance and cooperation among a group's females), the social strategies of male spider monkeys might reflect the intensity of male competition between rather than within groups over access to females. Thus, we would expect that male-competition over mates within a spider monkey group should not be as intense as in either unimale-multifemale primate societies or in multimale primate societies in which female philopatry and male dispersal are the norm. The low sexual dimorphism and small testis size seen in *Ateles* seem to support this hypothesis.

In this chapter, we draw together evidence from previous studies and from our own long-term research on white-bellied spider monkeys *(Ateles belzebuth)* in the western Amazon to evaluate the role that male aggression plays in spider monkey societies. We first describe aggression patterns seen in wild spider monkeys and focus on the phenomenon of female-directed aggression by males. After testing and rejecting alternative hypotheses for this behavior, we conclude that male aggression toward females is best interpreted as a unique form of cooperative and diffuse sexual coercion. That is, we suggest that rather than representing an individual strategy whereby a male, through direct coercion, gains immediate reproductive access to a female, the behavior of female-directed aggression instead benefits multiple (and possibly related) males by asserting their long-term control over females' reproduction and perhaps discouraging them from mating outside of the group. We suggest that this form of cooperative social control seen in spider monkeys—as well as in chimpanzees and in a handful of other male-philopatric mammals—reflects one facet of a relatively unique male reproductive strategy that is perhaps best characterized as "cooperative polygyny."

The Social System of Ateles

Among the New World primates, spider monkeys are noteworthy because of their remarkable convergence with chimpanzees in various features of their social systems. Despite marked differences in the intensity of direct male-male competition observed within groups and in the level of sexual dimorphism, both spider monkeys and chimpanzees live in large, fission-fusion societies that contain multiple adult males and females. Male spider monkeys, like male chimpanzees, are philopatric and form the core of the social group, whereas females in both species commonly emigrate from their natal groups as they reach reproductive age (Shimooka et al. in press; Di Fiore et al. in review). In addition, in both species, males develop strong, affiliative bonds with other males in their groups: males show frequent affiliative behaviors with one another (e.g., embraces, play), and, in many populations, groups of males will often travel together and cooperate with one another in territorial defense against males from adjacent groups (Watts and Mitani 2001).

Almost all studies addressing the social behavior of spider monkeys have also found differences in the ranging and grouping behavior of males and females (Shimooka 2003, 2005) as well as in their agonistic interactions with other group members (Klein 1972; Fedigan and Baxter 1984; Campbell 2003). Males have been observed to use the peripheral areas of their home range more often than females, especially those with dependent offspring (Chapman et al. 1989; Shimooka 2005), and to "patrol" their home range, particularly in areas bordering the ranges of adjacent groups (Shimooka 2005; Link and Di Fiore unpublished data), and to enter neighboring territories (Aureli et al. 2006).

Males also tend to travel farther each day than females, and all-male "bands" usually travel faster than mixed groups or female groups. Shimooka (2005) has suggested that the longer and faster travel rates of males, as well as their larger use of their home ranges, serve two main, nonexclusive functions: territorial defense and the monitoring of dispersed females within their groups. The co-ordinated behavior of these all-male "bands" thus seems to constitute a form of either mate-defense or resource-defense polygyny.

Given the pattern of male philopatry and female dispersal seen in spider monkeys, males within groups are presumed to be more closely related to one another than are females (Di Fiore 2003). Indeed, under most situations male-male competition within spider monkey social groups seems to be relaxed, as would be expected if males are related to one another (but see Campbell 2006; Valero et al. 2006), and a recent genetic study has shown that in at least one group of wild spider monkeys, most adult males were close relatives of one an-other (Di Fiore, in press; Di Fiore et al. in review). Direct competition over ac-cess to females has never been reported in wild spider monkeys, and the low frequency of male-male aggression, the lack of sexual dimorphism in body size in the genus, and the fact that several males within social groups sire offspring (Di Fiore, Link, and Spehar, unpublished data) are all consistent with the sug-gestion of reduced direct competition among males.

Patterns of Aggression in Spider Monkeys

Aggressive behavior in spider monkeys can be divided into three main cate-gories: intragroup low-intensity aggression, intragroup high-intensity aggres-sion, and intergroup aggression. Intragroup low-intensity aggression accounts for most cases of aggression observed in long-term studies of spider monkeys. Included in this category of aggression are behaviors such as chases, soft bites, grabbing, "grunt" vocalizations, staring and open mouth threats, and displace-ments. These behaviors are considered low-intensity aggression because they have never been reported to cause a serious injury to the target individuals. Responses to low-intensity intragroup aggression include retreats, submissive "fear grimaces," and high-pitched vocalizations ("squeals"). Often a target ani-mal is forced down a tree several meters onto the main trunk, or even to the ground. In almost all long-term studies of spider monkeys, the most common type of low-intensity intragroup aggression is that directed from males to fe-males (Table 7.1). Female-directed aggression is common in populations of

Table 7.1 Survey of aggression patterns seen in long-term studies of spider monkeys.

Taxon	Study Site	Population Type	Habitat Type	Direction of Aggression						# Cases	Reference
				From Males	To Males	From Females	To Females	M→F	F→M		
A. b. belzebuth	La Macarena, Colombia	Wild	Natural	60%	0%	38%	100%	60%	0%	50	1
A. b. belzebuth	San Francisco Zoo, USA	Captive		79%	9%	15%	91%	70%	0%	34	2
A. b. belzebuth	Yasuní National Park, Ecuador	Wild	Natural	82%	4%	18%	96%	78%	0%	124	3
A. b. chamek	Cocha Cashu, Peru	Wild	Natural	71%	13%	29%	87%	60%	2%	158	4
A. geoffroyi	Tikal, Guatemala	Wild	Natural	NR	NR	NR	NR	76%	0%	NR	5
A. geoffroyi	Los Inocentes, Costa Rica	Wild	Fragmented	65%	13%	35%	87%	56%	0%	78	6
A. geoffroyi	Dept. of Ethology, UNAM, Mexico	Captive		76%	37%	24%	63%	55%	14%	49	7
A. hybridus	Las Quinchas, Colombia	Wild	Fragmented	75%	0%	25%	100%	75%	0%	12	8

NR = not reported
1. Klein (1972). 2. Rondinelli and Klein (1976). 3. This study. 4. Symington (1987). 5. Fedigan and Baxter (1984). 6. McDaniel (1994). 7. Anaya-Heuertas and Mondragon-Ceballos (1998). 8. Link and Guerrero (unpublished data).

spider monkeys characterized by different levels of habitat and social disturbance (e.g., in fragmented as well as continuous forests and among reintro-
duced groups as well as in natural populations) that could potentially affect the
relatedness among males and thus their propensity for cooperative behavior
(Klein 1972; Fedigan and Baxter 1984; Symington 1987; van Roosmalen and
Klein 1988; McDaniel 1994; Campbell 2003; Link and Guerrero unpublished data). Male-to-female aggression is also common among captive spider
monkey (Eisenberg 1976; Rondinelli and Klein 1976; Anaya-Heuertas and
Mondragon-Ceballos 1998). Across all of these studies, the general pattern of
female-directed aggression is very consistent, involving stereotyped displays and
chases that are normally directed toward a single adult or subadult female by
one or more males. Although male-to-female aggression does not seem to be
directed toward causing physical injury to females, it is very clearly distressing
to the recipient.

The prevalence of intersexual aggression in spider monkeys is particularly
striking because aggressive interactions among same-sexed animals—especially
males—are extremely rare. Thus, compared with chimpanzees—where males
do aggressively compete with one another and can typically be ranked into a
dominance hierarchy, and where the degree of sexual dimorphism is considerable as a result—social relationships among spider monkeys males are far more
tolerant, and aggressive interactions are so rare that only long-term studies with
individually recognized animals have been able to comment at all about the nature of within-group male hierarchies (Symington 1987; Chapman 1990). Moreover, *Ateles* is one of the only primate genera in which males within social groups
have been reported to routinely form coalitions to direct aggression toward females (Eisenberg and Kuehn 1966).

The contexts in which female-directed aggression occurs have been used to
infer the ultimate mechanisms behind this behavioral pattern. Van Roosmalen
and Klein (1988) mentioned that aggression by males toward females was commonly observed at fruiting trees where the animals involved in the aggression
were feeding together. By contrast, Symington (1987) found that only 21% of
female-directed male aggression occurred within feeding contexts, while females directed aggressions toward other females in feeding trees twice as much.
Symington (1987) suggested that, although female–female aggression might
reflect intragroup feeding competition, males may be directing aggression toward females in order to induce urination, which could be used to indirectly assess their reproductive status (see also Slater et al. in press). Klein (1974) also

reported that a substantial proportion of aggression occurred shortly after spider monkey subgroups fused. This later pattern has also been reported for several other spider monkey populations (Aureli and Schaffner 2007; Link and Di Fiore unpublished data).

Campbell (2003) studied the aggressive behaviors of males toward females in relation to female reproductive state. Specifically, she tested whether aggression directed toward females by males was associated with sniffing behavior, as suggested by Symington (1987), and whether the timing of attacks correlated with the reproductive status of females, thus explicitly investigating whether aggression was being used as a form of sexual coercion. Campbell (2003) found no difference in the rates of either urination or defecation of females after being attacked versus when they were not attacked, suggesting that female-directed aggression was not intended to induce urination or defecation. In examining the relationship between male attacks and the reproductive status of females, Campbell (2003) found that females were attacked differentially during different phases of the reproductive cycle. Furthermore, for two females whose cycling profiles could be assessed, male attacks were more frequent outside rather than during the females' ovulatory periods, and the proportions of attacks occurring during the ovulatory versus the periovulatory periods did not differ from what was expected. In fact, only 1 out of 16 copulations observed was preceded by a male attack on the same day. Moreover, in both Campbell's (2003) and other studies, cycling females are not the only ones aggressed; a substantial proportion of aggression was directed toward lactating (and thus noncycling) adult females, as well as toward subadult females with whom males will probably never have reproductive opportunities, since young female spider monkeys emigrate upon reaching sexual maturity (see Shimooka et al. in press).

Recently, Slater et al. (in press) reported a correlation between the low-intensity aggression directed from males toward females and the female reproductive state. In deed, most aggression was directed towards females who were potentially cycling, although this aggression occurred within a relatively long window of time, suggesting that it was not precisely associated with the timing of ovulation. This study strongly suggests that male-to-female aggression is at least somewhat related to sexual context.

Two other classes of aggressive interactions have also been reported in spider monkeys, both of which are less common than the low-intensity aggressive interactions seem among groupmates. First, a number of authors have reported rare cases of severe—even lethal—aggression among the males of a group. In most cases, these incidents of high-intensity intragroup aggression have been executed jointly by several adult males toward another male, generally a subadult or juvenile (Fedigan and Baxter 1984; Campbell 2006; Valero et al. 2006). Both Fedigan and Baxter (1984) and Chapman et al. (1989) have noted that among *Ateles geoffroyi*, aggression directed toward juvenile males was almost two times as frequent as aggression toward juvenile females. Chapman et al. (1989) suggested that the female-biased adult sex ratio seen in many spider monkey groups may be due, in part, to the lower survival of young males as a result of being the targets of adult aggression. Also within this category of high-intensity aggression we would include the recent observations of infanticide and the first report of a forced copulation in spider monkeys (Gibson et al. 2007). In each of two long-term studies on spider monkeys, Gibson et al. (2007) have observed an infanticide event, and the victims were both infant males.

High-intensity intragroup aggression among males is likely to be associated primarily with situations where within-group competition over females outweighs the benefits of having a large number of males in a group. Such a social situation may occur either when the operational sex ratio within a group approaches 1:1 (Valero et al. 2006) or when territory or mate defense against males from adjacent groups is unnecessary—for example, on Barro Colorado Island, Panama, where only one social group exists and where several of the documented cases of lethal intragroup aggression among males have taken place (Campbell 2006).

Second, at a number of spider monkey study sites, researchers have described sometimes protracted aggressive encounters between several members of two adjacent spider monkey groups. Most of these intergroup encounters take place along the edges of a group's home range and are characterized by members of one or both groups rapidly approaching the other. Such encounters can escalate into chases and even include physical contact (e.g., bites), and they often end in long bouts of vocalizations from several members of each group spread from 50 to several hundred meters away. Males are much more actively involved in these intergroup encounters than females, and although no lethal event has been observed to date, serious "knife-cut" wounds have been seen on males after these encounters (Link and Di Fiore, unpublished data) suggesting these encounters could have a lethal outcome. Adult males have also been observed to cooperatively "patrol" their home range boundaries at several study sites. For example, van Roosmalen and Klein (1988) noted that adult male *Ateles paniscus* in Surinam cooperate in patrolling through their home range and jointly participate in "long-distance agonistic behaviors". In fact, van Roosmalen and Klein suggested that males might be defending a large territory that comprises the ranges of several females, a similar pattern to that observed in some groups of chimpanzees. More recently, during the course of a 16-month study of a single group of *Ateles belzebuth*, Inaba and Izawa (2000) recorded 48 intergroup encounters, 34 of which occurred at a mineral lick located in an area of boundary overlap with an adjacent group. During these encounters, male-male interactions were aggressive, whereas female-female interactions were more tolerant (Inaba and Izawa 2000). For this same population, Shimooka (2005) has described five patrolling events where an average of 5.8 (±1.1) males (out of seven in the group) were present, suggesting that most group males participate in this cooperative behavior. In another population of *Ateles belzebuth*, Link and Di Fiore (unpublished data) have also observed aggressive encounters between males of adjacent groups during patrolling events on the border of their study group's range. Symington (1987) has suggested that the number of males present during intergroup aggression might determine the outcome of such encounters. In a multisite comparison, Wallace (2008) has recently reported that the length of the "risky" border of a spider monkey group—that portion of the home range periphery shared immediately with a neighboring group—explains about 88% of the variation on the number of males resident in a group.

In summary, the three types of aggression observed in spider monkeys all point to the existence of a unique pattern of male social behavior in these primates. Male spider monkeys actively and cooperatively defend territories against males of other groups. Within groups, male social relationships are generally tolerant to affiliative (although the extent of male bonding is likely to be influenced by the relative strength of within- versus between-group competition over access to females), and males may cooperate with one another in aggression against female groupmates.

The Functions of Intersexual Aggression

Several nonexclusive explanations have been proposed for female-directed aggression among primates in general and among spider monkeys in particular (Muller et al., Chapter 1 in this volume). First, female-directed aggression may simply be a manifestation of direct *intragroup feeding competition*. If this were the case, we would expect (a) that attacks would occur more frequently during low fruit availability periods, (b) that most attacks should occur in feeding trees (particularly those with few food items), and (c) that both male and female group mates should be targets for male aggression. Second, targeting aggression toward females may qualify as a strategy of *direct sexual coercion* that males use to force or intimidate females into mating with them soon thereafter—at a cost to the female (Smuts and Smuts 1993). If this hypothesis were true, then (a) attacks should increase the chance of a copulation occurring subsequent to the aggression—and thus there should be a temporal association between attacks and mating behavior, and (b) attacks should be directed mostly at cycling rather than lactating, pregnant, or pre-reproductive age females. Finally, female-directed aggression may also reflect a delayed and/or diffuse form of sexual coercion—which we refer to as *social control*—in which males attack females in order to increase the chances that females will copulate with them in the future. With respect to this last option, we note that for taxa living in multimale groups it might also be adaptive for males to practice *cooperative social control*, if the potential for between-group competition over access to females is high relative to the level of within-group mating competition. Such a strategy of *cooperative social control* might be expected particularly for male-philopatric taxa such as spider monkeys and chimpanzees because of the likely closer genetic relatedness among males in a social group, though close relatedness among males is not theoretically necessary for *cooperative social control* to be adaptive. A large suite of predictions can be made based on the hypothesis that male-to-female aggression represents either individual or cooperative *social control*. First, we would not necessarily expect there to be a temporal association between female-directed aggression and subsequent mating behavior, nor would we expect male-to-female aggression to occur primarily in feeding trees, as we would predict for the *direct sexual coercion* and *intragroup feeding competition* hypotheses, respectively. Second, because social coercion should function akin to a form of male conditioning of females generally, we would not necessarily expect aggression to correlate with the reproductive status of females (e.g., cycling versus noncycling). Neither the *direct sexual coercion* nor the *intragroup feeding competition* hypothesis would easily account for the victims of male aggression to be females of various ages and reproductive states. Third, if *cooperative social control* is at play, then we would predict that most males within groups should practice female-directed aggression, regardless of their individual mating success, because it should improve the reproductive success of male relatives vis-à-vis males in other groups. Moreover, we would expect that males should at least sometimes cooperate to direct aggression against females. Neither of these behaviors would be easily consistent with *direct sexual coercion*. Finally, if attacks function to discourage females from mating with males of adjacent communities, then we might expect that they would occur more often in the boundary areas between groups, though this pattern might be obscured by the different tendencies of males versus females to visit border portions of the range. Table 7.2 outlines these predictions concerning the patterning of male-to-female aggression associated with each of these hypotheses.

In the remainder of this chapter, we present new data on patterns of aggression—particularly intersexual aggression—in two social groups of white-bellied spider monkeys *(Ateles belzebuth)* from Yasuní National Park in eastern Ecuador. We then evaluate the extent to which the alternative explanations for

Table 7.2   Predictions based on alternative explanations for female-directed aggression in spider monkeys.

Prediction	Hypothesis			Results
	IFC	DSC	CSC	
Both sexes are targets of aggression	Yes	No	No	No
FDA is evident predominantly during periods of fruit scarcity.	Yes	No	No	No
FDA is evident predominantly in feeding trees	Yes	No	No	No
FDA increases male's chance of mating	No	Yes	Maybe	Data not available
There is Temporal association of FDA and copulations/consortships	No	Yes	Maybe	Maybe
FDA targets predominantly cycling females	No	Yes	No	Maybe
All males practice FDA, regardless of mating success	No	No	Yes	Yes
Males cooperate in FDA	No	No	Yes	Yes
FDA is seen more in boundary areas of range	No	No	Yes	No

FDA = female-directed aggression; IFC = intersexual feeding competition, DSC = direct sexual coercion, CSC = cooperative social control

intersexual aggression outlined earlier in this chapter may apply in this population. We conclude that our observations are most consistent with an interpretation of female-directed aggression in spider monkeys as being an example of *cooperative social control*—that is, a diffuse form of sexual coercion practiced by all of the male members of a group. This behavior appears to be an integral part of the reproductive strategy of cooperative mate-and-resource-defense polygyny that is followed by male spider monkeys.

Methods

Given the infrequent nature of aggressive interactions among spider monkeys, collecting enough data to test hypothesis about the contexts and patterns of this behavior requires a long-term data set. Thus, in order to evaluate whether patterns of aggression are consistent with sexual coercion in these primates, we complement our own field data with qualitative accounts of aggression in spider monkeys from other sites.

We collected behavioral data on two groups of spider monkeys in the Yasuní National Park and Biosphere Reserve in lowland Ecuador. Group CAT-1 was studied from July 2002 through November 2004 at the Proyecto Primates Research Site (about 20 months of fieldwork), whereas group MQ-1 was studied from June 2005 to July 2005 and from August 2006 to December 2006 (about 7 months). Behavioral and ecological data were collected using focal animal sampling during partial to whole-day follows of all adult group members (see Link and Di Fiore 2006 and Spehar 2006 for a detailed overview of behavioral sampling methods). In addition, we recorded the size, composition, and location of the subgroup in which our focal individuals were traveling at 15-minute intervals throughout the day, and all social behaviors that were observed during the focal follows were recorded *ad libitum*. Aggressive behaviors in spider monkeys are sufficiently conspicuous that such *ad libitum* sampling within a small subgroup closely approximates "all-occurrence sampling" (Altmann 1974).

In this chapter, we focus on within-group "low-intensity" agonistic interactions, which are, by far, the most common type of agonistic behavior observed in our studies. At other sites, more escalated—and even lethal—aggression has occasionally been observed within social groups, primarily between males, but such events are extremely rare and have never been seen in our study groups. Within the category of "low-intensity" intragroup agonistic behavior, we analyze displacements and other forms of aggression separately. Low-intensity aggression in spider monkeys is stereotyped and generally consists of threats, chasing, and/or biting by the aggressor (who also commonly emits "grunt" vocalizations) followed by a sudden retreat by the victim, who drops down to a few meters above ground while making loud "squeal" vocalizations.

For each aggressive interaction and displacement, we recorded either the identities or the age-sex classes of the initiator(s) and recipient(s) of the interaction, identified the context in which interaction took place (e.g., whether it occurred within a feeding or nonfeeding situation), and mapped the locations of the incident within the group's home range. Recognized age-sex categories included adult males (AMs), adult females without dependent offspring (AFs), adult females with dependent offspring ≤2 years old (AFDs), and subadult females (SAFs). Given that spider monkeys' interbirth intervals are about 32 to 45 months and that pregnancy lasts approximately 7.5 months, we assumed that females with offspring ≤2 years old were not cycling during our studies.

For the purposes of our spatial analyses, we superimposed a 1-hectare grid on top of the set of location records noted during focal sampling for each study group, and we scored each 1-hectare cell entered by the group as either a "border" quadrat (i.e., one of the cells comprising the edge of the group's range) or an "interior" quadrat (i.e., cells from which the monkeys would have to cross a border quadrat to exit the range). We then compared the observed spatial patterning of cases of male-to-female aggression in boundary versus interior quadrats to an expected value derived from the proportion of location records falling into each of those cell types.

Results

Overview of Aggressive Interactions

In roughly 1740 hours of focal observation (1440 on CAT-1, 300 on MQ-1), we recorded a total of 124 overt low-intensity aggressive interactions (i.e., threats, chases, and bites) and 41 displacements (Figures 7.1a and 7.1b). Males were the main initiators of overt aggression, being responsible for 82% (N = 102) of all interactions. Females, on the other hand, were the targets of 96% of aggression, with 81% targeted toward AFs, 9% toward AFDs, and 10% toward SAFs. As reported in most studies on spider monkey social behavior, aggression was most frequently directed by males toward females, with male-to-female

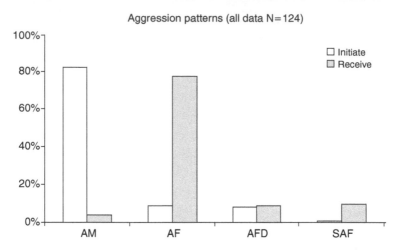

Figure 7.1a Initiators and recipients of 124 cases of overt aggression. AM = adult male, AF = adult female without a dependent offspring, AFD = adult female accompanied by a dependent offspring ≤2 years of age, SAF = subadult female.

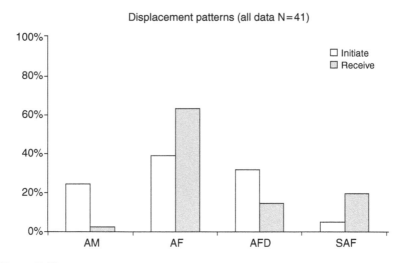

Figure 7.1b Initiators and recipients of 41 displacement interactions. Age-sex class abbreviations are as noted in Figure 7.1a.

aggression representing 78% of all within-group aggressive interactions (Table 7.1). Of these, 28% (N = 27) were performed by coalitions of two or more males. Males were never the targets of aggression from females, and male-male aggression was also extremely rare, accounting only for 4% (N = 5) of observed interactions.

The patterning of displacements was somewhat different from that of overt aggression. Displacements were more frequently initiated by females (76%, N = 31) and were directed almost exclusively toward females (98%, N = 40). Females were never observed to displace males. Thus, although male-to-female aggressive interactions tended to be overt, conspicuous, and commonly involved threats and chasing, female-female aggressive interactions tended to be more subtle.

Contexts of Male-to-Female Aggression

From a total of 88 incidents of male-to-female aggression in which we were also able to identify the location where the aggression took place, 33% (N = 29) occurred in feeding trees. Given the proportion of their active (i.e., nonresting) time that spider monkeys from our study populations spend feeding versus engaged in other activities, male-to-female aggression while feeding was actually less common than expected by chance ($Chi^2 = 6.4$, df = 1, p < 0.05). In addition, of the 29 cases of female-directed aggression occurring in feeding trees, in at least 7 cases both the male and female subsequently went back to the same tree to continue eating just after the aggression ended, and in 15 cases neither the male nor the female went back to the feeding tree. In two instances, the female went back to feed while the male did not, and in only three cases did the male return to feed without the female as would be expected if within-group competition were occurring. Taken as a whole, these results undermine the suggestion that male-to-female aggression reflects intersexual feeding competition.

Characteristics of Targeted Females and Male Aggressors

Out of a total of 97 incidents of overt aggression directed by males toward females in our two study groups, 87% (N = 84) were directed toward adult females without dependent offspring (AFs), whereas adult females with dependents (AFDs) were targeted only 9% of the time (N = 9) and subadult females (SAFs) only 4% of the time (N = 4) (Table 7.3). For both study groups, adult females without dependent offspring—that is, those who were potentially cycling—were the targets of male aggression more often than expected by chance, given their proportional representation in the group (CAT-1: $Chi^2 = 22.7$, df = 2, p < 0.01; MQ-1: $Chi^2 = 49.3$, df = 2, p < 0.01). Given that we did not collect hormonal data to monitor the ovarian cycles of our study animals, we cannot ascertain which of these females may actually have been cycling and thus able to conceive. Nonetheless, the tendency of males to target females without

Table 7.3   Distribution of 97 cases of female-directed aggression seen in this study.

	Group		
	Cat-1	MQ-1	ALL
AM to AF	50	34	84
AM to SAF	4	0	4
AM to AFD	5	4	9
Total	59	38	97

dependent offspring is clear, which is consistent with the hypothesis that female-directed aggression represents a form of either direct or indirect sexual coercion.

In both of our study groups, all adult males were seen to participate in at least some incidents of aggression directed toward females. However, in many cases we were unable to verify the individual identity (or identities) of aggressing males during the brief span of an agonistic event. Thus, we are unable to address the interesting question of whether certain males are more commonly aggressors than others.

Copulations and Aggression

Although we observed only eight copulations, for three of these the female involved was a recipient of male aggression within two days (before or after) of when the copulation was observed, and in each of these cases, more than one male participated in the aggression. For example, on one occasion the five adult males of CAT-1 directed sustained aggression toward a female that was observed copulating two days previously. For more than 90 minutes, the males directed branch-breaking displays toward her and chased her, often coordinating their efforts. Following this aggressive encounter, the female left her previous subgroup and immediately joined a subgroup containing only one of the males for the remainder of the day. Thus, although a robust relationship between female-directed aggression and either consortship formation or mating was not observed (but see Aureli and Schaffner 2007; Campbell and Gibson in press), these behaviors were associated with one another at least occasionally, lending some support to the *direct sexual coercion* hypothesis. The sexual behavior of spider monkeys is not easily observed in wild studies because it is a secretive behavior that generally occurs in consortships formed by a single adult male and adult female (Campbell and Gibson in press). Such consortships are easily missed in a fission-fusion social system in which observers can only be with a small fraction of a group at any one time. Thus, even a weak association of sexual behavior and male-to-female aggression is likely to be significant.

Spatial and Temporal Patterning of Male-to-Female Aggression

Only 24% of incidents of male-to-female aggression occurred in border quadrats for the CAT-1 group, and only 14% occurred in border quadrats for the MQ-1 group, whereas the proportion of location records falling into border areas for these groups was 45% and 21%, respectively. Thus, contrary to one of the predictions of the *social control* hypothesis, female-directed aggression was not more common in border areas than in interior portions of the range (CAT-1: $Chi^2 = 1.32$, df=1, NS; MQ-1: $Chi^2 = 0.19$, df=1, NS).

Also contrary to a key prediction of the *intragroup feeding competition* hypothesis, we found no clear association between season of the year and the frequency of incidents of male-to-female aggression, at least over a 14-month period of observation on the CAT-1 group by a single observer (Figure 7.2).

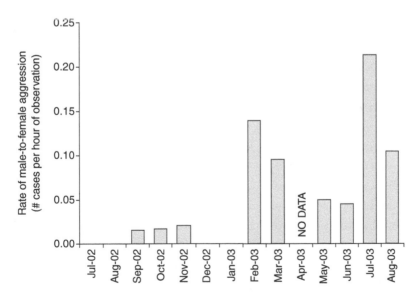

Figure 7.2 Temporal patterning of cases of male-to-female aggression in one study group across a 14-month period.

Although we do not have quantitative data on fruit availability or rainfall patterns at the site across this time period, variation in the rate of male-to-female aggression was not qualitatively linked to either of these factors. First, Yasuní is far less seasonal than many other neotropical forests, in terms of month-to-month variation in average daily temperatures, rainfall, and fruit availability. Second, the highest rates of male-to-female aggression are seen both when fruit abundance is typically high (February/March) and low (July/August).

Discussion

In this study, as in studies of spider monkeys at other sites, we found a very clear pattern of regular aggression directed from males toward females. Overall, our results, which are summarized in Table 7.2, support the idea that male-to-female aggression in spider monkeys functions as a form of sexual coercion and is an important component of the reproductive strategy of males. We consider that the "social control" executed by male spider monkeys represents a diffuse form of sexual coercion because (1) it is directed not only toward cycling females but other females of various ages as well, and (2) targeted females are clearly distressed when aggressed against, which might influence their willingness to mate with other potential males from outside their social group. Indeed, such "social control" might be a key factor influencing the pattern observed across spider monkeys where adult females only mate with their group's males (Gibson, personal communication; Link, Di Fiore, and Spehar unpublished data). Although most female-directed aggression does not appear to regularly increase males' direct access to mating opportunities immediately following attacks (Campbell 2003; Symington 1987), it may do so at least occasionally, thus providing support for the *direct sexual coercion* hypothesis as well. In fact, Aureli and Schaffner (2007) found that low-intensity aggression is most often targeted to females that are potentially cycling, suggesting that one function of this kind of aggression is related to the sexual behavior of spider monkeys. The possibility for direct sexual coercion among spider monkeys has also been demonstrated by a recent observation of forced copulation in *Ateles belzebuth chamek* (Gibson et al. 2007).

Nonetheless, the lack of a more robust relationship between female-directed aggression and subsequent mating by aggressing males—coupled with the fact that all males perform such behavior, sometimes cooperatively and sometimes targeting noncycling females—suggests that male-to-female aggression is best considered to be a more diffuse and indirect method of coercion. We thus interpret female-directed aggression as one component in a broader reproductive strategy of *social control* that males execute both individually and cooperatively, and in some instances it may also function as direct sexual coercion. Such a strategy—in which a group of males uses aggression and intimidation to control the reproductive behavior of females—would make sense within male-philopatric societies like those of spider monkeys, where the genetic relatedness among group males should be higher than that among females. In fact, preliminary data on genetic relatedness among individuals in at least one of our study groups suggests that males are much more closely related to one another, on average, than females (Di Fiore, in press; Di Fiore et al. in review). Moreover, if between-group competition over access to reproductive females is sufficiently strong, it could still benefit unrelated males to cooperate with one another to improve their ability to defend a large range—and reproductive access to the set of females occupying that range—from males of neighboring groups.

Several alternatives to either direct or indirect sexual coercion have been proposed to explain consistent patterns of female-directed aggression in primates and other mammals, including *male "herding," intergroup feeding competition,* and *intragroup feeding competition.* Here we briefly describe why these alternatives do not seem appropriate for explaining female-directed aggression in spider monkeys.

First, in several primate species (e.g., Hamadryas baboons, *Papio hamadryas hamadryas*), aggression from males to females has been interpreted as a form of *male "herding,"* whereby individual males keep close control over the social relationships of a harem of females. However, unlike the situation within the one-male units of Hamadryas baboons, the social associations among group members within a spider monkey community are extremely fluid, with the composition of foraging and ranging parties varying over short spatiotemporal intervals. Moreover, given the large home ranges of spider monkeys groups (average across groups and species = 279 ha, N = 15 groups: Wallace 2008), group members may spend long periods of time (even several days) without meeting one another. Thus, *male "herding"*—in the sense of using physical attacks to discourage females from straying far from a dominant, harem-holding male—is not a sufficient explanation for female-directed aggression in spider monkeys, where males and females are often ranging separately.

Second, the fact that aggressive interactions between social groups mainly involve males, coupled with the observation that parties of males making "raids" or "patrols" into neighboring territories do not spend time feeding, argues strongly against *intergroup feeding competition* as an explanation for female-directed aggression (Aureli et al. 2006; Link and Di Fiore unpublished data).

Finally, the contexts and typical pattern of male-to-female aggression in spider monkeys do not support the hypothesis that males are displacing females as a response to *intragroup feeding competition.* Symington (1987) found that competition for feeding resources was more intense between females and that male-female aggression occurred mostly in other social contexts. Our results, too, indicate that aggression from males to females within feeding contexts occurs at a lower rate than that expected by chance alone, and only in a small proportion of these aggressions did the male return to feed and prevent the attacked female from doing so, as would be expected if competition over resources was at play. Moreover, attacks were most often directed toward a single female, even when large parties were present at the feeding source where the aggression took place. The fact that female-directed aggression occurs throughout the year and not primarily during periods of marked food scarcity (which are rare in Yasuní) also undermines the feeding competition hypothesis, although we note that we do not have the data to test this prediction quantitatively. Similarly, the expected outcome of aggression in a feeding context would be that of the aggressor displacing the victim from the feeding tree and then replacing the displaced individual in the disputed resource. Clearly, our data does not support this pattern.

Social Control in Spider Monkeys

Symington (1987) was one of the first to suggest a potential role of spider monkey aggression as a form of sexual coercion, as she observed that males mostly targeted their aggressive displays toward "cycling" females (without small offspring). During our study, we also found that most male attacks were directed toward females whom we assume could potentially be "cycling"—those without dependent offspring more than 2 years old—thus providing some support for the idea that male-female aggression could be acting as a form of direct sexual coercion. However, in our study, female-directed aggression seldom resulted in males obtaining immediate access to mating opportunities (but see Gibson et al. 2007). In fact, we only observed one case of male-to-female aggression resulting in the formation of a consortship, and that involved multiple males directing repeated aggressions toward a female (who was observed copulating two days earlier) for 90 minutes before she split from her subgroup and formed a new subgroup with one of the aggressors. Slater et al. (in press) reported that females often left their original subgroups with an aggressor after receiving low-intensity aggression (about 41% of aggressions), suggesting that this pattern is far more common in other populations than we have observed in our study. Thus, although aggression directed toward females by male spider monkeys may occasionally initiate a consortship or "provoke" a mating opportunity and thus constitute direct sexual coercion, we would argue that this coercion is generally more indirect. That is, rather than providing an immediate reproductive benefit, female-directed aggression constitutes a more diffuse form of sexual coercion: males are regularly aggressive toward the females of their social groups in order to maintain control over them and potentially increase their future mating opportunities. Social control is plausibly interpreted as a strategy by which all or several males in a community coordinate efforts to control reproductive access to a group of females.

The competitive regime that male spider monkeys face over access to females parallels that proposed by Wrangham (1980) for female-bonded groups living under contexts of high between-group competition for food. In this case, however, competition occurs between groups of (presumbably related) males over access to the females occupying a given range. Male spider monkeys jointly confront males from other communities, while within their social groups they assert "control" over females, both individually and cooperatively, through repeated low-intensity aggression. Such cooperative behavior could have evolved as a male mating strategy whereby an individual's fitness is increased through the defense of a range used by several females and by mutually monopolizing access to these females. In fact, Symington (1987:167) has already suggested that male spider monkeys do not act cooperatively in order "to gain immediate access to receptive females . . . but in order to gain long term reproductive benefits by maintaining the integrity of the community range."

Spider monkeys, we would argue, face an interesting socio-ecological situation in which the competition among males between groups to recruit, monitor, and monopolize reproductive access to females is greater than the mating competition they face within groups. That is, it seems that under the contexts of high competition between social groups, spider monkeys have evolved a tolerant social system through which males cooperatively defend their ranges and face low levels of direct contest competition over females within their groups. Such a system could be reinforced by the higher relatedness of males within social groups, bringing a potential role of kin selection to the shaping of male sexual strategies. In fact, preliminary data on reproductive behavior and paternity in the Yasuní spider monkeys have revealed (1) that several males copulate with each female during a single reproductive cycle, (2) that multiple males sire offspring within the group, even offspring of the same cohort, and (3) that successive infants of the same female may be sired by different males. All of these findings confirm that mating and reproductive skew is not as high as that in other primate societies (Di Fiore, Link, and Spehar, unpublished data).

One criticism of "cooperative" social control might be that it is subject to the problem of collective action: why should individual males not simply let other males incur the costs of conditioning or disciplining females and then collect the benefits of female "group fidelity"? Two points are relevant here. First, if spider monkey groups are indeed composed of related males, then the issue of whether a collective action problem in fact exists comes down to the actual calculus of indirect fitness. Given that the cost to males of directing aggression toward females is likely to be very low, the benefit accruing to male relatives in terms of improving their mating access to females need not be very high for the behavior to be nonetheless beneficial from an inclusive fitness standpoint. Second, even if males are not closely related, cooperative social control could still evolve as one component in a constellation of male cooperative behavior, providing the intensity of between-group competition over access to females is sufficiently high that aggression between groups can result in the loss of male group members and, ultimately, the differential survival of groups. Dissolution of social groups through the extermination of all of a group's males has been reported for chimpanzees (Nishida et al. 1985; Goodall 1986), and in spider monkeys intergroup encounters are largely resolved in favor of groups with a larger number of males present (Inaba 2002; Wallace 2008; Link and Di Fiore unpublished data). Under these conditions, male-male collaboration and support might be so important to group persistence (and thus to the survival and reproduction of individual group members) that cooperation, even among unrelated males, has evolved and is maintained through the differential success of groups with more versus fewer cooperating males—a possible example of a true "group selection" effect. Thus, the collection action problem does not seem to be sufficiently serious to reject the idea that female-directed aggression can be both an individual reproductive strategy and a cooperative one that might be favored by kin selection, true group selection, or both.

The pattern of low-intensity female-directed aggressions seen in spider monkeys has not been reported for other ateline primates (howler monkeys: genus *Alouatta;* woolly monkeys: genus *Lagothrix;* or muriquis: genus *Brachyteles*), even though all of them have been subjects of long-term behavioral studies. Although greater male philopatry than female philopatry is presumably ancestral within the ateline clade—and despite the fact that ateline males are generally more cooperative than females within social groups—male sexual strategies seem to differ substantially between these genera (Di Fiore and Campbell 2007). For example, in howler monkeys, the male-male competitive regime is intense. One or a few males within groups typically monopolize reproductive access to a small group of females, and infanticide is a common male behavior upon takeover of a group by a new male or coalition of males. Relationships among male muriquis and woolly monkeys, on the other hand, are far more tolerant. In both species, males have been observed copulating with a single female without overt aggression among them or toward females. Compared to the other atelines, male spider monkeys have small testes for their body size, suggesting that sperm competition is less relevant than in other genera (e.g., *Lagothrix* and *Brachyteles*). Similarly, body size dimorphism is less pronounced (and secondary sexual characters less developed) in spider monkeys than in either *Lagothrix* or *Alouatta,* again suggesting a lower direct male-male competition among *Ateles.* Thus, it seems that male reproductive strategies and social behavior in spider monkeys are unique even within the atelines.

Recently, van Schaik et al. (2004) have suggested that sexual coercion as an effective male reproductive strategy is only seen among primates in Old World monkeys and apes (catarrhines) and not among either lemurs (Lemuroidea) or New World monkeys (platyrrhines). Although we would agree that direct sexual coercion (e.g., forced copulations, attacks on females for rebuffing a male's advance) are seldom seen outside of catarrhines, our results nonetheless suggest that males in at least one platyrrhine genus direct most of their aggression toward females in a manner that is consistent with a more diffuse form of sexual coercion. Particularly interesting, we think, is the fact that this female-directed aggression is often performed by multiple males, which we suggest reflects a cooperative or mutualistic male reproductive strategy that is adaptive under conditions of high between-group competition over access to females. Such a situation, we speculate, may be generally true among taxa that—like spider monkeys—live in male-bonded, fission-fusion societies, including bottlenose dolphins where male aggression toward females, including coalitionary aggression, is also seen (Connor et al. 1992a, b; Scott et al. 2005). Further comparative study is needed to evaluate whether comparable patterns of cooperative social control and cooperative polygyny characterize can be found in additional animal taxa.

Acknowledgments

We are very grateful to the Ecuadorian government, especially officials of the Ministerio de Ambiente, for their continued interest in our primate research and for permission to conduct this research. Thanks are due also to the Waorani communities of Guiyero and Timpoka for permission to work in their traditional lands. Drs. Friedemann Koester and Laura Arcos Terán of the Pontificia Universidad Catolica de Ecuador and Drs. David Romo and Kelly Swing of the Universidad San Francisco de Quito have provided critical scientific and logistical support for our research over many years of fieldwork in the Yasuní region. Many volunteers, students, and colleagues—including Wampi Ahua, Renee Bauer, Erin Fleming, Jamille Heer, Paul Mathewson, Kristin Phillips, Yukiko Shimooka, Monica Ramirez, Dylan Schwindt, and Scott Suarez—spent long hours with us in the forest, following spider monkeys and assisting with data collection; we are most grateful for their help and companionship. This chapter has benefited from conversations with and/or comments from Filippo Aureli, Christina Campbell, Nicole Gibson, Martin Muller, Colleen Schaffner, and two anonymous reviewers. Funding for this project was generously provided by the National Science Foundation (AD, AL), the Wenner-Gren Foundation for Anthropological Research (AD), the L.S.B. Leakey Foundation (AD, SNS), New York University (AD, AL, SNS), and the New York Consortium in Evolutionary Primatology (AD, AL, SNS). The research described here was done in full agreement with Ecuadorian legislation and was approved by the IACUC committee of New York University.

References

Altmann, J. "Observational Study of Behavior: Sampling Methods." *Behaviour* 49 (1974): 227–262.

Anaya-Heuertas, C., and R. Mondragon-Ceballos. "Social Behavior of Black-Handed Spider Monkeys *(Ateles geoffroyi)* Reared as Home Pets." *International Journal of Primatology* 19 (1998): 767–784.

Aureli , F., and C. M. Schaffner. "Aggression and Conflict Management at Fusion in Spider Monkeys." *Biology Letters* 3 (2007): 147–149.

Aureli, F., C. M. Schaffner, J. Verpooten, K. Slater, and G. Ramos-Fernandez. "Raiding Parties of Male Spider Monkeys: Insights into Human Warfare?" *American Journal of Physical Anthropology* 131 (2006): 486–497.

Campbell, C. J. "Female-Directed Aggression in Free-Ranging *Ateles geoffroyi.*" *International Journal of Primatology* 24 (2003): 223–237.

———. "Lethal Intragroup Aggression by Adult Male Spider Monkeys *(Ateles geoffroyi).*" *American Journal of Primatology* 68 (2006): 1197–1201.

Campbell, C. J., and K. N. Gibson. "Spider Monkey Reproduction and Sexual Behavior. In *Spider Monkeys: Behavior, Ecology and Evolution of the Genus* Ateles, ed. C. J. Campbell. Cambridge: Cambridge University Press (2008).

Chapman, C. A. "Association Patterns of Spider Monkeys: The Influence of Ecology and Sex on Social Organization." *Behavioral Ecology and Sociobiology* 26 (1990): 409–414.

Chapman, C. A., L. M. Fedigan, L. Fedigan, and L. J. Chapman. "Post-Weaning Resource Competition and Sex Ratios in Spider Monkeys." *Oikos* 54 (1989): 315–319.

Connor, R. C., R. A. Smolker, and A. F. Richards. "Two Levels of Alliance Formation among Male Bottlenose Dolphins *(Tursiops* sp.)." *Proceedings of the National Academy of Sciences, USA* 89 (1992a): 987–990.

———. "Dolphin Coalitions and Alliances." In *Coalitions and Alliances in Humans and Other Animals,* eds. A. H. Harcourt, and F. B. M. de Waal, pp. 415–443. Oxford: Oxford University Press, 1992b.

Di Fiore, A. "Molecular Genetic Approaches to the Study of Primate Behavior, Social Organization, and Reproduction." *Yearbook of Physical Anthropology* 46 (2003): 62–99.

———. "Genetic Approaches to the Study of Dispersal and Kinship in New World Primates." In *South American Primates: Comparative Perspectives in the Study of Behavior, Ecology, and Conservation,* eds. P. A. Garber, A. Estrada, J. C. Bicca-Marques, E. Heymann, and K. B. Strier. New York: Springer (2008).

Di Fiore, A., and C. J. Campbell. "The Atelines: Variation in Ecology, Behavior, and Social Organization." In *Primates in Perspective,* eds. C. J. Campbell, A. Fuentes, K. C. MacKinnon, M. Panger, and S. K. Beader, pp. 155–185. New York: Oxford University Press, 2007.

Di Fiore, A., A. Link, S. N. Spehar, and C. A. Schmitt. "Dispersal Patterns in Sympatric Woolly and Spider Monkeys: Integrating Molecular and Observational Data." (in review).

Eisenberg, J. F. "Communication Mechanisms and Social Integration in the Black Spider Monkey, *Ateles fusciceps robustus,* and Related Species." *Smithsonian Contributions to Zoology* 213 (1976): 1–108.

Eisenberg, J. F., and R. E. Kuehn. "The Behavior of *Ateles geoffroyi* and Related Species." *Smithsonian Miscellaneous Collections* 151 (1966): 1–63.

Fedigan, L. M., and M. J. Baxter. "Sex Differences and Social Organization in Free-Ranging Spider Monkeys *(Ateles geoffroyi).*" *Primates* 25 (1984): 279–294.

Gibson, K. N., L. G. Vick, A. C. Palma, F. M. del Rocío Carrasco, D.M, G. Ramos-Fernández. "Intra-community infanticide and a forced copulation in spider monkeys: a multi-site comparison between Cocha Cashu, Peru and Punta Laguna, Mexico". *American Journal of Primatology* 70 (2007): 1–5.

Goodall, J. *The Chimpanzees of Gombe: Patterns of Behavior.* Cambridge, Mass.: The Belknap Press of Harvard University Press, 1986.

Inaba, A., and K. Izawa. "Ecological and Sociological Studies in Wild Spider Monkeys MB-1 Group at La Macarena, Colombia." In *Adaptive Significance of Fission Fusion Society in Ateles,* ed. K. Izawa, pp. 21–36. Sendai, Japan: Miyagi University of Education, 2000.

Klein, L. L. *The Ecology and Social Behavior of the Spider Monkey, Ateles belzebuth.* Unpublished Ph.D. thesis, Berkeley, Calif., 1972.

———. "Agonistic Behavior in Neotropical Primates." In *Primate Aggression, Territoriality, and Xenophobia,* ed. R. L. Holloway, pp. 77–122. New York: Academic Press, 1974.

Lehmann, J., and C. Boesch. "To Fission or to Fusion: Effects of Community Size on Wild Chimpanzee *(Pan troglodytes verus)* Social Organisation." *Behavioral Ecology and Sociobiology* 56 (2004): 207–216.

Link, A., and A. Di Fiore. "Seed Dispersal by Spider Monkeys and Its Importance in the Maintenance of Neotropical Rain-Forest Diversity." *Journal of Tropical Ecology* 22 (2006): 335–346.

Link, A., A. C. Palma, A. Velez, and A. G. de Luna. "Costs of Twins in Free-Ranging White-Bellied Spider Monkeys *(Ateles belzebuth belzebuth)* at Tinigua National Park, Colombia." *Primates* 47 (2006): 131–139.

McDaniel, P. S. "The Social Behavior and Ecology of the Black-handed Spider Monkey *(Ateles geoffroyi)."* Ph.D. thesis, St. Louis, Mo. 1994.

Morin, P. A., J. J. Moore, R. Chakraborty, L. Jin, J. Goodall, and D. S. Woodruff. "Kin Selection, Social Structure, Gene Flow, and the Evolution of Chimpanzees." *Science* 265 (1994): 1193–1201.

Muller, M. N., S. M. Kahlenberg, M. Emery Thompson, and R. W. Wrangham. "Male Coercion and the Costs of Promiscuous Mating for Female Chimpanzees." *Proceedings of the Royal Society of London B* 274 (2007): 1009–1014.

Nishida, T., M. Hiraiwa-Hasegawa, T. Hasegawa, and Y. Takahata. "Group Extinction and Female Transfer in Wild Chimpanzees in the Mahale National Park, Tanzania. *Zietschrift für Tierpsychologie* 67 (1985): 284–301.

Rondinelli, R., and Klein, L. L. "An Analysis of Adult Social Spacing Tendencies and Related Social Interactoins in a Colony of Spider Monkeys *(Ateles Geoffroyi)* at the San Francisco Zoo." *Folia Primatologica* 25 (1976): 122–142.

Scott, E. M., J. Mann, J. Watson-Capps, B. L. Sargeant, and R. C. Connor. "Aggression in Bottlenose Dolphins: Evidence for Sexual Coercion, Male-Male Competition, and Female Tolerance through Analysis of Tooth-Rake Marks and Behaviour." *Behaviour* 142 (2005): 21–44.

Shimooka, Y. "Seasonal Variation in Association Patterns of Wild Spider Monkeys *(Ateles belzebuth belzebuth)* at La Macarena, Colombia." *Primates* 44 (2003): 83–90.

————. "Sexual Differences in Ranging of *Ateles belzebuth belzebuth* at La Macarena, Colombia." *International Journal of Primatology* 26 (2005): 385–406.

Slater, K. Y., C. M Schaeffner, and F. Aureli. "Female-Directed Male Aggression in Wild Spider Monkeys *(Ateles geoffroyi yucatanensis):* Evidence of Sexual Coercion?" *International Journal of Primatology.* In press.

Smuts, B. B., and R. W. Smuts. "Male Aggression and Sexual Coercion of Females in Nonhuman Primates: Evidence and Theoretical Implications." *Advances in the Study of Behaviour* 22 (1993):1–63.

Spehar, S. N. "The Function of the Long Call in White-Bellied Spider Monkeys *(Ateles belzebuth)* in Yasuní National Park, Ecuador." Ph.D. thesis, New York University, 2006.

Symington, M. M. Ecological and Social Correlates of Party Size in the Black Spider Monkey, *Ateles paniscus chamek.* Unpublished Ph.D. thesis, Princeton University, 1987.

Trivers, R. L. "Parental Investment and Sexual Selection." In *Sexual Selection and the Descent of Man: 1871–1971,* ed. B. G. Campbell, pp. 136–179. Chicago: Aldine Publishing Co., 1972.

Valero, A., C. M. Schaffner, L. G. Vick, F. Aureli, and G. Ramos-Fernandez. "Intragroup Lethal Aggression in Wild Spider Monkeys." *American Journal of Primatology* 68 (2006): 732–737.

van Roosmalen, M. G. M., and L. L. Klein. "The Spider Monkeys, Genus *Ateles.*" In *Ecology and Behavior of Neotropical Primates,* eds. R. A. Mittermeier, A. B. Rylands, A. F. Coimbra-Filho, and G. A. B. da Fonseca, pp. 455–537. Washington, D.C.: World Wildlife Fund, 1988.

van Schaik, C. P., G. R. Pradhan, and M. A. van Noorddwijk. "Mating Conflict in Primates: Infanticide, Sexual Harassment, and Female Sexuality." In *Sexual Selection in Primates: New and Comparative Perspectives,* eds. P. M. Kappeler and C. P. van Schaik, pp. 131–150. Cambridge: Cambridge University Press, 2004.

Watts, D., and J. Mitani. "Boundary Patrols and Intergroup Encounters among Wild Chimpanzees." Behaviour 138 (2001): 299–327.

Wrangham, R. W. "Feeding Behaviour of Chimpanzees in Gombe National Park, Tanzania." In *Primate Ecology: Studies of Feeding and Ranging Behaviour in Lemurs, Monkeys, and Apes,* ed. T. H. Clutton-Brock, pp. 504–538. London: Academic Press, 1977.

————. "An Ecological Model of Female-Bonded Primate Groups." *Behaviour* 75 (1980): 262–300.

8

Male Aggression against Females and Sexual Coercion in Chimpanzees

Martin N. Muller, Sonya M. Kahlenberg,
and Richard W. Wrangham

Aggressive competition among males is a prominent feature of chimpanzee social life, and descriptions of violent—occasionally lethal—conflicts over both
dominance status and territory abound in the literature (Muller and Mitani
2005). Male aggression against females has received considerably less attention,
but it is clear from published reports that such aggression occurs regularly at all
long-term chimpanzee study sites (Nishida and Hiraiwa-Hasegawa 1985;
Goodall 1986; Smuts and Smuts 1993; Watts 1998; Matsumoto-Oda and Oda
1998; Watts and Mitani 2001; Muller 2002; Newton-Fisher 2006; Stumpf and
Boesch 2006; Muller et al. 2007). There are few published data on the rates of
intersexual aggression in chimpanzees, but in our own site of Kanyawara, in
Uganda's Kibale National Park, females are as likely as males to be victims of
male aggression (Figure 8.1). This behavior occasionally takes severe forms, such
as prolonged beatings with fists, feet, or branches. Such surfeits of male brutality
have not been seen everywhere (e.g., Budongo: Newton-Fisher 2006), but male
chimpanzees attack and wound females more frequently than do many primate
males, including those of their closest relative, the bonobo (Hohmann and Fruth
2003; Paoli, Chapter 16 in this volume).

Most research on male-female aggression in chimpanzees has focused on the
victimization of females from neighboring communities (e.g., Wolf and Schulman 1984; Goodall 1986; Williams et al. 2004). Although females can be hurt
or occasionally killed by extracommunity males, such aggression is rare. Aggression toward community females has been examined primarily in the context of
forced consortships, in which a male compels a cycling female to accompany

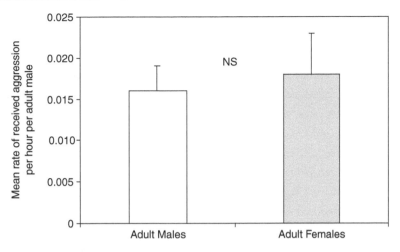

Figure 8.1 Kanyawara females are as likely as males to be victims of male aggression. Bars show mean dyadic rates of male aggression for males (N = 11) and females (N = 7) (Mann-Whitney test: P < 0.0001). All dyads were observed together for at least 25 hours (focal data from 1998).

him to a peripheral part of the territory, where rival males are excluded from mating (Goodall 1986; Smuts and Smuts 1993). Goodall (1986) was the first to suggest that regular, apparently unprovoked attacks on anestrous, cycling females might represent male intimidation and might specifically function to dissuade future resistance to the establishment of consortships. This hypothesis was later expanded and refined by Smuts and Smuts (1993). The study of intersexual aggression within communities has been widely neglected since Smuts and Smuts's seminal paper, however, and published data are scant.

In this chapter we draw on a 10-year data set from Kanyawara to examine patterns of male aggression against females within a wild chimpanzee community. There are four main objectives. First, we describe the major contexts of male-female aggression at Kanyawara and present data on the forms and frequency of the behavior. Second, we discuss functional hypotheses for male-female aggression in contexts that appear unrelated to sexual coercion. Third, we present evidence that much male-female aggression in chimpanzees represents sexual coercion, and we suggest an important distinction between direct and indirect mechanisms of coercion. Fourth, we examine the relationship between male coercion and female choice in chimpanzees, and we question some of the assumptions underlying recent studies of female choice.

Methods

The Kanyawara chimpanzee community in Kibale National Park, Uganda has been studied continuously since 1987. Struhsaker (1997) provides a detailed description of the forest. The data in this chapter came from two sources. For focal aggression rates, we used observations collected by the first author between January and December 1998. To examine longer-term patterns of aggression, we used 10 years of all-occurrence sampling data (Altmann 1974) collected between January 1994 and December 2003 by long-term Ugandan field assistants working closely with university-based researchers (graduate students, postdoctoral researchers, or one of the authors).

All-occurrence sampling of aggression is generally possible at Kanyawara, because the boisterous nature of chimpanzee agonism renders it highly conspicuous to observers. Nevertheless, it is clear that the long-term data underestimate true rates of aggression, because some interactions are obscured by vegetation. Muller et al. (2007) demonstrated that for male aggression against females, dyadic rates calculated from the long-term data correlate strongly with actual rates from focal data.

We focused on three categories of male aggression: *Charging displays* involved exaggerated locomotion, piloerection, and branch-shaking directed at specific females. *Chases* were recorded when a male pursued a fleeing female, who was generally screaming. All incidents of contact aggression were recorded as *Attacks;* these included hits, kicks, or slaps delivered in passing, as well as extended episodes of pounding, dragging, and biting (Muller and Wrangham 2004a). Mild threats, including hand waves and head tips (Bygott 1974), also occurred, albeit at low rates. These are not considered here because they are sometimes ignored by the recipient and are therefore easily overlooked by observers.

Prior to 1998, there were minor inconsistencies in the classification of aggression by different long-term observers. Specifically, some observers recorded aggressive acts as *charges,* but they were clearly *attacks* because physical contact was described. This situation was remedied in 1998, when M.N.M. and R.W.W. standardized data collection on aggression. Consequently, all analyses requiring discrimination between contact and noncontact forms of aggression exclude data collected prior to 1998. Dates of data collection for individual analyses are noted in the text.

During this study, adult males were observed in association with adult females (i.e., at least one male and one female were in a group during the 15-minute party composition scan) for 12,829 hours between 1994 and 2003. Over this period, 1295 independent acts of male-female aggression were recorded in the long-term data. Aggression involved 13 males and 23 females, representing every adult in the community.

Frequency and Nature of Male Aggression against Females

Focal data indicate that intersexual aggression regularly occurs among Kanyawara chimpanzees, with individual females suffering charges, chases, or physical attacks from individual males at a mean rate of 0.017 times per hour (Muller et al. 2007). A female traveling in a party with five adult males could thus be expected to receive aggression, on average, once per day (i.e., 12 hours of observation), a rate indistinguishable from that of male aggression received by males. This figure is an average across all adult females from all parties in 1998 and, as such, masks considerable variation. As detailed later in this chapter, both noncycling and nulliparous females receive less male aggression than cycling mothers do.

Long-term data indicate that most episodes of male-female aggression occurred without physical contact. Directed charges and chases accounted for 65% of aggressive episodes, compared to 35% involving actual attacks (data from 1998–2003, n = 997 independent aggressive acts). This proportion is nearly identical to what Goodall (1986) reported for male-female aggression during a one-year period at Gombe (aggression without physical contact = 66%; aggression with physical contact = 34%; n = 154 total aggressive acts).

Attacks on females varied dramatically in their intensity. Some consisted of perfunctory kicks or slaps that likely did little physical harm, but elicited clear signs of distress, such as whimpering or screaming. Other attacks were more vicious, incorporating extended episodes of hitting, kicking, biting, dragging a female on the ground, lifting a female and slamming her into the ground, jumping up and down on a female's back, and often, a combination of two or more of these behaviors. Readers who have never observed such an attack in the wild, especially those who are familiar with the milder forms of aggression predominant in most primate male repertoires, may not fully appreciate the ferocity of these assaults. A sobering description of one incident appears in Box 8.1.

Most male-female aggression was dyadic, but in 9% of observations more than one male simultaneously directed aggression at a single female (115 of 1276 incidents in which the number of male aggressors was unambiguous;

Box 8.1 Males attack an anestrous female at Kanyawara: June 22, 2002

Sonya M. Kahlenberg

Makoku and Twig, two young, low-ranking males, were grooming each other on the ground. Suddenly, and for no apparent reason, Makoku charged toward Outamba, a high-ranking, middle-aged female who was resting nearby with her offspring. Makoku grabbed Outamba and hit her hard with his fists, upon which Twig immediately joined in the attack. Outamba screamed loudly and tried to shield her infant, Tacugama, from the aggression by holding him tightly to her belly. Her older offspring, Kilimi and Tenkere, fled, screaming in distress. After being assaulted for a full minute, Outamba managed to free herself from the males and run away. Moments later, however, a high-ranking male called Johnny caught Outamba and pinned her down by standing on her back. Outamba remained silent, face-down on the ground, with Tacugama tucked underneath her. For the next two minutes, Johnny alternated between stomping on Outamba's back and hitting her with his fists. Another high-ranking male, Tofu, joined in, delivering punches and pulling Outamba's hair. When the males stopped briefly, Outamba escaped to a nearby tree. As she scrambled up the tree, Johnny caught her leg and pulled her down to the ground where she landed hard. The two males continued to beat her, and Johnny grabbed a small sapling that was still attached to the ground and used it to repeatedly flail her. After about two minutes, the males lost interest, and Outamba again escaped to the nearest tree. Soon after the males moved away, Outamba came to the ground and was reunited with Kilimi and Tenkere, who examined and groomed their mother's multiple bleeding wounds. Observers noted blood on the ground and additional drops on the leaves of the sapling used as a weapon.

data from 1994–2003). The modal coalition involved two males, with a maximum of five. Coalitionary aggression was more likely than individual aggression to involve physical assault (53% of coalitionary aggression [57 of 107 incidents] vs. 33% of individual aggression [296 of 897 incidents]; chi-square test: $\chi^2 = 16.6$, df $= 1$, p $= 0.00004$; data from 1998 to 2003).

Detailed data on female responses to male aggression were available for 1055 independent events (82% of the total; 1994–2003 data). Females showed submission to male aggression in 96% of cases by immediately fleeing, emitting sounds of distress (i.e., whimpers, screams) or submissive (i.e., pant-grunts, pant-barks) vocalizations (see Nishida et al. 1999 for descriptions of these vocalizations), and/or presenting their hindquarters for male inspection. Females were described retaliating against (i.e., chasing or attacking) adult males 5% of the time, but, in marked contrast to male aggression, female reprisals never involved more than a quick hit or slap and were usually accompanied by female submissive behavior (54% of retaliatory acts).

Third parties rarely intervened to break up conflicts or aid targeted females (4% of total independent aggressive acts, n $= 45$ interventions). When intervention did occur, supporters were normally adult males (62%) or close kin (28%: immature offspring $= 20\%$, adult offspring $= 4\%$, mothers $= 4\%$). Adult females came to the support of unrelated adult females in only four cases (9% of observed interventions).

Females varied significantly in the amount of aggression that they received from males, and female fecundity (i.e., probability of conception) was a strong predictor of received aggression (Muller et al. 2006). We focused on two independent measures of fecundity: estrous state and parity. Estrous state was assessed by the presence of maximal sexual swellings, which in chimpanzees are estrogen-dependent markers of the follicular phase (Graham 1981). We treated parity as a separate indicator of fecundity because, in Kanyawara, parous females had higher probabilities of conception than nulliparous females (copulations per conception: parous females <500, nulliparous females >1000) (Wrangham 2002).

Long-term data indicate that parous females were subject to significantly higher rates of male aggression when cycling and maximally swollen than during periods of lactational amenorrhea (Figure 8.2). Furthermore, a cross-sectional comparison showed that median rates of received male aggression for maximally swollen parous females (0.049 ± 0.010 times/h) were significantly higher than those for maximally swollen nulliparous females (0.015 ± 0.008 times/h) (Muller

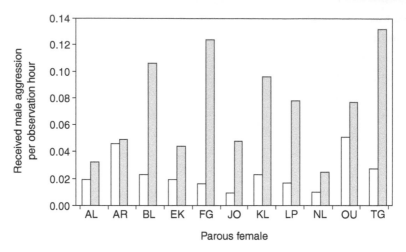

Figure 8.2 Male aggression against females intensifies in reproductive contexts. Parous females received significantly higher median rates of male aggression when they were in estrus (shaded bars) than during periods of lactational amenorrhea (white bars; Wilcoxon signed-rank test: N = 11, P = 0.003). Estrous periods included observations when females maintained maximally tumescent sexual swellings, but were not pregnant. Periods of lactational amenorrhea included observations from the birth of an infant until the female resumed full sexual swellings. Females were observed for at least 100 hours in each reproductive state.

et al. 2007). This comparison is based on absolute rates of received aggression. Rates of received aggression by male/female dyad showed the same pattern (Muller et al. 2007). Four females who began the study as nulliparous also experienced increased levels of male aggression when they mated as parous females (nulliparous median: 0.016 ± 0.010 times/h; parous median: 0.055 ± 0.025 times/h). This comparison is based on absolute rates of received aggression. Rates of received aggression by male/female dyad again showed the same pattern (nulliparous median: 0.000 ± 0.001 times/h, 10–11 males [median = 11] per female; parous median: 0.008 ± 0.003 times/h, 7–11 males [median = 9.5] per female; Muller et al. 2007).

Noncoercive Functions of Male-Female Aggression

Smuts and Smuts (1993) define sexual coercion as "use by a male of force, or threat of force, that functions to increase his chances that a female will mate with him at a time when she is likely to be fertile, and to decrease the chances that she will mate with other males, at some cost to the female." The association between female fecundity and received male aggression in Kanyawara suggests that some aggression there functions as sexual coercion, a hypothesis that we address in detail later in this chapter. Male aggression is not a unitary phenomenon, however, and some expressions of it may have functions other than sexual coercion. We therefore begin by examining the evidence for aggression in contexts apparently unrelated to sexual motivation.

Resource Competition

Overt competition over plant foods is uncommon among chimpanzees, though much of the aggression exhibited by females occurs in this context (38% in 1978 Gombe data: Goodall 1986; 80% in 1998 Kanyawara data: Muller 2002). An adult male can normally supplant a female from a feeding spot simply by approaching her or, occasionally, by delivering a mild threat (e.g., an arm wave or soft cough). Thus, when overt intersexual feeding competition does take place, it rarely escalates. The exception occurs when meat is the resource being consumed. Competition over meat provokes frequent squabbling among chimpanzees, and male-female aggression in feeding contexts likely varies with the frequency of successful hunting. For example, Goodall (1986) reported that during a one-year period at Gombe, approximately 12% of male-female aggression took place during feeding. Meat eating accounted for approximately 10% of that total (estimated from Figure 12.4c in Goodall:349). In one year at Kanyawara (1998), on the other hand, little hunting took place, and aggression in feeding contexts accounted for less than 1% of male-female aggression. In neither case does feeding competition explain a substantial proportion of male aggression against females.

Dominance Acquisition

Dominance striving appears to motivate much of the aggression among male chimpanzees (Bygott 1974; Goodall 1986; Muller 2002). Thus, it is possible that some fraction of male-female aggression reflects intersexual struggle for dominance. The ontogeny of male aggression suggests that this is indeed the case for adolescent males, but not for adults.

Starting in their juvenile years (ages 5–8), male chimpanzees frequently harass community females by flailing branches or throwing sticks and stones at them, behaviors that females generally ignore (Pusey 1990; Nishida 2003). As males enter adolescence (9–15 years), this teasing behavior is replaced by threats, charging displays, and attacks. These reactions initially incite female retaliation that, in turn, triggers male escalation. Eventually, however, such exchanges invariably result in female submission, formally acknowledged by the female's pant-grunt vocalization (Pusey 1990; Nishida 2003).

Adolescent males primarily target females outside of estrus, and rates of aggression against females peak during late adolescence and early adulthood (Figure 8.3). Given that adolescent males do not target females when they are most likely to conceive, their aggression appears to be unrelated to sex. Instead, female-directed aggression by young males is likely associated with dominance striving. Nishida (2003) reported that adolescent males in Mahale were much more aggressive toward females who retaliated against harassment than females who did not fight back. He concluded that heightened male-female aggression during male adolescence initiates the process of female domination that precedes heightened aggression toward low-ranking males and a young male's integration into the adult male status hierarchy. This process is regularly observed at Kanyawara, but does not account for the male-female aggression reported later in this chapter, which is limited to acts by fully adult males.

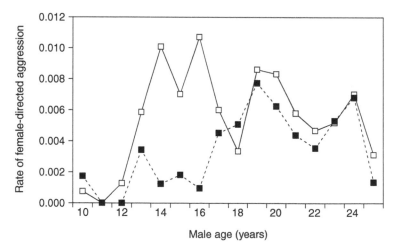

Figure 8.3 Male aggression against females as a function of male age and female reproductive state. Aggression against anestrous (i.e., without sexual swellings) and estrous (i.e., with maximal swellings) females is depicted by open and black points, respectively. Points are mean dyadic rates across males of a given age (1–4 males/age, median = 3). All dyads were observed together for ≥ 25 hours.

Is it possible that some intersexual aggression reflects continuing conflict over dominance status during adulthood? Given that adult male chimpanzees in the wild are universally dominant to adult females, and females never threaten an adult male's rank (Goodall 1986; Takahata 1990; Kahlenberg 2006), this would seem unlikely. However, because of their fission-fusion social system and the dispersed nature of female ranging, chimpanzee males encounter females living in the central part of their territory much more often than females living on the periphery (Emery Thompson and Wrangham 2006). Therefore, it is possible that some proportion of male-female aggression functions to resolve indistinct dominance relationships with rarely encountered females. If this were a frequent cause of aggression, then peripheral females would be expected to receive higher rates of aggression from males than central females. However, at Kanyawara, the opposite pattern is evident, with central females receiving significantly higher rates of male aggression (Figure 8.4). Thus, dominance striving is unlikely to account for this behavior.

Signaling to Rival Males

Because male dominance striving is such a prominent feature of chimpanzee society, it is possible that an important function of male-female aggression is to

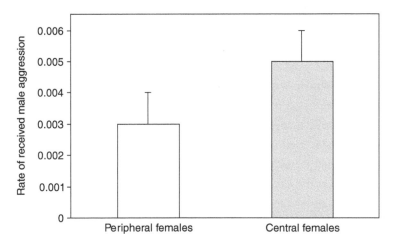

Figure 8.4 Females living in the central (N = 12) region of the Kanyawara home range receive more aggression from males than females living in the peripheral region (N = 10; Mann-Whitney test: P = 0.023). Bars show medians of dyadic medians ± SE. All dyads were observed together ≥ 50 hours (8–13 males per female, median = 11). Central/peripheral designations follow Emery Thompson and Wrangham (2006).

signal a male's fighting ability to other males. Kitchen et al. (Chapter 6 in this volume) refer to this idea as the "male-male competition hypothesis." Rodseth and Novak (Chapter 12 in this volume) advocate a version of the hypothesis for chimpanzees, suggesting that "a male sometimes targets a female to make a political point—not because she represents any serious challenge to him, but precisely because she is politically insignificant and can thus serve as a safe (yet believable) proxy for a real political rival."

A major theoretical problem for Rodseth and Novak's interpretation is that for such signaling to have any benefit for the aggressor, male rivals must attend to its content. However, it is not evident why males should acknowledge a signal that is not costly to produce. Chimpanzees can confidently assume that every adult male in their community is dominant to every female and that a male's ability to dominate females is unrelated to his status in the male hierarchy. Thus, it is unlikely that females could ever serve as a "believable" proxy for a political rival. Perhaps a more probable suggestion is that a male might target a female who has a special (e.g., close kin) relationship with another male in order to provoke that male as part of a status challenge (cf. Wittig and Boesch 2003). However, considering that adult males intervened in only 29 of 1125 episodes of male-female aggression, it seems unlikely that they regularly interpret intersexual aggression as a threat.

A weaker version of the male-male competition hypothesis suggests that males target females not as proxies for their rivals, but simply because they are likely to scream and flee, adding flair to a male's charging display. In this formulation, females would be a chimpanzee equivalent of the empty kerosene cans that the alpha male Mike famously crashed together during charging displays to terrorize larger males at Gombe (Goodall 1986), or the humans that alpha male Charlie dragged in the same context (Wrangham personal observation). But males rarely drag females, preferring to pummel them in place. This suggests that a male's primary motivation is to terrorize his victim rather than to maximize the effect of his attack as a signal to bystanders.

This hypothesis suffers from the same empirical problem that afflicts the stronger version: namely, it fails to explain why male aggression should be directed so often at females, and specifically at fecund females. If males are seeking a convenient proxy for their aggression, or simply attempting to make noise, then any low-ranking victim should suffice, from juvenile males to subadult females to infants. The fact that cycling, parous females are the most frequent victims of male aggression suggests that male motivation is more complex. It might be argued

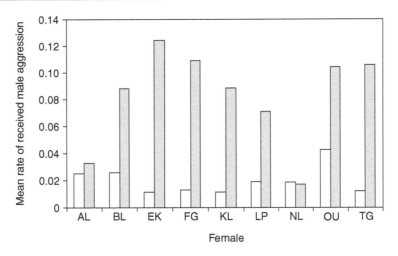

Figure 8.5 Even in large male parties (>6 adult males present), females receive more male aggression when estrous (shaded bars) than when lactating (white bars) (Wilcoxon signed-rank test: P=0.01). Bars indicate overall rates. Females were observed ≥50 hours in each reproductive state.

that estrous females tend to attract large groups of males and that male aggression toward such females can therefore be explained by an audience effect, in which the benefit of advertising one's fighting ability increases in the presence of multiple rivals. However, this would suggest that females should receive high rates of aggression whenever they are in large male parties, regardless of their estrous state. As illustrated by Figure 8.5, this is not the case at Kanyawara.

Male Policing

Another possible function of male-female aggression in chimpanzees is to limit female-female aggression (Watts et al. 2000). At Kanyawara, conflicts occasionally break out among females, and when males are present, they aggressively intervene in these conflicts around 23% of the time (Kahlenberg et al. 2008a). Males typically direct aggression toward the antagonist (loser support) or toward both conflict participants (impartial intervention), and all male interventions abruptly end female conflicts. Females at Kanyawara receive significantly less aggression in the presence of males than when traveling in all-female parties (Figure 8.6; Kahlenberg et al. in press a), which may partly be explained by this male "policing." Although the ultimate explanation for male policing is unclear, it may represent an attempt to retain female immigrants in the commu-

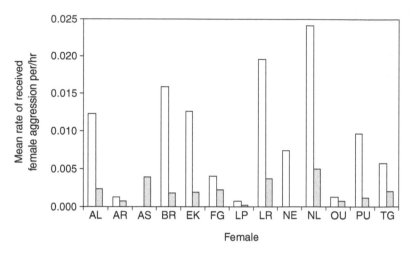

Figure 8.6 Females receive more intrasexual aggression in female-only parties (white bars) than in mixed-sex parties (shaded bars) (Wilcoxon signed-rank test: N=13, P=0.006). Y-values are aggressive acts received from females/hour in association with at least one other female. Mixed-sex parties were those with ≥1 adult male present at the time of aggression. Female-only parties had no adult males at the time of aggression. Females were observed for ≥50 hours in each condition. Data from 1997–2003.

nity, since they are frequent targets of resident female aggression (Nishida 1989; Kahlenberg et al. 2008b). Overall, male policing appears to account for a small but regular proportion of male-female aggression—approximately 7% in one year at Kanyawara (data from 2005, n=248 total events).

Male Aggression as Sexual Coercion

Several lines of evidence suggest that, at Kanyawara, chimpanzee males use aggression as a coercive mating tactic, making females more likely to mate with them and/or less likely to mate with other males (Muller et al. 2007). First, females experience higher rates of male aggression during periods of maximal swelling, when conception is most likely to occur (Emery Thompson 2005), than during lactational amenorrhea. This is true for both absolute rates and dyadic rates (i.e., corrected for party size) of received male aggression. Second, parous females, who are more attractive to males (Muller et al. 2006), receive higher rates of male aggression during estrus than do nulliparous females (Muller et al. 2007). Third, individual males exhibit increased copulation rates with the parous females toward whom they are most aggressive (Figure 8.7).

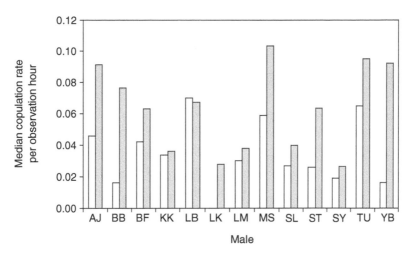

Figure 8.7 We examined median dyadic rates of aggression for each of 13 males with 15 parous females. For each male we calculated median copulation rates with females who received above (>) or below (≤) the median amount of aggression for that male. All dyads were observed together for at least 25 hours when the female was in estrus (2–14 females per male, median = 11). Males copulated at significantly higher rates with the parous females toward whom they were more aggressive (shaded bars) than those toward whom they were less aggressive (white bars; Wilcoxon signed-rank test: n = 13 males, P = 0.002).

The association between rates of copulation and received aggression at Kanyawara is consistent with sexual coercion, but could theoretically have resulted from other processes as well. We therefore tested two alternatives to the coercion hypothesis (Muller et al. 2007). First, it is possible that females were not actually coerced, but preferred to mate with high-ranking males, who happened to be more aggressive. This hypothesis is undermined by our observation that the correlation between aggression and copulation rates was not limited to high-ranking males. Twelve of 13 males at Kanyawara showed elevated copulation rates with the females toward whom they were most aggressive. Furthermore, although focal data do indicate that high-ranking males are generally more aggressive, this relationship is driven by a correlation between rank and male aggression against males. Our data show no correlation either between male dominance rank and total aggression directed at females, or between rates of male aggression against males and male aggression against females. These patterns support the coercion hypothesis and challenge the concept of male-female aggression as an incidental byproduct of male-male competition.

A second potential alternative explanation for the association between mating and aggression is that males were generally aggressive toward any female in direct proximity with them and that females necessarily approached males in order to copulate (Soltis et al. 1997). We tested this hypothesis by analyzing long-term aggression and mating data from nine nulliparous females. If aggression is the inevitable result of close male proximity, then nulliparous females should exhibit the same correlation between mating and aggression found in parous females. If, however, the relationship results from coercion, then no correlation between aggression and mating should exist for nulliparous females because they are less attractive to males, and males show little interest in monopolizing them (Wrangham 2002; Muller and Wrangham 2004a; Muller et al. 2006). Males in our study did not copulate at increased rates with the nulliparous females toward whom they were most aggressive, and we found no association between the amount of aggression directed by males at estrous nulliparous females and the number of times males copulated with those females (Muller et al. 2007).

A final criterion that must be met for male-female aggression to be considered coercive is that there must be a cost to the behavior, such that females would be better off not experiencing it (Smuts and Smuts 1993). Although receiving aggression may appear to be inherently undesirable, it is possible that it might benefit females in some way, for example, by allowing them to test the quality of potential mates (Smuts and Smuts 1993; Szykman et al. 2003). If the evolutionary benefits of such testing outweigh the costs of the aggression itself, then the behavior might not be considered coercive, as it would not generate an evolutionary conflict between males and females (Watson-Capps, Chapter 2 in this volume). This difficulty has led some investigators to conclude that sexual conflict can never be inferred from observations of behavior (Pizzari and Snook 2003). Although this is an important problem, particularly for the studies of invertebrates that comprise the bulk of sexual conflict research, the complex social cognition exhibited by nonhuman primates appears to offer partial mitigation. Chimpanzees, for example, recognize conspecifics as individuals, remember prior social interactions, and form long-term social relationships in relatively stable communities. If female resistance represents a strategy to select "more persistent and thus better quality partners" (Pizzari and Snook 2003), then females are apparently repeatedly testing males who are familiar to them from hundreds of previous interactions. If the costs of such testing are significant, there should be strong selection on females to reduce them by utilizing information from previous interactions in their mating decisions. Therefore, despite the difficulties of establishing sexual conflict from behavioral observations alone, it is useful to ask whether male aggression imposes substantial costs on female chimpanzees.

One obvious cost females incur from male aggression is physical injury. This is sometimes the direct result of contact with male fists, feet, or canines. However, incidental damage also occurs, as females regularly fall from or within trees while attempting to evade male aggressors. During one severe attack at Kanyawara, a female's lower lip was split down the midline to the chin, leaving her lower teeth exposed and the severed tissue hanging. Another female had a fist-size chunk of tissue torn from her sexual swelling, which prevented her from defecating normally for many months. Less serious afflictions, including cuts and swollen areas on female sexual skins, limbs, and faces, are seen regularly but are difficult to quantify. Many injuries go unnoticed because the chimpanzees' dark hair obscures both external wounds and the bruises indicative of internal wounds.

Soft tissue damage is probably the most common result of male attacks, but skeletal wounding appears to occur at remarkably high rates in some populations. Novak and Hatch (Chapter 13 in this volume) examined a Liberian chimpanzee sample and found "localized depression fractures or punctures" that appeared to have resulted from "powerful blows or bites" in more than 46% of young females and 73% of older females, in locations primarily on the back of the skull. It is not possible to confidently attribute all such injuries to male aggression, but given the comparatively low rates of female-female aggression at most sites, they are suggestive.

Because it is difficult to accurately assess rates of wounding in chimpanzees, we have employed noninvasive urine sampling to quantify levels of stress hormone (glucocorticoid) production as a potential measure of the costs of male aggression. Although acute glucocorticoid secretion represents an adaptive response, it also constitutes a physiological cost, as energy must be redirected from processes such as reproduction and growth to meet the demands of the stressor (Sapolsky 2002). Chronic activation of the stress response incurs additional costs, as sustained glucocorticoid exposure is associated with protein breakdown, muscle wasting, immunosuppression, gastric ulcers, and atherosclerosis (Rabin 1999; Sapolsky 2002).

In accordance with the idea that male aggression imposes physiological costs on females, we found that female cortisol (the primary glucocorticoid in chimpanzees) excretion at Kanyawara was elevated in reproductive contexts and that such increases were correlated with female fecundity in a similar manner as rates of aggression. Accordingly, cycling parous females exhibited significantly higher levels of urinary cortisol than cycling nulliparous females, and parous females showed elevated levels of cortisol excretion during estrous periods compared to periods of lactational amenorrhea (Muller et al. 2007). In contrast, nulliparous females showed no difference between average levels of glucocorticoid excretion on estrous versus nonestrous days.

It is difficult to demonstrate conclusively that male aggression was responsible for the observed elevations in female cortisol because energetic stress resulting from increased travel or feeding competition can also affect cortisol excretion (e.g., Muller and Wrangham 2004b). However, the contrast between parous and nulliparous females suggests that aggression was likely the relevant stressor. Even when they are not in estrus, nulliparous females spend more time than parous females in parties with males, incurring associated travel and feeding costs (Wrangham 2000). Yet nulliparous females maintained significantly lower cortisol levels than parous females. Furthermore, estrous nulliparous females at Kanyawara mate with males at rates equivalent to those of estrous parous females (Wrangham 2002; Emery Thompson 2005), yet they did not show the marked increase in cortisol excretion during periods of maximal swelling that parous females did. The salient difference between nulliparous and parous females appears to be their attractiveness to males (Muller et al. 2006), which results in heightened male coercion.

Mechanisms of Sexual Coercion

Because of evidence that male aggression against females both constrained female mate choice and imposed costs on females, we conclude that, at Kanyawara, such aggression frequently functioned as sexual coercion. However, this conclusion says little about the actual means of coercion. In theory, coercive aggression could increase male copulation rates through at least two mechanisms: by overcoming female resistance (direct coercion) and by limiting female promiscuity (indirect coercion/coercive mate guarding). A complete account of sexual coercion in chimpanzees entails consideration of these alternatives.

Indirect Coercion (Sequestration, Herding, and Punishment)

A chimpanzee male will sometimes accompany a female to a peripheral part of their territory for several days to more than a month, and during this time he enjoys exclusive mating access. Several researchers have stressed that male initiation of such "consortships" is often coercive (Goodall 1986; Smuts and Smuts 1993; Clutton-Brock and Parker 1995). In fact, almost all previous discussions of sexual coercion in this species have centered on this mechanism. Goodall (1986, p. 453) noted that if a female is reluctant to follow a male trying to initiate a consortship, he will sometimes employ "a fair amount of brutality" in the form of aggressive displays and attacks. This strategy appears to be effective, since males at Gombe, unless "crippled or very old," are typically able to force females into consorting with them (Goodall 1986).

We classify forced consortships as indirect coercion (specifically as *sequestration:* Muller et al., Chapter 1 in this volume), because they appear designed to restrict a female's access to other males rather than to overcome female mating reluctance. That is, females sometimes resist consorting with males whom they would normally accept as copulation partners in a multimale mating environment. The evolutionary significance of forced consortships is not known, however. Even at Gombe, where consortships appear to be most common, no more than 20% of conceptions occur in consortships (Constable et al. 2001) and only a fraction of these are forced (Goodall 1986). At other sites, such as Taï, forced consortships have not been observed (Boesch and Boesch-Achermann 2000). Only a few consortships of any sort have been recorded in 20 years of observation at Kanyawara (Wrangham 2002). Kyleb Wild describes an unsuccessful attempt at forced consortship from Kanyawara in Box 8.2 (Wild 2006).

When mating occurs in multimale groups, chimpanzee males commonly attempt to maintain exclusive access to estrous females by aggressively guarding them (Tutin 1979; Hasegawa and Hiraiwa-Hasegawa 1983). This is typically a one-male activity, but males are also known to cooperatively guard females at some sites (Gombe: Tutin 1979; Kanyawara: Wrangham personal observation; Ngogo: Watts 1998). Mate guarding is characterized by frequent aggression directed by the guarding male toward rival males to separate them from the estrous female. Guarding males also frequently direct aggression at the estrous female, particularly when attempts to prevent other males from copulating are unsuccessful (Watts 1998). Thus, the most likely mechanism for sexual coercion in chimpanzees is mate guarding, which includes both *herding* (aggression directed toward females to induce immediate separation from rival males and to restore proximity to the guarding male) and *punishment* (physical retribution against female association or copulation with other males).

Like sequestration, herding and punishment are classified as indirect coercion

Box 8.2: A failed attempt to force a consortship at Kanyawara: November 21, 2006

Kyleb Wild

Dominant female Outamba had recently shown the first signs of estrus since her previous birth two years earlier. Although she had not yet reached maximal swelling, she was beginning to attract the attention of some adult males. On this day Yogi, the lowest ranking adult male in the community, was the only adult male traveling in Outamba's party.

In the late morning Yogi displayed toward Outamba and some of the other females. Outamba initially moved away casually, while making submissive pant-grunts, but as Yogi's displays continued, she screamed and tried to avoid him. She then cautiously approached and presented her estrous swelling for Yogi's inspection. When he ignored her gesture, Outamba again began to scream. Yogi abruptly and severely attacked her by hitting, kicking, and dragging her across the ground. Her offspring remained nearby making pant-barks, while the rest of the party moved away from the conflict. The attack injured Outamba's wrist, and she limped slightly for hours afterward.

In a conciliatory gesture, Outamba submissively approached Yogi and placed her hand in his mouth. She again presented her swelling, and this time Yogi inspected it, and briefly groomed her. He then moved a few meters away and Outamba followed him. Yogi shook a nearby tree branch while looking at Outamba, a common signal for females to approach, and Outamba responded by moving closer. Yogi moved a few meters away, shook another branch, and Outamba followed. This sequence was repeated many times over the next seven hours: Yogi soliciting with branch shakes or slaps on the ground, and Outamba following for short distances with her family before stopping to rest, feed, or groom. By nightfall, Outamba and her offspring had followed Yogi a few hundred meters from the other chimpanzees in the area.

The next morning vocalizations from other community members were heard in the distance, but neither Outamba nor Yogi responded. Outamba eventually left her night nest and approached Yogi, making submissive vocalizations. When other chimpanzees were again heard in the distance, Outamba began loudly screaming. Yogi severely attacked her, pounding on her with fists and feet. Outamba continued to scream, and attempted to hit and bite Yogi before fleeing rapidly toward the calls. Yogi followed in close pursuit. After running for several hundred meters, Outamba encountered a group of adult males. The first three males she met displayed and chased her while she screamed and tried to avoid them. Kakama, a high-ranking male, hit Outamba once, and she sat next to him, screaming briefly and pant-grunting. Kakama touched her back reassuringly; then she presented to him, and he inspected her swelling. Three more males, including the alpha, then arrived, making charging displays. Yogi briefly charged toward Outamba but then moved away. Two of the arriving males hit Outamba while she tried to avoid them. Outamba presented her swelling to Big Brown, another high-ranking male, who did not respond to her. The entire group then moved on with Yogi trailing far behind. Yogi made no more solicitation attempts.

because their intent is not to overcome female mating reluctance, but to limit female promiscuity. Because herding and punishment are generally accompanied by male aggression against rival males, they are expected to involve primarily high-ranking males. Consistent with the idea that much male aggression against females represents mate guarding, high-ranking males at Kanyawara exhibit higher rates of aggression toward parous estrous females than do low-ranking males (Figure 8.8).

Direct Coercion (Harassment, Intimidation, and Forced Copulation)

Direct coercion involves the use of force to overcome female resistance to mating. In the short term, this comprises *harassment* (repeated attempts to mate that induce eventual female submission) and, in the longer term, *intimidation* (physical retribution against female refusals to mate). Because direct coercion

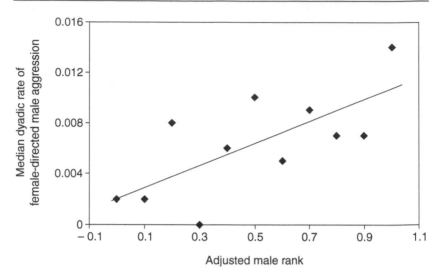

Figure 8.8 High-ranking males direct more aggression towards estrous parous females than do low-ranking males (linear regression: $r^2 = 0.49$, $p = 0.016$). Raw data and details of rank calculation and adjustment methods are given in Kahlenberg (2006). Rates are dyadic (2–21 females per rank, median = 8). Data from 1994–2003.

entails female resistance, it is expected to involve primarily nonpreferred, or low-ranking, males. As previously noted, the fact that higher-ranking males show elevated rates of aggression toward estrous females suggests that sexual coercion in chimpanzees is often indirect. However, the fact that middle- and low-ranking males at Kanyawara, like high-ranking males, copulated most frequently with the females toward whom they were most aggressive, suggests that direct coercion does occur.

The most extreme form of direct coercion, *forced copulation*, rarely occurs in chimpanzees. Tutin (1979), for example, recorded only two instances out of 1137 observed copulations at Gombe. By contrast, in some orangutan populations more than 50% of copulations are forced (Knott, Chapter 4 in this volume). The comparatively low rate in chimpanzees appears to reflect relatively weak female preferences (or rather aversions). In other words, females almost never show extreme resistance to male courtship. For this reason they tend to mate with every male in their community during an estrous cycle (Wrangham 2002). Consistent with this interpretation, Goodall (1986) noted that, at Gombe, forced copulations generally involve close maternal relatives (brothers or sons), for whom females would be expected to maintain strong aversions, owing to the potentially deleterious effects of inbreeding (Pusey and Wolf 1996).

Long-Term Strategies

Although male aggression against females at Kanyawara intensified during estrous periods, it was not restricted to that context. Cycling but anestrous females and lactating females also regularly experienced aggression from males, most of which could not be accounted for by obvious factors like feeding competition or male policing. What is the function of this behavior? Goodall (1986:482) suggested that such acts function to instill a "fearful respect" in females so that they will comply with male demands in future mating situations. If true, then this behavior would constitute a form of *conditioning aggression* designed not to increase a male's immediate mating success, but to manipulate female sexuality over the long term.

The conditioning hypothesis is difficult to test, and in this chapter we have focused primarily on aggression that takes place when females have maximal sexual swellings. However, data from Kanyawara do suggest an important role for long-term processes. We recently discovered, for example, that over a period of 10 years, individual females at our site initiated periovulatory copulations most frequently with the males who had been most aggressive toward them throughout their cycles (i.e., during both estrous and nonestrous periods; Muller et al. in preparation). This finding is clearly consistent with a conditioning strategy and is of particular importance because female-initiated, periovulatory copulations have been used uncritically as an indicator of female preference in chimpanzees (Stumpf and Boesch 2005, 2006). The long-term effects of male aggression on female sexual behavior represent a critical area for future research on the mechanisms of coercion in chimpanzees.

Male Coercion and Female Choice

Distinguishing direct from indirect forms of coercion is complicated by the need to interpret female preferences. Such preferences are difficult to discern in chimpanzees because females mate promiscuously, actively soliciting copulations from multiple males. This complication extends to other Old World primates that maintain exaggerated sexual swellings, as these appear to function as erotic stimuli, promoting male interest, so as to facilitate multimale mating (Hrdy 1981; Zinner et al. 2004).

Although much evidence suggests that multimale mating functions to confuse paternity, and thereby decrease the risk of infanticide by males (Hrdy 1979; van Schaik and Janson 2000; Wolff and Macdonald 2004), van Schaik et al. (2000) note that females who bias paternity toward high-ranking males gain additional protection for their offspring by inducing those males to invest in defense against infanticide. Van Schaik et al. predict that females should mate promiscuously early in the follicular phase in order to give all males a nonzero probability of conception, but that during the periovulatory period, when conception is most likely to occur, females should adopt a preference for high-ranking males (reviewed in Clarke et al., Chapter 3 in this volume).

Chimpanzees present an interesting test case for the van Schaik model because mating is highly promiscuous, and infanticide by adult males within the social group is an important risk for females (Nishida and Kawanaka 1985; Clark Arcadi and Wrangham 1999; Murray et al. 2007). Stumpf and Boesch (2005, 2006) recently evaluated these predictions in two communities of wild chimpanzees living in Taï National Park, Ivory Coast. They quantified female preferences by establishing rates of proceptivity (female-initiated sexual behavior) and resistance (ignoring or avoiding male solicitations) across male-female dyads. They then looked to see whether rates of proceptivity and resistance changed during periods when females were most fecund.

Consistent with the van Schaik model, females at Taï appeared to mate more selectively during periods when they were more likely to conceive. Specifically, rates of female proceptivity were lower, and rates of female resistance higher, during the periovulatory period (POP). Males who were generally approached by females for copulations were approached more frequently during the POP. Males whose sexual advances were generally resisted by females were resisted more consistently during the POP. Females were more proceptive toward males who were high-ranking or soon became high ranking (Stumpf and Boesch 2005).

Although Stumpf and Boesch (2005, 2006) interpret their findings as strong evidence for female choice in chimpanzees, two flaws in their methodology render this conclusion untenable. First, the sample of males in the study was small, with three in the first community and four in the second. Thus, females had a limited pool of mates to choose from, and competition among males for access to females was presumably relaxed in comparison to most sites. Second, and more importantly, the behavioral measures of preference in the study were inadequate because they did not control for the potential effects of male coercion. Specifically, Stumpf and Boesch (2006, p. 750) use the word "preference" to indicate the male "with whom a female would like to mate." However, their behavioral measure of preference is simply how often a female approaches a male for copulation. The unsupported assumption is that females are free to solicit copulations from (or resist the advances of) males at little or no cost, and that female sexual initiations are thus a direct indication of unconstrained preference.

To see how male aggression, or merely the threat of aggression, might constrain female proceptivity, however, consider the following scenario. An estrous female chimpanzee arrives at a fruiting tree in which a group of males is feeding. As she advances, the female directs submissive pant-grunts toward the males; her grunts peak in intensity as she enters the tree and approaches the alpha male. She hurriedly presents her sexual swelling for his inspection, and a brief copulation ensues. The preceding is a common occurrence at our field site and undoubtedly at others, and under the definitions employed by Stumpf and Boesch, would qualify as female proceptive behavior—an indicator of female "preference." However, more data are required to decide whether the female rushed to greet the alpha out of choice (because she preferred him to other males) or from coercion (because she risked violence if she failed to acknowledge his presence).

Measures of female resistance are similarly problematic. Stumpf and Boesch assume that resistance denotes negative preference, but it is often unclear whether a female actively dislikes the prospect of mating with a particular male or is simply afraid that to do so will invite punishment from a higher-ranked suitor. In a detailed study at Mahale, Matsumoto-Oda (1999) observed that copulations with high-ranking males increased significantly during the periovulatory period. This suggested to her that females preferred high-ranking males at this time. However, she could not exclude the alternative hypothesis that high-ranking males were guarding females more closely near ovulation. In support of the latter idea, male solicitations at Mahale were significantly more likely to succeed when higher-ranking males were absent (Matsumoto-Oda 1999). Similarly, almost all female solicitations of adult males failed when higher-ranked males were nearby.

An additional complication is that females may be resistant to male solicitations not because they find a male objectionable, but because they are wary of approaching a potential aggressor. This difficulty is well illustrated by a re-markable sequence of photographs taken by David Bygott in Gombe National Park (Figure 8.9). These show Humphrey, an aggressive, high-ranking male, courting a nulliparous female, Gilka. Although Gilka presents her estrous swelling to Humphrey, she is clearly reluctant to approach him, and she tries to keep a small tree between them. Humphrey threatens Gilka, first shaking a palm frond and then a small bush. She screams and clings to the tree, refusing to aban-don its dubious shelter. Eventually, Humphrey's patience is exhausted, and he at-tacks, striking her on the back. This sequence would clearly be categorized as female resistance in Stumpf and Boesch's methodology. How many observers, though, would feel comfortable in concluding that Gilka had a strong aversion to mating with Humphrey rather than the alternative that she was simply afraid to approach him?

One means of addressing these issues is to test for a correlation between rates of male aggression and rates of female proceptivity and resistance. Stumpf and Boesch (2006) noted that "males who were generally more aggressive toward fe-males did not obtain higher mating success with any female" in their study. However, this measure is insufficient because it does not consider individual, male-female dyads. The relevant question is not whether males who were "gen-erally more aggressive" had higher mating success, but whether males who were particularly aggressive toward individual females were able to influence the be-havior of those females. The Kanyawara data on conditioning aggression alluded to above are consistent with such an interpretation.

Another potential test of the van Schaik hypothesis involves focusing on the behavior of high-ranking males, and particularly the alpha, during the pe-riovulatory period. If females are eager to bias paternity toward the alpha male, and the alpha male is concerned with paternity certainty, then their interests should be aligned during the fecund period. Thus, the alpha male should be less aggressive toward females as ovulation approaches. High rates of aggres-sion in the period immediately preceding ovulation would suggest a conflict of interest and undermine the idea that females behave so as to concentrate pa-ternity in the alpha.

We are currently analyzing long-term data from Kanyawara to test this pre-diction and have thus far found little support for the original van Schaik model. The alpha male at Kanyawara shows no decrease in aggression toward females during the periovulatory period, suggesting a clear conflict of interest. But why should a female's interests diverge from those of the alpha? One solution is

Figure 8.9 (a) Humphrey, an aggressive high-ranking male, sits and shakes a palm frond as he courts Gilka, a young adult female. (b) Gilka presents her estrous swelling while staying on the opposite side of a tree.

Figure 8.9 (c) Humphrey, with hair and penis erect, moves around the tree; Gilka tries to keep the tree between them. (d) Humphrey flails the palm frond more vigorously; Gilka screams and clings to the tree but does not fully present for copulation. (Gilka's abnormally swollen brow and nose are the result of a fungal infection.)

Figure 8.9 (e) Humphrey tries again, shaking the branch of a bush; Gilka stays on her side of the tree. (f) Humphrey, still with erection, stands and grabs hold of Gilka, who crouches screaming.

Figure 8.9 (g) Humphrey begins his attack, slapping Gilka on the back. Photos and captions by David Bygott, Gombe National Park, 1970.

adumbrated by Clarke et al. (Chapter 3 in this volume), who note that, in a fission-fusion species like chimpanzees, females frequently travel alone or in small groups and regularly encounter potentially infanticidal males in the absence of the alpha. Thus, even high-ranking males may not be able to offer reliable protection from infanticide. Under such circumstances, the benefits of paternity confusion may be paramount. If true, then the only potential benefit to females of biasing paternity toward high-ranking males in fission-fusion species would be "good genes." Stumpf and Boesch's model implicitly assumes that such benefits outweigh the risk of infanticide inherent in any attempt to concentrate paternity in a single male. We consider this an open question that should prove a fruitful area for future investigations of male coercion and its interaction with female choice.

Summary

Female chimpanzees are frequent victims of male aggression, which ranges from mild threats to prolonged and vicious attacks. Some of this aggression occurs in the context of feeding competition, male "policing" of female disputes, and—in the case of adolescent male aggression—competition for status. Much aggression appears to function as sexual coercion, however, as fecundable females are the most frequent victims, rates of male aggression correlate with copulation rates across adult-male/parous-female dyads, and aggression is costly for females. Forced copulations are rare among chimpanzees, primarily because females rarely exhibit extreme resistance to male solicitations. Direct coercion in the form of harassment and intimidation appears to occur, however, when low-ranking males attempt to overcome female reluctance to mate in the presence of higher-ranked, mate-guarding suitors. High-ranking males regularly practice coercive mate guarding, which is characterized as a male counterstrategy to female attempts at paternity confusion. At some sites males also engage in forced consortships, but the extent to which this strategy is successful is not clear. Preliminary evidence from Kanyawara suggests that much intersexual aggression in chimpanzees reflects a long-term male strategy of behavioral conditioning, through which males attempt to influence the future mating decisions of females. Such conditioning aggression calls into question previous work on female choice in chimpanzees, which had assumed a priori that male aggression did not influence female mating behavior.

Acknowledgments

Research at Kanyawara was funded by the U.S. National Science Foundation (including grants 0416125, 9729123 and 9807448), the L. S. B. Leakey Foundation, the National Geographic Society, and the Wenner-Gren Foundation. We thank the Uganda Wildlife Authority and Makerere University Biological Field Station for supporting long-term research at Kanyawara, and the field staff of the Kibale Chimpanzee Project for behavioral data collection. Finally, we are grateful to Melissa Emery Thompson for helpful comments on the chapter and to David Bygott for his photographs of Gombe chimpanzees.

References

Altmann, J. "Observational Study of Behavior: Sampling Methods." *Behaviour* 49 (1974): 227–267.

Boesch, C., and H. Boesch-Achermann. *The Chimpanzees of the Taï Forest: Behavioral Ecology and Evolution.* Oxford: Oxford University Press, 2000.

Bygott, J. D. *Agonistic Behaviour and Dominance in Wild Chimpanzees.* Ph.D. Dissertation. Cambridge: Cambridge University 1974.

Clark Arcadi, A., and R. W. Wrangham. "Infanticide in Chimpanzees: Review of Cases and a New Within-Group Observation from the Kanyawara Study Group in Kibale National Park." *Primates* 40 (1999): 337–351.

Clutton-Brock, T. H., and G. A. Parker. "Sexual Coercion in Animal Societies." *Animal Behavior* 49 (1995): 1345–1365.

Constable, J. L., M. V. Ashley, J. Goodall, and A. E. Pusey. "Noninvasive Paternity Assignment in Gombe Chimpanzees." *Molecular Ecology* 10 (2001): 1279–1300.

Emery Thompson, M. "Reproductive Endocrinology of Wild Female Chimpanzees *(Pan troglodytes schweinfurthii):* Methodological Considerations and the Role of Hormones in Sex and Conception." *American Journal of Primatology* 67 (2005): 137–158.

Emery Thompson, M., and R. W. Wrangham. "Comparison of Sex Differences in Gregariousness in Fission-Fusion Species: Reducing Bias by Standardizing for Party Size." In *Primates of Western Uganda,* eds. N. Newton-Fisher, H. Notman, V. Reynolds, and J. Paterson, pp. 209–226. New York: Springer, 2006.

Goodall, J. *The Chimpanzees of Gombe: Patterns of Behavior.* Cambridge, Mass.: Harvard University Press, 1986.

Graham, C. E. "Menstrual Cycle of the Great Apes." In *Reproductive Biology of the Great Apes: Comparative and Biomedical Perspectives,* ed. C. E. Graham, pp. 1–43. New York: Academic Press, 1981.

Hasegawa, T., and M. Hiraiwa-Hasegawa. "Opportunistic and Restrictive Matings Among Wild Chimpanzees in the Mahale Mountains, Tanzania." *Journal of Ethology* 1 (1983): 75–85.

Hohmann, G., and B. Fruth. "Intra- and Inter-Sexual Aggression by Bonobos in the Context of Mating." *Behaviour* 140 (2003): 1389–1413.

Hrdy, S. B. "Infanticide Among Animals: A Review, Classification, and Examination of the Implications for the Reproductive Strategies of Females." *Ethology and Sociobiology* 1 (1979): 13–40.

Hrdy, S. B. *The Woman That Never Evolved.* Cambridge, Mass.: Harvard University Press, 1981.

Kahlenberg, S. M. "Female-Female Competition and Male Sexual Coercion in Kanyawara Chimpanzees." Ph.D. Dissertation. Harvard University, Cambridge, 2006.

Kahlenberg, S. M., M. Emery Thompson, M. N. Muller, and R. W. Wrangham. "Immigration Costs for Female Chimpanzees and Male Protection as an Immigrant Counterstrategy to Intrasexual Aggression." *Animal Behaviour* 76 (2008a): 1497–1509.

Kahlenberg, S. M., M. Emery Thompson, and R. W. Wrangham. "Female Competition over Core Areas in *Pan troglodytes schweinfurthii,* Kibale National Park, Uganda." *International Journal of Primatology* 29 (2008b): 931–947.

Matsumoto-Oda, A. "Female Choice in the Opportunistic Mating of Wild Chimpanzees *(Pan troglodytes schweinfurthii)* at Mahale." *Behavioral Ecology and Sociobiology* 46 (1999): 258–266.

Matsumoto-Oda, A., and R. Oda. "Changes in the Activity Budget of Cycling Female Chimpanzees." *American Journal of Primatology* 46 (1998): 157–166.

Muller, M. N. "Agonistic Relations among Kanyawara Chimpanzees." In *Behavioural Diversity in Chimpanzees and Bonobos,* eds. C. Boesch, G. Hohmann, and L. Marchant, pp. 112–124. Cambridge: Cambridge University Press, 2002.

Muller, M. N., M. Emery Thompson, and R. W. Wrangham. "Male Chimpanzees Prefer Mating with Old Females." *Current Biology* 16 (2006): 2234–2238.

Muller, M. N., S. M. Kahlenberg, M. Emery Thompson, and R. W. Wrangham. "Male Coercion and the Costs of Promiscuous Mating for Female Chimpanzees." *Proceedings of the Royal Society B: Biological Sciences.* 274 (2007): 1009–1014.

Muller, M. N., and J. C. Mitani. "Conflict and Cooperation in Wild Chimpanzees." *Advances in the Study of Behavior* 35 (2005): 275–331.

Muller, M. N., and R. W. Wrangham. "The Reproductive Ecology of Male Hominoids." In *Reproductive Ecology and Human Evolution,* ed. P. T. Ellison, pp. 397–427. New York: Aldine, 2001.

———. "Dominance, Aggression and Testosterone in Wild Chimpanzees: A Test of the 'Challenge Hypothesis'." *Animal Behaviour* 67 (2004a) 113–123.

———. "Dominance, Cortisol and Stress in Wild Chimpanzees *(Pan troglodytes schweinfurthii)*." *Behavioral Ecology and Sociobiology* 55 (2004b): 332–340.

Murray, C. M., E. Wroblewski, and A. E. Pusey. "New Case of Intragroup Infanticide in the Chimpanzees of Gombe National Park." *International Journal of Primatology* 28 (2007): 23–37.

Newton-Fisher, N. "Female Coalitions against Male Aggression in Wild Chimpanzees of the Budongo Forest." *International Journal of Primatology* 27 (2006): 1589–1599.

Nishida, T. "Social Interactions between Resident and Immigrant Female Chimpanzees." In *Understanding Chimpanzees,* eds. P. G. Heltne and L. A. Marquardt, pp. 68–89. Cambridge, Mass.: Harvard University Press, 1989.

———. "Harassment of Mature Female Chimpanzees by Young Males in the Mahale Mountains." *International Journal of Primatology* 24 (2003): 503–514.

Nishida, T., and M. Hiraiwa-Hasegawa. "Responses to a Stranger Mother-Son Pair in the Wild Chimpanzee: A Case Report." *Primates* 16 (1985): 1–13.

Nishida, T., T. Kano, J. Goodall, W. C. McGrew, and M. Nakamura. "Ethogram and Ethnography of Mahale Chimpanzees." *Anthropological Science* 107 (1999): 141–188.

Nishida, T., and K. Kawanaka. "Within-Group Cannibalism by Adult Male Chimpanzees." *Primates* 26 (1985): 274–284.

Pizzari, T., and R. R. Snook. "Sexual Conflict and Sexual Selection: Chasing Away Paradigm Shifts." *Evolution* 57 (2003): 1223–1236.

Pusey, A. E. "Behavioural Changes at Adolescence in Chimpanzees." *Behaviour* 115 (1990): 203–246.

Pusey, A. E., and M. Wolf. "Inbreeding Avoidance in Animals." *Trends in Ecology and Evolution* 11 (1996): 201–206.

Rabin, B. S. *Stress, Immune Function, and Health.* New York: Wiley-Liss, 1999.

Sapolsky, R. M. "Endocrinology of the Stress-Response." In *Behavioral Endocrinology,* eds. J. B. Becker, S. M. Breedlove, D. Crews, and M. M. McCarthy, pp. 409–450. Cambridge, Mass.: MIT Press, 2002.

Smuts, B. B., and R. W. Smuts. "Male Aggression and Sexual Coercion of Females in Nonhuman Primates and Other Mammals: Evidence and Theoretical Implications." *Advances in the Study of Behavior* 22 (1993): 1–63.

Soltis, J., F. Mitsunaga, K. Shimizu, Y. Yanagihara, and M. Nozaki. "Sexual Selection in Japanese Macaques I: Female Mate Choice or Male Sexual Coercion?" *Animal Behaviour* 54 (1997): 725–736.

Struhsaker, T. T. *Ecology of an African Rain Forest.* Gainesville: University Press of Florida Press, 1997.

Stumpf, R., and C. Boesch. "Does Promiscuous Mating Preclude Female Choice? Female Sexual Strategies in Chimpanzees *(Pan Troglodytes Verus)* of the Tai National Park, Cote d'Ivoire." *Behavioral Ecology and Sociobiology* 57 (2005): 511–524.

———. "The Efficacy of Female Choice in Chimpanzees of the Tai Forest, Cote d'Ivoire." *Behavioral Ecology and Sociobiology* 60 (2006): 749–765.

Szykman, M., A. L. Engh, R. C. Van Horn, E. E. Boydston, K. T. Scribner, and K. E. Holekamp." Rare Male Aggression Directed toward Females in a Female-Dominated Society: Baiting Behavior in the Spotted Hyena." *Aggressive Behavior* 29 (2003): 457–474.

Takahata, Y. "Adult Males' Social Relationships with Adult Females." In *The Chimpanzees of the Mahale Mountains: Sexual and Life History Strategies,* ed. T. Nishida, pp. 133–148. Tokyo: University of Tokyo Press, 1990.

Tutin, C. E. G. "Mating Patterns and Reproductive Strategies in a Community of Wild Chimpanzees." *Behavioral Ecology and Sociobiology* 6 (1979): 39–48.

van Schaik, C. P., and C. H. Janson, eds. *Infanticide by Males and Its Implications.* Cambridge: Cambridge University Press, 2000.

van Schaik, C. P., G. R. Pradhan, and M. A. van Noordwijk. "Mating Conflict in Primates: Infanticide, Sexual Harassment and Female Sexuality." In *Sexual Selection in Primates: New and Comparative Perspectives,* eds. P. Kappeler and C. P. van Schaik, pp. 131–150. Cambridge: Cambridge University Press, 2004.

Watts, D. P. "Coalitionary Mate Guarding by Male Chimpanzees at Ngogo, Kibale National Park, Uganda." *Behavioral Ecology and Sociobiology* 44 (1998): 43–55.

Watts, D. P., F. Colmenares, and K. Arnold. "Redirection, Consolation, and Male Policing: How Targets of Aggression Interact with Bystanders." In *Natural Conflict Resolution,* eds. F. Aureli and F. B. M. de Waal, pp. 281–301. Berkeley: University of California Press, 2000.

Watts, D. P., and J. C. Mitani. "Boundary Patrols and Intergroup Encounters in Wild Chimpanzees." *Behaviour* 138 (2001): 299–327.

Wild, K. Unpublished field notes. November 21, 2006.

Williams, J. M., G. W. Oehlert, J. V. Carlis, and A. E. Pusey. "Why Do Male Chimpanzees Defend a Group Range?" *Animal Behaviour* 68 (2004): 523–532.

Wolf, K., and S. R. Schulman. "Male Response to 'Stranger' Females as a Function of Female Reproductive Value among Chimpanzees." *American Naturalist* 123 (1984): 163–174.

Wolff, J. O., and D. W. Macdonald. "Promiscuous Females Protect Their Offspring." *Trends in Ecology and Evolution* 19 (2004): 127–134.

Wrangham, R. W. "On the Evolution of Ape Social Systems." *Social Science Information* 18 (1979): 335–368.

———. "Why Are Male Chimpanzees More Gregarious Than Mothers? A Scramble Competition Hypothesis." In *Primate Males*, ed. P. M. Kappeler, pp. 248–258. Cambridge: Cambridge University Press, 2000.

———. "The Cost of Sexual Attraction: Is There a Tradeoff in Female *Pan* between Sex Appeal and Received Coercion?" In *Behavioural Diversity in Chimpanzees and Bonobos*, eds. C. Boesch, G. Hohmann, and L. Marchant, pp. 204–215. Cambridge: Cambridge University Press, 2002.

Zinner, D. P., C. L. Nunn, C. P. van Schaik, and P. M. Kappeler. "Sexual Selection and Exaggerated Sexual Swellings of Female Primates." In *Sexual Selection in Primates: New and Comparative Perspectives*, eds. P. Kappeler and C. P. van Schaik, pp. 71–89. Cambridge: Cambridge University Press, 2004.

9

Sexual Coercion in Dolphin Consortships: A Comparison with Chimpanzees

Richard C. Connor and Nicole L. Vollmer

Until we published our observations of sexual coercion in Indian Ocean bottlenose dolphins in the early 1990s, they enjoyed, at least in the public view, a reputation of being gentle creatures almost incapable of aggression. When infanticide was reported in populations from Scotland and Virginia a few years later (Patterson et al. 1998; Dunn et al. 2002), it became clear that conflicts between the sexes over reproduction were as important in shaping the reproductive strategies of male and female dolphins as they are in many other mammals.

What struck us immediately was how similar the pattern of sexual coercion in dolphins was to that found in common chimpanzees. In both species, individual females are coerced into accompanying males on consortships for extended periods. This was intriguing because, like chimpanzees, the bottlenose dolphins *(Tursiops* sp.*)* that inhabit Shark Bay, Western Australia are large-brained, highly social mammals that exhibit complex within-group male alliances in the context of a highly dynamic fission-fusion grouping pattern (Connor et al. 2000; Connor and Mann 2006). It seemed astonishing that animals living in such disparate habitats might have converged to such a degree. On the other hand, there were also striking differences between chimpanzee and dolphin societies, including the frequency and nature of sexual coercion. We might expect that those very different habitats they live in would speak to this issue.

Therefore, we begin this chapter by describing the Shark Bay bottlenose dolphin society, as it is the social context of sexual coercion that invites comparison with chimpanzees. We then describe the pattern of sexual coercion in the dolphins, considering not only why it occurs, but why females must tolerate it and possible counterstrategies they employ. The consequence of sexual coercion, being monopolized, exacerbates another set of costs from males, the risk of infanticide, and provokes another female counterstrategy, multiple cycling. We then explore contrasts between Shark Bay and the well-studied population of bottlenose dolphins in Sarasota, Florida. This comparison suggests that a tailored comparative study of mating associations at the two sights would be highly informative regarding the ecology of sexual coercion in dolphins. In the second half of this chapter, we document similarities and differences between the pattern of coercive consortships in dolphins and chimpanzees, and then discuss why this style of coercion is less common among chimpanzees. Our review suggests that an important factor favoring coercive consortships is the rate at which estrous females are encountered by males. The comparison is useful because it directs attention to a little-considered variable.

Social Group Size and Fission-Fusion Grouping Dynamics in the Shark Bay Dolphins

Bottlenose dolphins and some other inshore delphinids exhibit the classic "chimp-like" or "atomistic" fission-fusion grouping pattern in which individuals associate in small parties that change frequently in size and composition as individuals and subgroups join and leave.

In Shark Bay this fission-fusion grouping occurs in an open social network (Connor et al. 1992b, 2000). In a > 600 km² study area, we have not discovered any social divisions reminiscent of the intergroup boundaries that define the chimpanzee community, the baboon "troop," or any other semiclosed primate group (Connor and Mann 2006). Rather, we find in the dolphins a mosaic of overlapping home ranges of variable size that extend throughout the study area and likely beyond; for example, the range of dolphin "A" overlaps with B but not C, while the range of B overlaps with C. This combination of strong and well-differentiated social bonds and an open social network is unusual and has interesting cognitive implications above and beyond that sometimes posited for fission-fusion societies, where interactions impacting social relationships often occur "off-camera" with respect to a given individual (Smolker et al. 1992; Connor 2007). Whereas primates interact with all members of their social group and likely know the kin and dominance relations of others, dolphins will often have incomplete knowledge about the social relationships of their associates because they overlap little or not at all with individuals their associates may know well (Connor and Mann 2006; Connor 2007). Although we cannot rule out the existence of a community boundary beyond our study area, such a finding would not change the basic fact that individuals in the southern part of our study range are unfamiliar with individuals in the north, even though they share many associates.

Philopatry and Sex-Specific Bonds in Bottlenose Dolphins

In terrestrial mammals members of one sex, usually female, remain in their natal area while the other sex disperses. Their low cost of locomotion and concomitant large ranges, in combination with the fact that they do not crèche their infants, allow dolphins and whales to violate the terrestrial "one sex disperses" rule (Connor 2000). Given that both sexes remain in their natal range, the important variable becomes the extent to which males and females remain with their mothers. At one extreme we find both sexes remaining with their mothers in the resident or "fish-eating" killer whales *(Orcinus orca)* off southwestern Canada. Bottlenose dolphins exhibit another extreme, where females, but not males, continue to associate strongly with their mothers as adults, even though both sexes continue to use their natal ranges (Connor et al. 2000). In Shark Bay, many males continue to use their natal range as adults, but we cannot rule out emigration by some males or even females for that matter.

Male–Male Bonds: Nested Alliances

The strongest associations among adult dolphins in Shark Bay are found between males that form alliances. Bottlenose dolphin alliances are clearly associated with mating tactics. Two to three males cooperate in "first-order" alliances to form and maintain consortships with individual females, and "second-order alliances," composed of teams of first-order alliances, cooperate in conflicts with other males over females. Second-order alliances range in size from 4 to 14 or more males. Associations between males in first-order alliances can be strong and durable, with males exhibiting pairwise coefficients of association near 80–100 (on a scale of 0–100) each year for up to two decades (Connor and Mann 2006). In contrast, some males form trios and pairs with a number of other males, switching partners on a regular basis.

One possible correlate of pair and trio stability is the size of their second-order alliance. The very strongly bonded males tend to associate in small groups; one typically finds two first-order alliances together. At the other extreme, in one 14-member "super-alliance," trio membership varied as males often switched partners between consortships (Connor et al. 1999). Superalliance males differed in the stability of their first-order alliances, but even the most stable trio exhibited coefficients of association below that of males that associate in smaller groups (Connor et al. 2001). Further analysis of relationships within the superalliance revealed a complex social structure, with males exhibiting strong preferences in their choice of which males to partner with for consortships. Interestingly, trio stability in the super-alliance was correlated with days observed consorting females (Connor et al. 2001).

Relatedness may be another key factor that impacts the size and stability of male groups. Males in stable alliances were related more often than expected by chance, whereas males in the super-alliance were not, even if partner preferences were included in the analysis (Krutzen et al. 2003).

Female–Female Bonds: Variation and Foraging Specializations

Female dolphins exhibit moderately strong associations in many cases, but perhaps the most interesting feature of female-female associations is their variability. Some females are clearly much more social than others, as measured by the number of social affiliates and the strength of particular female-female associations (Smolker et al. 1992; Connor and Mann 2006; Connor et al. in prep). This is likely related to the pronounced variation in individual foraging strategies among Shark Bay females. Such variation within a social network is not uncommon in marine mammals (see Connor 2001) but is rare in terrestrial mammals, including primates. The well-known "sponge-carrying" (Smolker et al. 1997; Krutzen et al. 2005) is a striking example of a foraging specialization whose participants are relatively asocial, likely because of the distribution of targeted prey (i.e., patch size) and the time required to forage successfully in this manner (Connor et al. 2000a). Variation between groups in foraging tactics also provides a possible explanation for variation in the size of second-order alliances among males.

The duration and strength of many male-male associations conceal an alliance "fickleness" as long-term alliance relationships are sometimes terminated and new ones formed (see below; Connor et al. 1992b; Connor and Mann 2006). Thus, female-female associations in dolphins may not be as strong as male-male associations, but they may, at least compared to males, be relatively immune to shifts of strategic winds. Such stability might be expected if female associations are based more on kinship, a topic currently under investigation.

Coercive Consortships of Females by Male Bottlenose Dolphins

We have no evidence of direct sexual coercion in dolphins, including forced copulation or other behaviors *directly* associated with male attempts to mate (e.g., harassment or intimidation: Clutton-Brock 1995; Muller et al. Chapter 1 in this volume). We have documented indirect sexual coercion in the form of consortships initiated and maintained by aggression (Connor et al. 1992, 1996). Evidence of coercion in consortships was reported in over half the consortships by nonprovisioned males observed by Connor et al. (1992a, b) and Watson-Capps (2005). However, the percentage of consortships that are coerced is likely much higher, and we cannot exclude the possibility that it may approach 100%. Our uncertainty in this domain is due to the simple fact that establishing that a consortship is coerced is difficult, requiring observation of any of several brief events at or near the surface indicating that the males are controlling the movements of, or "herding," the female. These events include the males biting the female, chasing her during a capture or after she bolts from them, and hitting her and charging at her or producing threat vocalizations called "pops," in air, where they can be heard and localized by observers in boats (Connor and Smolker 1996). In one case, a consortship was observed for 11 hours over a two-day period; in the space of a few seconds, one of the males surfaced beside the female, popped at her, head-jerked, then charged and hit her (Connor et al. 1996).

The aggression observed in consortships is clearly a male tactic designed to sequester and intimidate females into staying with the males and not simply because males tend to be aggressive to females in their group. This is demonstrated by the long chases in some captures, the extensive chases to retrieve females who have bolted from the males, the fact that females who have received aggression continue to remain with males though other activities such as resting and foraging, and the "pop" vocalization, which is clearly a "come hither" signal not heard systematically in any other context (Connor et al. 1992a, b; 1996; Connor and Smolker 1996; Vollmer and Connor in prep).

Male dolphins use the cover of their second-order alliance partners to protect their consortships. Especially during the mating season, males in consortships are almost always with or near other members of their group. We have also observed males escorting their females to locales that are used infrequently by rival males, suggesting that males may sometimes actively avoid other male groups.

The duration of some dolphin consortships greatly exceeds the 5– to 7–day period predicted from captive data on estrogen levels, and a few exceed the estimated length of an estrus cycle (30 days). Taking a tip from chimpanzee studies, we suggested that some consortships are established in *anticipation* of female estrus (Connor et al. 1996). Goodall (1986) reported that 21% of consortships by male chimpanzees were initiated when the female was flat.

Consortships are clearly an important mating tactic by male dolphins in Shark Bay, but we cannot say it is the only one. Connor et al. (1996) reported associations between estrus females and several male alliances that resembled the opportunistic mating parties of chimpanzees (Goodall 1986). However, this occurred in a highly unusual circumstance, where the three provisioned females remained in a limited area (offshore of the provisioning beach). All three females had recently lost their calves due to a pollution event at the beach and had resumed cycling at the same time. Connor et al. (1996) also suggested that old males, who consort females infrequently, may employ alternative tactics and may even be preferred by females. However, the low frequency with which they consort females may also reflect a decline in their competitive ability.

Conflict is expected within alliances because they consort with one female at a time and paternities cannot be shared. We do not know how mating opportunities or paternities are distributed within a pair or a trio. Observations of mating are not very useful in this regard. Bottlenose dolphins may rival or even exceed bonobos in the extent to which sex occurs in nonreproductive contexts (Connor et al. 2000; Mann 2006). Shark Bay bottlenose dolphins, from infants to adults, engage in frequent social sex (e.g., male-male mounting is very common), so observations of mounting during consortships are of dubious value for estimating paternity (and intromission itself is difficult to observe). The distribution of paternities from a preliminary genetic study hints at some skew in mating success within the more stable pairs and trios (Krutzen et al. 2004).

Although first-order alliances depend on second-order alliance partners to help defend their consortships from rival males, conflict is expected in these relationships as well. Bottlenose dolphins negotiate their social relationships by petting and stroking each other with their flippers (Connor et al. 2000). Synchrony may also be used to signal affiliation in dolphins (Figure 9.1). Two

Figure 9.1 Two Shark Bay males perform a synchronous display on opposite sides of a female. Photo by Richard Connor.

or three allied males often swim side-by-side breaking the surface in unison. When two allied first-order alliances are together, synchrony is (not surprisingly) found most often between members of the same pair or trio (Connor et al. 2006). Occasionally, males from different first-order alliances surface synchronously. Between-alliance synchrony occurs significantly more often when the males are excited around female consorts. These findings suggest that synchrony reflects social tension as well as social bonds in a manner similar to the well-known genital-genital rubbing of bonobos (Connor et al. 2006).

Rates of Aggression during Consortships

Obtaining an accurate estimate for rates of male dolphin aggression toward females is difficult for several reasons. First, the dolphins spend much of their time beneath the surface and out of sight. There is no reason to think that aggression toward females occurs only at the surface; our hydrophones clearly indicate otherwise. Second, sea state can significantly impede our ability to detect surface aggression, especially behaviors such as pops, which are much more likely to be heard in calm conditions. Third, many high-intensity bouts of behavior involve the dolphins splashing and racing around, with frequent group direction changes, making it difficult for observers to keep up and to see who is doing what to whom. Male-male aggression occurs often enough during consortships that we cannot assume that all acts are directed toward the female. Finally, some behaviors, even when observed clearly, are ambiguous as to whether they are aggressive. Mounting, chasing, and "goosing," whereby one individual directs its rostrum into the genital area of another, may occur in aggressive but also in more playful or affiliative contexts (Connor et al. 2000).

Given these limitations, what is the observed rate of aggression, including threats, toward females by males during consortships? Most of the consortships reported by Connor (1992a, b) involved 8 males who associated in three alliances. We examined rates of aggression during 113 hours (on 30 days) of focal observations on these 8 males during consortships with 15 different females. We considered only behaviors that were clear threats or acts of aggression, including hits, biting, head-jerks, charging at the female, and pops. Given that these behaviors occur as briefly observed events that may occur in a sequence, we counted bouts of aggression rather than each event (pop or charge). We defined a bout as including all events occurring within five minutes of each other; a period that covers about two to three surfacing bouts. By this criterion, only 5 of 35 intervals between aggressive events recorded during the same follow were between 5 and 10 minutes (average 48 minutes). The rate of aggression directed at consorted females was 0.51 bouts per hour, or about one bout every two hours. This falls to 0.48 bouts/hour if you eliminate the four captures, to include only rates during "established" consortships. These values are somewhat higher than Watson-Capps (2005) found during 126 hours of observing cycling females in consortships (0.36 events/hour, or once every 2.8 hours).

An estimate of how much we underestimate rates of aggression by focusing exclusively on surface observations can be obtained from underwater recordings during follows. The majority (75%) of aggressive bouts in our sample included only observations of males popping while resting at the surface. Vollmer and Connor (in prep) compared the rates that pops were heard underwater during consortships versus follows when males were not considered to be consorting females. During 49.1 hours of continuous recordings during 10 consortships, a total of 81 bouts of pops were heard for an average rate of one bout every 36 minutes of recording, or 1.65 bouts/hour.

The underwater pop data are problematic because we cannot say which dolphins produced the pops, but they do suggest that rates of aggression and threats toward female consorts, especially the pop vocalization, may be considerably higher than the most generous estimates based on surface observations. More accurate estimates of pop rates may be obtained by combining focal observations with continuous recording using a hydrophone array, which will improve our ability to identify which male or alliance pops are coming from. Some readers will think the rates of underwater pops should be much higher still, given the rate pops are heard at the surface. But a typical context in which we hear pops is preceding a change in the direction of group travel, and the males often rest at the surface, where pops are audible in air, before initiating a change in direction.

Many factors could impact rates of aggression by male dolphins toward females during consortships. One possibility has already been suggested: pops may be heard more often if males are engaging in behavior that includes more direction changes. Watson-Capps (2005) found that rates of aggression were higher toward younger females who may require more "training." Less desirable males might be expected to behave aggressively to maintain a consortship. Range overlap should impact rates of aggression significantly. Range size varies considerably for males and females. Consider the case where a female's range is contained entirely within the range of a consorting alliance versus the case where only a slight overlap exists. In the latter case, if the female escapes to part of her range beyond the males' range, the males are much less likely to recapture her. We predict that the frequency of popping by males will be correlated with the degree of range overlap between the consorting alliance and the female.

Coercion as a Male Mating Tactic: Costs and Benefits

The costs and benefits of coercive consortships in dolphins must be evaluated against the alternative: mate guarding, which requires that males behave aggressively toward others only to prevent mating between the female and other males. Controlling a female's movements is obviously more costly than simply following her. The payoff for this additional cost must be found in an increased chance to inseminate her, primarily by reducing the risk that other males usurp the female. Especially during the mating season, two or more alliances in a second-order alliance will have a female consort. It may often behoove males to keep females away from more powerful groups in their home range, and they may want to avoid leaving their normal range to follow a female to areas where they might encounter unfamiliar male groups. If male dolphins simply followed females, they would often be led away from each other to each female's core foraging area, where they would be vulnerable to attack by other second-order alliances.

Shark Bay versus Sarasota

Another bottlenose dolphin research site, Sarasota Bay, Florida, offers an interesting contrast to Shark Bay. In Sarasota, adult males form pairs, but trios are unknown and some males travel alone (Wells 2003). Connor et al. (2000) suggested that this difference might be due to differences in population density. A lower population density translates into a lower encounter rate among rival males in competitive circumstances, reducing selection for alliance formation (Connor and Whitehead 2005). Data found in Watson-Capps (2005) support this hypothesis; Watson-Capps reports a density in Shark Bay of 2.4 dolphins/km^2, compared with 1.0 dolphins/km^2 in Sarasota. The lower encounter rate in Sarasota between males in the company of females allows males to consort in pairs or alone and may explain the lack of a second-order alliance structure in Sarasota. However, the lower encounter rates reduce selection not only for male alliance formation, but also for herding relative to guarding. We expect that, relative to Shark Bay, selection for herding is reduced by two factors: a reduced likelihood of encountering rival males and the lack of a need to remain near second-order alliance partners. A comparative study of coercion during consortships in the two populations would be revealing.

Counterstrategies by Female Dolphins

Coerced consortships impose costs on females from a risk of injury (Connor et al. 1992a, 1996; Scott et al. 2005), reduced foraging time (Watson-Capps 2005), and loss of mate choice options (Connor et al. 1996). Females might prefer to mate with a different male, or, more likely, females might prefer to mate with many males (Connor et al. 1996). Females can avoid mating with particular males by rolling away from them (see Watson-Capps 2005), but males might be able to coerce mating by intimidation or by cooperating to physically control the females. Given the paucity of copulations observed, we cannot say if male dolphins engage in either or both of these tactics. Females often pet or stroke males shortly after receiving aggression; this might be an example of submissive behavior to appease aggressive males (Watson-Capps 2005).

Like males, females use petting and stroking in affiliative interactions, but we also find an affiliative contact behavior that, among adults, is almost exclusive to interactions between females. In "contact swimming," one female rests her flipper against the side of another, slightly behind and below her dorsal fin, so that the two females appear "glued together" for periods of seconds to 20 minutes (Richards 1996; Connor et al. 2006). Most occurrences of contact swimming were recorded in male-biased groups, often in cases where one of the females was being herded or harassed. Given this context, it is tempting to speculate that contact swimming might be associated with female-female cooperation against aggressive males. This would not be surprising given that sexual size dimorphism is minimal in the Shark Bay population. However, cooperative female aggression has been observed against male juveniles only (Connor et al. 1992b). Such coalitions may be important to females, given the costs that juvenile males may inflict on adult females. Female coalitions against adult males would have to be strong enough, in most cases, to defeat not just the consorting males but the entire second-order alliance. The lack of female-female coalitions against males led Connor et al. (2006) to suggest that contact swimming is a general affiliative signal. While contact swimming is seen more often when one participant is the target of male aggression and harassment, it occurs in other contexts as well (e.g., two females traveling alone: Richards 1996).

Although joint aggression by females against adult males has not been observed in Shark Bay, several possible cases of cooperative deception by females against adult males have been seen. Connor and Mann (2006) describe a case where females may have cooperated to conceal a consorted female, allowing her to escape from her captors.

Female bottlenose dolphins in captivity ovulate two to seven times during the year they conceive, and this is reflected in the multiple attractive periods over several months exhibited by female dolphins in Shark Bay (Connor et al. 1996). Given the costs of coercive consortships to female dolphins (Watson-Capps 2005), multiple cycling must have important benefits to females. Connor et al. (1996) suggested that multiple female cycling (including possibly nonconceptive cycles) in dolphins is a tactic to confuse paternity and reduce the risk of infanticide—which was unknown in dolphins at the time but has been discovered subsequently in at least two locations (Patterson et al. 1998; Dunn et al. 2002) but not Shark Bay. Dolphin infanticide was previously overlooked, it seems, because the killing blows often fail to leave obvious external injuries and possible direct observations may have been interpreted benignly (e.g., "calf tossing": Dunn et al. 2000). Connor, Read, and Wrangham (2000) pointed out that infanticide might be widespread in odontocetes, given that breeding is often diffusely seasonal, that lactation exceeds the gestation period, and, in at least some species, that there is a year-round association between males and females (see Van Schaik and Kappeler 2001). The risk of encountering unfamiliar males may strongly favor multiple cycling in fission-fusion species that lack such year-round associations, especially in the open social network of bottlenose dolphins.

Coerced Consortships in Dolphins and Chimpanzees

The aggressive herding of females that we observe in Shark Bay shows interesting similarities to chimpanzee consortships. In fact, initial recognition that consortships are coerced in Shark Bay owes much to an observation from Gombe. In 1986 a provisioned male dolphin called Snubby, accompanied by a female consort, approached the beach in Shark Bay where he and several other dolphins were hand-fed. The unhabituated female refused to approach closely, so Snubby swam to a point halfway between the female and a person offering a favorite fish, and there he remained, whirling back and forth toward the female and the fish bucket. That same year Jane Goodall's book, *The Chimpanzees of Gombe,* appeared with a strikingly similar account of the provisioned male Goliath rushing back and forth between bananas and an unhabituated female consort.

Coercive consortships are one of three mating tactics described in chimpanzees (Tutin 1979). Most mating is *opportunistic* and occurs in multimale groups with estrous females. *Possessive* mating also occurs in multimale groups, but individual males maintain exclusive mating access to an estrous female by directing aggression at both the female and other males in the group. This strategy, performed by high-ranking males near the time of ovulation, is more successful than is suggested by mating frequency alone (Muller and Mitani 2005). Watts (1998) reported possessive mate guarding by cooperating male pairs in the unusually large community of Ngogo in Kibale. Coercive *consortships* are always performed by one male and are not the most common mating tactic in any population of chimpanzees (Muller and Mitani 2005). A key variable impacting the rate at which male chimpanzees threaten females during consortships will be risks from encountering other males. Watts (1998) reported a relatively high rate of three to four aggressive events per hour directed at females guarded by pairs of males. We would expect much lower rates of aggression during chimpanzee consortships after females are led away from other community males.

In both dolphins and chimpanzees, it is easier to establish a consortship when a female is alone, and finding a female alone is less likely when she is maximally attractive. A swollen female chimpanzee will likely be found in a multimale group, and an attractive female dolphin is likely to be in a consortship with another alliance. In chimpanzees, consortships are established and protected by escorting females to peripheral areas of the community range, significantly reducing the chance of encountering other males. The male chimpanzee strategy of guarding, especially around the time of ovulation, appears to be a successful tactic for high-ranking males only. While chimpanzees herd females to keep them away from other males, dolphins herd females to keep them close to other males (their first- and second-order alliance partners). The benefit that chimpanzees derive from herding—avoiding rival males—is less obvious for dolphins, as rival males may be encountered almost anywhere in a male dolphin's range.

Coerced Consortships in Context: The Convergence of Social and Life History Attributes of Dolphins and Chimpanzees

The comparison between dolphins and chimpanzees would not be that significant if coerced consortships were the only shared attribute, for aggressive herding is found in a range of mammals (Smuts and Smuts 1993). In our view, the comparison is attractive because chimpanzees and dolphins also share similar life histories and the consortships occur in a similar social milieu. Given these overall similarities, it becomes interesting to consider why the convergence in male mating tactics is not more complete.

Strong Male-Male Bonds in a Fission-Fusion Grouping Pattern

Some of the earliest studies of wild dolphins drew comparisons with chimpanzee societies (e.g., Wursig 1978). Bottlenose dolphins and common chimpanzees exhibit similar fission-fusion grouping patterns. The average group size reported by Smolker at al. (1992) in Shark Bay (4.8) is in the range of typical party sizes for chimpanzees (4–10) (Muller and Mitani 2005). In both chimpanzees and Shark Bay bottlenose dolphins, the strongest same-sex associations are between adult males. Unlike chimpanzees, however, female dolphins are philopatric as well (Connor et al. 2000a).

Like the Shark Bay dolphins, male chimpanzees form nested alliances. Within communities, male chimpanzees form coalitions in competition over rank (Goodall 1986). Males of the same community cooperate against males from other communities, apparently to provision and protect their resident females by expanding the size of their territory (Williams et al. 2004). Having multiple-nested male alliances *within* a social network is more human than chimplike and has interesting implications for dolphin social cognition (Connor et al. 1992b; Connor and Mann 2006; Connor 2007).

Chimpanzee and Bottlenose Dolphin Life History

Bottlenose dolphins and chimpanzees also have similar life histories (Table 9.1). This is not surprising given that both species have large brains and there is an association between life history and brain size (van Schaik and Deaner 2003). This convergence is worth reviewing in the present discussion because we expect convergent male mating tactics to be impacted by the life history and reproductive schedule of females as well as the pattern of grouping and social bonds. In both species, young are dependent for three to five years, but in some cases even longer (seven to eight years in the Shark Bay dolphins: Connor et al. 2000). The shortest interbirth interval is three years in Shark Bay, but most are in the four- to six-year range. A female with a three-year interval may begin to become attractive to males when her calf is 1.5, but in most cases renewed periods of attractiveness do not commence until the calf is 2.5 or older (Connor et al. 1996). Postpartum amennorhea in chimpanzees lasts 2.0 to 4.5 years. Male chimpanzees may live into their 40s and females over 50 in exceptional cases.

Table 9.1 Life history characteristics of chimpanzees and bottlenose dolphins.

	Chimpanzee	SB Bottlenose Dolphin
Age of first reproduction (f)	13–15	11–12
Age of weaning	3–4 (range?)	3–5 (range 2–8)
Postpartum amennorhea	2–4.5	1.5–3
Interbirth interval	5–6	4–6
Maximum longevity ($\female,\male$)	40s, 50s	35–40+?

We do not have good maximum longevity data from Shark Bay; the oldest known female was 35 when she died after a stingray spine penetrated her heart. We would not be surprised to see male and female dolphins in Shark Bay over 40, but 50 years would be very surprising. Female Atlantic bottlenose dolphins *(Tursiops truncatus)* in Sarasota are known to live over 50 years (Wells 2003), but the Sarasota animals are 20 to 25% longer than the Shark Bay dolphins and are much heavier with greater sexual-size dimorphism. The two dolphin populations also differ in some basic life history attributes. Although general reproductive parameters are similar in Shark Bay and Sarasota dolphins, the extremes are telling. The youngest female to give birth in Shark Bay was 11 years old, but three females in Sarasota gave birth at ages 6–7. Females in the 6–7 age range in Shark Bay are clearly juveniles. In Sarasota, the minimum interbirth interval is two years compared to three years in Shark Bay (Connor et al. 2000; Wells 2003).

Explaining Differences in Coerced Consortships between Chimpanzees and Dolphins

The similarities that we find between the Shark Bay bottlenose dolphins and common chimpanzees are certainly interesting, but we want to consider why the convergence is not more complete: why are male chimpanzees not even more like dolphins who use consortships as their primary mating strategy?

For a male chimpanzee to form a successful consortship, he must be able to keep a female away from other males. His chances of doing so successfully will be enhanced if he initiates a consortship when no other males are around. Since estrous females are usually in the company of males, we often find, as already mentioned, consortships initiated with flat females. We may speculate that coerced consortships would be more common in chimpanzees if there were more females in estrus simultaneously dispersed in the community range—that is, if estrus females were *dispersed in space but not in time.* In such a case, a male would more often encounter solitary estrous or near-estrous females, given the greater probability that other community males would be occupied with females elsewhere. However, there are few estrous females simultaneously available within a chimpanzee community. Watts (1998) reports observing, in the largest chimpanzee community known, estrous females on 96/282 observation days (34%). On most days only one estrous female was observed, and no day had more than 3 estrous females. This contrasts sharply with the dolphin situation where it was not uncommon in the mating season to find three or four females being consorted by males in one large 14-member second-order alliance (Connor et al. 1999, 2001). During our most recent study (2001–2006), it was not unusual to observe 5 to 10 consortships during the course of a day during the mating season. Clearly, there are more dispersed estrous females available in the range of a male dolphin in Shark Bay.

A number of factors might increase or decrease the probability of a male encountering an estrous female and being able to form a successful consortship with her. Here we evaluate 12 factors to see which are likely to contribute to the chimpanzee/dolphin difference in consortship frequency. We consider the adult sex ratio, five factors that impact the population-wide operational sex ratio (interval between cycling periods, infant mortality, duration of cycling periods, duration of attractiveness within a cycle, and mating seasonality), four factors that, for a given operational sex ratio, impact the chance of a male encountering an unencumbered estrous female (female dispersion, population density, home range, and day range), and finally, female tactics and the number of males in a consortship. Asterisks refer to the factors in this list that are likely important.

1. Adult sex ratio. *A relatively female-biased sex ratio in dolphins would increase encounter rates between males and estrous females.*

Wells (2003) reports a ratio among adults in Sarasota of close to 1:2 in favor of females. Mann and Sargeant (2003) reported that 46% of 358 sexed non-calves in Shark Bay are males, suggesting an adult sex ratio closer to 1:1. Goodall (1986, Table 5.2) reports a sex ratio ranging from just under 1:2 to over 1:3 during 1972–1983, counting "peripheral" and old females. The sex ratio in the Taï chimpanzee community during the stable period from 1982 to 1987 ranged from over 1:2 to just over 1:3 (Boesch and Boesch-Achermann 2000, Table 2.2). Thus, there typically appear to be fewer females per male among dolphins than among chimpanzees. This contradicts the idea that more consortships occur in dolphins because they have relatively more females. We conclude that this explanation is *unlikely* to account for the chimpanzee dolphin difference in use of consortships as the primary mating tactic.

Factors 2–6 influence the population-wide operational sex ratio by affecting the ratio of mature males to "attractive" females in the population that might be candidates for consorting.

*   *   *

2. Interval between cycling periods (gestation plus postpartum amenorrhea). *A shorter interval between cycling periods would increase the proportion of mature females in estrus at a given time.*

Chimpanzee gestation is 228 days (0.62 years), and the duration of postpartum amenorrhea ranges from 2.0 to 4.4 years (Knott 2001) for a total interval between attractive periods of 2.5–5 years. Dolphins have a one-year gestation period, and most females become attractive when their calf is 2.5 years old, for a total of 3.5 years (Connor et al. 1996). Chimpanzees continue to exhibit swellings during pregnancy (Wallis and Goodall 1993), but it is not clear if dolphins have attractive periods during pregnancy. These estimates do not suggest a significant difference between dolphins and chimpanzees in this variable. We conclude that this explanation is *unlikely* to account for the chimpanzee-dolphin difference in use of consortships as the primary mating tactic.

3. Infant mortality. *A higher infant mortality would shorten the average interval between cycling periods, increasing the proportion of females in estrus.*

Mann et al. (2000) estimated the infant mortality among dolphin infants by age 3 at 44%. Nishida et al. (2003) report that 50% of infants at Mahale died by age 5. More generally, the figures in Hill et al. (2001) from five chimpanzee study sites indicate 35–40% mortality by age 3 in chimpanzees. These estimates do not indicate a prominent difference between dolphins and chimpanzees in this variable. We conclude that this explanation is *unlikely* to account for the chimpanzee-dolphin difference in use of consortships as the primary mating tactic.

4. Duration of the cycling period. *A longer period of cycling would increase the proportion of estrous females.*

Knott (2001) describes the "waiting time to conception" in chimpanzees as the interval between the resumption of cycling and the next conception. This interval was highly variable between sites, ranging from 4.7 months at Gombe to 26.9 months at Taï. Whether females cycled continuously during the longer intervals is not mentioned. Similar figures have not been calculated for the Shark Bay dolphins, but a "typical" female who begins being consorted when her surviving calf is 2.5 and conceives that season will cycle for 4 to 6 months. Some continue to be consorted or are consorted again the next season, which would increase the average significantly. It seems unlikely that this variable is important, given that the dolphin "mean" is likely to fall within the chimpanzee range. We conclude that this explanation is *unlikely* to account for the chimpanzee-dolphin difference in use of consortships as the primary mating tactic.

5. Duration of attractiveness. *If females spend a larger proportion of the cycling period in an "attractive" state, a higher proportion of females would be attractive at any given time.*

The period of maximal swelling in a chimpanzee is 10 to 12 days in a 32–36 day cycle. Connor et al. (1996) estimated attractive periods as being the length of time females were consorted with no intervening intervals of >7 days. Most attractive periods fell within the 5–7 day span predicted from captive studies of estrogen levels, but a number were in the 1–2 week category (and some consortships lasted much longer than that). Assessment of this variable is difficult because the probability of conception usually occurs in the last two days before detumescence in chimpanzees and high-ranking males are more likely to mate guard during this time. It is unknown if male dolphins increase levels of competition near ovulation, but differences in the intensity of herding and competition for females suggest the possibility. Again, striking differences between the dolphins and chimpanzees in the duration of attractiveness are not apparent. We conclude that this explanation is *unlikely* to account for the chimpanzee-dolphin difference in use of consortships as the primary mating tactic.

*6. Mating seasonality. *Greater seasonality will produce time periods in which a greater proportion of females are attractive, increasing consortship opportunities, relative to the aseasonal case where attractive females are more dispersed temporally.*

Conceptions are more seasonal in the Shark Bay dolphins than in chimpanzees. Mann et al. (2000) reported that 73% of births occur during the three-month period October–December, which is also the period the conceptions occurred given the 12-month gestation period. Cycling is seasonal as well, even taking into account that the same females that conceive in the October–December period in Shark Bay often begin cycling (i.e., start to be consorted by males) during June–July of the same year (Connor et al. 1996).

Mating, and the number of estrous females, are seasonal in chimpanzees, even if births are not. However, the degree of seasonality is considerably less than is found in the Shark Bay dolphins. In Gombe, the highest percentage of conceptions in any three-month period was 33%, and a value of 75% was reached only by considering a seven-month "season" (Wallis 2002). We conclude that this explanation is *plausible and may account, partially,* for the chimpanzee-dolphin difference in use of consortships as the primary mating tactic.

For a given population operational sex ratio, the following factors (7–10) will influence a male's opportunity to form consortships with unencumbered estrous females.

7. Female dispersion. *A high level of female dispersion (i.e., spread out and traveling alone) will increase the consortship opportunities for males relative to a population where females are clumped in one or a few large groups.*

This category is somewhat difficult to address because we are interested in the dispersion of estrous females, but in most cases the estrous females were discovered by male conspecifics before they were found by researchers. Therefore, given that consortships are often initiated before or in the early stages of swelling for chimpanzees (and given that a similar phenomenon may occur in dolphins), we compare the sociability of anestrous females. For female sociability to explain the dolphin-chimpanzee difference, it must be true that chimpanzee females are less dispersed spatially than dolphins, but this is not the case. Anestrous female chimpanzees are, in most populations, mostly solitary. Goodall (1986) reported an average party size of 1.6 for five anestrous females in Gombe. In Taï, party sizes are in general larger than in other populations, and some females may have strong associations with one other female (Boesch and Boesch-Achermann 2000). In Shark Bay the word for female associations is, again, variability. However, the solitary extreme in dolphins is more like the chimpanzee norm; even the "solitary" sponge-carrying females occasionally rest and travel in moderate to large female groups. Many female dolphins have moderately strong associations with other females and are often found resting and traveling in female groups (Smolker et al. 1992). In sum, the data suggest a difference in the opposite direction to that required to explain the chimpanzee-dolphin difference in consortship formation. We conclude that this explanation is *unlikely* to account for the chimpanzee-dolphin difference in use of consortships as the primary mating tactic. We note that estrous females might shift their movements in ways that reduce dispersion, but we consider this in 11. Female mating tactics.

8. Density. *A higher density of dispersed females in a male's home range will increase his encounter rate with estrous females.*

Boesch and Boesch-Achermann (2000) reported chimpanzee densities across four research sites ranging from 2.2 to 4.28 individuals/km^2 (median = 3.06) [page 107, Table 5.11]. The value of 2.4 dolphins/km^2 given by Watson-Capps (2005) for Shark Bay fits within this range. We conclude that this explanation is *unlikely* to account for the chimpanzee-dolphin difference in use of consortships as the primary mating tactic.

*9. Home range. *For a given density, a larger home range will increase the number of estrous females in a male's range.*

The average territory size in chimpanzees is 5 to 30 km^2. Males will cover the entire community range, but individual females typically occupy smaller ranges within the community range. The Shark Bay social network spans an area >600 km^2, but individual home ranges are smaller. Watson-Capps (2005) reported highly variable home range sizes for adult females (7 to 101 km^2), but the mean (48 km^2) was considerably higher than a large chimpanzee territory. Male dolphin home range sizes are larger still, averaging 102 km^2 (Randic 2008). We conclude that this explanation is *plausible and may account, partially,* for the chimpanzee-dolphin difference in use of consortships as the primary mating tactic.

*10. Day range. *For a given density, a larger day range will increase a male's encounter rate with estrous females.*

Another striking difference between dolphins and chimpanzees emerges when we consider day range. Adult chimps travel between 2.0 and 4.5 km/day, depending on the sex and community studied (Pontzer and Wrangham 2004). The few estimates available of day range in bottlenose dolphins are all at least an order of magnitude higher than chimpanzee day ranges, which are typical distances dolphins travel every 1 to 2 *hours*. Bottlenose dolphins in Shark Bay cease movement occasionally for a few seconds to, in rare cases, a few minutes and the rest of the time they are moving. Their typical travel speed is 2 to 3 km/hr, and, unlike chimpanzees, they continue to travel at night. We conclude that this explanation is *plausible and may account, partially,* for the chimpanzee-dolphin difference in use of consortships as the primary mating tactic.

*11. Female mating tactics.

Female mating tactics might differ between the species in ways that significantly affect the number of unencumbered estrous females available for consortships. This could occur if male chimpanzees and dolphins differ in their ability to persuade females to accompany them on consortships, or if females change their movement patterns at the onset of estrus. Perhaps female chimpanzees can more easily avoid consortships than dolphins. In spite of their clear dominance over females, single male chimpanzees may be less successful at coercing females into forming consortships than dolphins working as a team in alliances of two to three individuals. Also, female chimpanzees might disfavor consortships more (relative to alternative mating tactics) and thus allot more effort to avoiding consortships or pursuing other tactics that effectively reduce the number of "unencumbered females" available for consortships. Estrous female chimpanzees are often attracted to aggregations of males, whereas such behavior has not been documented in Shark Bay. We conclude that this explanation is *plausible and may account, partially,* for the chimpanzee-dolphin difference in use of consortships as the primary mating tactic.

The final factor, the number of males involved in consorting a single female, influences not the operational sex ratio but the "operational consortship ratio."

*12. How many males does it take to tango? *For a given sex ratio, a higher average number of males in a consortship will increase the number of consortship opportunities for males (but obviously not the opportunities to conceive).*

Consortships in chimpanzees and dolphins always involve a single female, who is guarded by one male chimpanzee but always two to three male dolphins (Connor et al. 1992a, b, 1996). The average number of males consorting females in Shark Bay is 2.9 (Connor 2007). Thus for every estrous female, three consortship "positions" are available. We conclude that this explanation is *plausible and may account, partially,* for the chimpanzee-dolphin difference in use of consortships as the primary mating tactic.

Summary: The Key Differences between Dolphins and Chimpanzees

Of the 12 factors reviewed in this chapter, five emerge as plausible explanations for the greater frequency and importance of consortships in dolphins compared to chimpanzees: mating seasonality, home and day range, female tactics, and number of males in a consortship. The much greater mating seasonality increases the clustering *in time* of spatially dispersed females in Shark Bay. At a roughly similar population density, the larger dolphin home range will increase the number of estrous females that a male may find, while the larger day range will increase the rate at which those females are encountered. Consortships might be less common in chimpanzees if female chimpanzees are more able to avoid them or, for reasons that are not presently clear, disfavor consortships more strongly than dolphins. Finally, the fact that there are three consortship slots per female dolphin compared to one slot for an estrous female chimpanzee greatly increases the relative number of consortship opportunities available for the dolphins. In the next and final section, we relate several of these differences to a fundamental difference between the aquatic dolphin and the terrestrial chimpanzee: the cost of locomotion.

Costs of Locomotion, Population Density, Encounter Rates, and Sexual Coercion in Dolphins

Compared to terrestrial mammals, dolphins enjoy relatively low costs of locomotion (Williams 1999). More specifically, Williams found that although the total cost of transport was similar in both habitats, the dolphins' elevated maintenance costs imply that the proportional increase in energy required to move from point A to point B is less for dolphins than for terrestrial mammals. The impact of cheap travel on the social lives of cetaceans has been explored in detail elsewhere (Connor et al. 2000a, b; Connor 2000; Whitehead 2003). Here we focus on the impact on consortships and coercion. First, assume that other factors are equal but dolphins have larger home ranges than chimpanzees. For the same population density and sex ratio, a male will overlap with more females than his chimpanzee counterpart. This increases consortship opportunities. Now consider travel time and velocity; the relatively "high speed" and continuous movements of dolphins will increase the rate at which they encounter females with whom they overlap in their home range. But a high encounter rate extends to males as well and may explain why Shark Bay males are "forced" to consort females in alliances rather than alone (Connor and Whitehead 2005; Whitehead and Connor 2005). This impact of a higher encounter rate, favoring male alliances, has the further effect of increasing the number of consortship opportunities by increasing the "slots" per female compared to chimpanzees. Our findings also suggest that a comparison of chimpanzee populations would be interesting. There is considerable variation among chimpanzee populations in the percentage of inseminations that occur during consortships. Consortships are considered rare in Kibale National Park in Uganda but account for 25% of conceptions in Gombe. Presently, the only hypothesis offered to explain this variation centers on reduced female dispersal and inbreeding avoidance at Gombe. Gombe females may choose to accompany males in consortships to avoid mating with high-ranking relatives (Muller and Mitani 2005). However, these populations also vary considerably in many of the 12 factors considered here that impact operational sex ratio and encounter rates of males with unencumbered estrous females.

References

Boesch, C., and H. Boesch-Achermann. *The Chimpanzees of the Taï Forest.* Oxford: Oxford University Press, 2000.

Connor, R. C. "Group Living in Whales and Dolphins." In *Cetacean Societies: Field Studies of Whales and Dolphins,* eds. J. Mann, R.Connor, P. Tyack, and H. Whitehead, pp. 199–218. Chicago: University of Chicago Press, 2000.

———. "Individual Foraging Specializations in Marine Mammals: Culture and Ecology." *Behavioral and Brain Sciences* 24 (2001): 329–330.

———. "Dolphin Social Intelligence: Complex Alliance Relationships in Bottlenose Dolphins and a Consideration of Selective Environments for Extreme Brain Size Evolution in Mammals." *Philosophical Transactions of the Royal Society: Biological Sciences.* 362 (2007): 587–602.

Connor, R. C., M. R. Heithaus, and L. M. Barre. "Super-Alliance of Bottlenose Dolphins." *Nature* 371(1999): 571–572.

———. "Complex Social Structure, Alliance Stability and Mating Access in a Bottlenose Dolphin 'Super-Alliance'." *Proceedings of the Royal Society of London: Biological Sciences* 268 (2001): 263–267.

Connor, R. C., and J. Mann. "Social Cognition in the Wild: Machiavellian Dolphins?" In *Rational Animals?,* eds. S. Hurley, and M. Nudds, pp. 329–367. Oxford: Oxford University Press, 2006.

Connor, R. C., A. J. Read, and R. W. Wrangham, "Male Reproductive Strategies and Social Bonds." In *Cetacean Societies: Field Studies of Whales and Dolphins,* eds. J. Mann, R. C. Connor, P. L. Tyack, and H. Whitehead, pp. 247–269. Chicago: University of Chicago Press, 2000b.

Connor, R. C., A. F. Richards, R. A. Smolker, and J. Mann. "Patterns of Female Attractiveness in Indian Ocean Bottlenose Dolphins." *Behaviour* 133 (1996): 37–69.

Connor, R. C., and R. A. Smolker. "'Pop' Goes the Dolphin: A Vocalization Male Bottlenose Dolphins Produce during Consortships." *Behaviour* 133 (1996): 643–662.

Connor, R. C., R. A. Smolker, and A. F. Richards. "Two Levels of Alliance Formation among Male Bottlenose Dolphins (*Tursiops* sp.)." *Proceedings of the National Academy of Sciences* 89 (1992a): 987–990.

———. "Dolphin Alliances and Coalitions." In *Coalitions and Alliances in Animals and Humans,* eds. A. H. Harcourt, and F. B. M. de Waal, pp. 415–443. Oxford: Oxford University Press, 1992b.

Connor, R. C., R. S. Wells, J. Mann, and A. Read. "The Bottlenose Dolphin: Social Relationships in a Fission-Fusion Society." In *Cetacean Societies: Field Studies of Whales and Dolphins,* eds. J. Mann, R. C. Connor, P. L. Tyack, and H. Whitehead, pp. 91–126. Chicago: University of Chicago Press, 2000a.

Connor, R. C., and H. Whitehead. "Alliances II: Rates of Encounter during Resource Utilization: A General Model of Intrasexual Alliance Formation in Fission-Fusion Societies." *Animal Behavior* 69 (2005): 127–132.

Dunn, D. G., S. G. Barco, D. A. Pabst, and W. A. McLellan. "Evidence for Infanticide in Bottlenose Dolphins of the Western North Atlantic." *Journal of Wildlife Diseases* 38 (2002): 505–510.

Goodall, Jane. *The Chimpanzees of Gombe: Patterns of Behavior.* Cambridge, Mass. Belknap Press, 1986.

Hill, K., C. Boesch, J. Goodall, A. Pusey, J. Williams, and R. Wrangham. "Mortality Rates among Wild Chimpanzees." *Journal of Human Evolution* 40 (2001): 437–450.

Kappeler, P. M., and C. P. van Schaik. "Evolution of Primate Social Systems." *International Journal of Primatology* 23 (2001): 707–740.

Knott, C. D. "Female Reproductive Ecology of the Apes." In *Reproductive Ecology and Human Evolution,* ed. P. T. Ellison, pp. 429–463. New York: Aldine de Gruyter, 2001.

Krützen, M., L. M. Barré, R. C. Connor, J. Mann, and W. B. Sherwin. "O Father, Where Art Thou? Paternity Assessment in an Open Fission-Fusion Society of Wild Bottlenose Dolphins (*Tursiops* sp.) in Shark Bay, Western Australia." *Molecular Ecology* 13 (2004): 1975–1990.

Krützen, M., J. Mann, M. Heithaus, R. Connor, L. Bejder, and B. Sherwin. "Cultural Transmission of Tool Use in Bottlenose Dolphins." *Proceedings of the National Academy of Sciences.* 105 (2005): 8939–8943.

Krützen, M, W. B. Sherwin, R. C. Connor, L. M. Barré, T. Van de Casteele, J. Mann, and R. Brooks. "Contrasting Relatedness Patterns in Bottlenose Dolphins (*Tursiops* sp.) with Different Alliance Strategies." *Proceedings of the Royal Society of London, Biological Sciences.* 270 (2003): 497–502.

Mann, J. "Sociosexual Behaviour among Indian Ocean Bottlenose Dolphins and the Development of Male-Male Bonds." In *Homosexual Behaviour in Animals: An Evolutionary Perspective*, eds. P. Vasey, and V. Sommer, pp. 107–130. Cambridge: Cambridge University Press, 2006.

Mann, J., R. C. Connor, L. M. Barre, and M. R. Heithaus. "Female Reproductive Success in Bottlenose Dolphins (*Tursiops* sp.): Life History, Habitat, Provisioning, and Group Size Effects." *Behavioral Ecology* 11 (2000): 210–219.

Mann J., and B. Sargeant. "Like Mother, Like Calf: The Ontogeny of Foraging Traditions in Wild Indian Ocean Bottlenose Dolphins (*Tursiops* sp.)." In *The Biology of Traditions: Models and Evidence*, eds. D. Fragaszy, and S. Perry, pp. 236–266. Cambridge: Cambridge University Press, 2003.

Muller, M. N., and J. C. Mitani. "Conflict and Cooperation in Wild Chimpanzees." *Advances in the Study of Behavior* 35 (2005): 275–331.

Nishida, T. "Alpha Status and Agonistic Alliance in Wild Chimpanzees *(Pan troglodytes schweinfurthii)*." *Primates* 24 (1983): 318–336.

Nishida, T., N. Corp, M. Hamai, T. Hasegawa, M. Hiraiwa-Hasegawa, K. Hosaka, K. D. Hunt, N. Itoh, K. Kawanaka, A. Matsumoto-Oda, J. C. Mitani, M. Nakamura, K. Oikoshi, T. Sakamaki, L. Turner, S. Uehara, and K. Zamma. "Demography, Female Life History, and Reproductive Profiles among the Chimpanzees of Mahale." *American Journal of Primatology* 59 (2003): 99–121.

Patterson, I., R. J. Reid, B. Wilson, K. Grellier, H. M. Ross, and P. M. Thompson. "Evidence for Infanticide in Bottlenose Dolphins: An Explanation for Violent Interactions with Harbour Porpoises?" *Proceedings of the Royal Society of London, B* 265 (1998) 1–4.

Pontzer, H., and R. W. Wrangham. "Climbing and the Daily Energy Cost of Locomotion in Wild Chimpanzees: Implications for Hominoid Locomotor Evolution." *Journal of Human Evolution* 46 (2004): 315–333.

Randic, S. Spatial "Analysis of the Distribution and Home ranges of Male Bottlenose Dolphins in Shark Bay, Australia." Master's thesis, University of Massachusetts, Dartmouth, 2008.

van Schaik, C. P., and R. O. Deaner "Life History and Brain Evolution." In *Animal Social Complexity*, eds. F. B. M. de Waal, and P. L. Tyack, pp. 1–25. Cambridge, Mass.: Harvard University Press, 2003.

Smolker R. A., A. F. Richards, R. C. Connor, J. Mann, and P. Berggren. "Sponge-Carrying by Indian Ocean Bottlenose Dolphins: Possible Tool-Use by a Delphinid." *Ethology* 103(1997): 454–465.

Smolker, R. A., A. F. Richards, R. C. Connor, and J. Pepper. "Association Patterns among Bottlenose Dolphins in Shark Bay, Western Australia." *Behaviour* 123 (1992): 38–69.

Smuts, B. B., and R. W. Smuts. "Male Aggression and Sexual Coercion of Females in Nonhuman Primates and Other Mammals: Evidence and Theoretical Implications." *Advances in the Study of Behavior* 22 (1993): 1–63.

Tutin, C. E. G. "Mating Patterns and Reproductive Strategies in a Community of Wild Chimpanzees *(Pan troglodytes schweinfurthii)*." *Behavioral Ecology and Sociobiology* 6 (1979): 29–38.

Wallis, J. "Seasonal Aspects of Reproduction and Sexual Behavior in Two Chimpanzee Populations: A Comparison of Gombe (Tanzania) and Budongo (Uganda)." In *Behavioural Diversity in Chimpanzees and Bonobos,* eds. C. Boesch, G. Hohmann, and L. F. Marchant, pp. 181–191. Cambridge: Cambridge University Press, 2002.

Wallis, J., and J. Goodall. "Anogenital Swelling in Pregnant Chimpanzees of Gombe National Park." *American Journal of Primatology* 31(1993): 89–98.

Watson-Capps, J. "Female Mating Behavior in the Context of Sexual Coercion and Female Ranging Behavior of Bottlenose Dolphins in Shark Bay, Western Australia." Ph.D. Dissertation, Georgetown University, 2005.

Watts, D. "Coalitionary Mate Guarding by Male Chimpanzees at Ngogo, Kibale National Park, Uganda." *Behavioral Ecology and Sociobiology* 44 (1998): 43–55.

Wells, R. S. "Dolphin Social Complexity: Lessons from Long-Term Study and Life History." In *Animal Social Complexity,* eds. F. B. M. de Waal, and P. L. Tyack, pp. 32–56. Cambridge, Mass.: Harvard University Press, 2003

Whitehead, H. *Sperm Whales: Social Evolution in the Ocean.* Chicago: University of Chicago Press, 2003.

Whitehead, H., and R. C. Connor. "Alliances I: How Large Should Alliances Be?" *Animal Behavior* 69 (2005): 117–126.

Williams, J. M., G. W. Oehlert, J. V. Carlis, and A. E. Pusey. "Why Do Male Chimpanzees Defend a Group Range?" *Animal Behaviour* 68 (2004): 523–532.

Williams, T. M. "The Evolution of Cost Efficient Swimming in Marine Mammals: Limits to Energetic Optimization." *Philosophical Transactions of the Royal Society: Biological Sciences* 354(1999): 193–201.

Wursig, B. "Occurrence and Group Organization of Atlantic Bottlenose Porpoises *(Tursiops truncatus)* in an Argentine Bay." *Biological Bulletin* 154 (1978): 348–359.

10

Male Aggression toward Females in Hamadryas Baboons: Conditioning, Coercion, and Control

Larissa Swedell and Amy Schreier

> She is following him, but apparently not closely enough, because he keeps threatening her and neckbiting her. She screams and kecks when she approaches him. She is about three meters behind him but apparently that's not close enough because he then turns around and threatens her, then runs after her and tries to neckbite her while she screams and holds onto a rock below him. He can't get to her very well and sort of bites her face instead. A while later he walks to the top of the cliff and she doesn't follow; he then threatens her, runs toward her, chases her, grabs her by her head hair, pulls her up to him and neckbites her. They both sit down, then she moves away from him again.
>
> *[Excerpt from field notes following takeover of Who by Ike on December 9, 1996 (Swedell 2000, 2006)]*

One of the most striking elements of hamadryas baboon social organization is the difference in female behavior before and after a takeover, a social transition in which one "leader male" appropriates a female from another. The day before the takeover described in the excerpt, "Syl" had five females, none of whom had ever been seen interacting with another male. One of the females was "Who," described above, and another was "Bea." The day after the takeover, any observer might have concluded that Syl had been wiped from Bea's memory. Syl sat only 3 m away, but Bea did not even look at him. Instead, she kept tightly close to her new leader male, Fel, following less than 1 m behind him at all times. While Who (above) did not follow Ike as readily as Bea did Fel, neither female was ever again seen to interact with Syl, their former leader male. In this chapter, we explore the proximate mechanisms behind this abrupt, seemingly inexplicable, shift in behavior by hamadryas females.

Hamadryas baboons *(Papio hamadryas hamadryas)* are unusual among mammals in that males physically coerce females into semipermanent social units (Kummer 1968a, 1968b). As the smallest and most stable social unit in hamadryas society, one-male units (OMUs) persist over time to the extent that males can defend the females in their OMU from other males. Females may thus remain with leader males for periods of several months to several years. Male challengers attempt aggressive takeovers in which, if successful, a male herds a female out of her previous OMU and incorporates her into his own unit. If a leader male is injured, dies, or is otherwise unable to defend his females, they may be taken over by multiple males and incorporated into several different OMUs.

Hamadryas are also unusual with respect to the multilayered social system that distinguishes them from other baboons (Kummer 1968b). In addition to OMUs, three other levels of social structure have been defined. The *band*, which functions as the ecological unit akin to the troop or group in other monkey societies, consists of several OMUs, their associated follower males, and a number of solitary males (i.e., those unattached to OMUs). Within bands, *clans* are more subtle groupings of OMUs and solitary males that associate most frequently and within which males may be related (Abegglen 1984; Schreier and Swedell 2007). Finally, *troops* are amalgamations of two or more bands at a sleeping site. Kummer and Kurt (1963) and Kummer (Kummer 1968a, b) first described this system in detail, and it has since been further studied and described by Abegglen (1984) and Swedell (2006), among others.

The relationship between a hamadryas leader male and each of his females, which likely represents the strongest social bond in hamadryas society, is the cohesive force maintaining the OMU social structure and is reinforced through reciprocal affiliative behavior and herding by leader males (Kummer 1968a, 1995; Colmenares et al. 2002; Swedell 2006). As in mountain gorillas, hamadryas males are both the main protectors and the main aggressors of their females, and females likely benefit from their association with a protective male because it increases the survival prospects of their offspring (Swedell 2006; Swedell and Saunders 2006; Robbins, Chapter 5 in this volume). The relationship between a hamadryas male leader and a female can be described as a "permanent consortship" in that males associate closely with and mate guard their females at all times (Bergman 2006; Swedell and Saunders 2006). This contrasts with the pattern seen in other baboons, macaques, and chimpanzees, in which males consort and mate guard only when females are in estrus (DeVore 1965; Huffman 1987; Morin 1993).

Bonds among males—expressed mainly among solitary males and between leader males and their followers—are likely the most enduring relationships in hamadryas society (Abegglen 1984). Unlike other male-bonded primates such as chimpanzees, however, relationships among hamadryas males are character-ized by a suppression of male-male competition, described by Kummer (1968b, 1971, 1973; Kummer et al. 1974) as "respect" for another male's "possession" of females, in which males do not regularly contest other males for access to estrous females. Although challenges and takeovers do occur, the timing of these events appears to be related to leader male tenure and fighting ability rather than to fe-male reproductive condition (Swedell 2000, 2006). Bonds among males—and male behavior in general—appear to drive hamadryas baboon social organiza-tion as a whole: males remain in their natal clans or bands; form long-lasting, af-filiative relationships with one another; and effectively control movement of females within and among OMUs (Kummer 1968a; Sigg et al. 1982; Abegglen 1984; Colmenares 1991, 1992).

By contrast, bonds among hamadryas females are weaker, most likely because females are repeatedly separated from one another during transfers among OMUs. Moreover, leader males typically herd females away from other OMUs and threaten them when they attempt to engage in extra-unit interactions, re-sulting in minimal contact across OMU boundaries (Kummer 1968a, 1968b; Colmenares et al. 1994; Swedell 2002, 2006).

The formation of a hamadryas one-male unit appears to occur in at least three ways. The first strategy, usually seen in subadult males, is to forge an affil-iative relationship with a young juvenile female who is still in her natal OMU. By developing a grooming relationship with this female and maintaining spatial as-sociation with her over time, a male gradually extricates her from her natal OMU and establishes his own "initial unit" with her as his first female (Kummer 1968a; Abegglen 1984). The second strategy, usually seen in adult males, is an opportunistic one in which a male challenges a leader male for one of his females (see Figure 10.1). This can occur at any time but appears to be most often suc-cessful after a leader male has been injured (Swedell 2000, 2006). A third strat-egy involves a male attaching himself to an OMU as a "follower" male, which may lead to the eventual inheritance of the females from the leader male when he is older and unable to defend them (Kummer 1968a; Abegglen 1984). It has been suggested that, in such cases, the leader and follower may be close relatives (Kummer 1968a; Abegglen 1984; Colmenares 1992). Whether follower males obtain females more often than solitary males, however, is not yet known, nor do we know the relative frequency of each of the above male strategies (Swedell et al. 2008).

As in many other primates with marked sexual dimorphism in body and ca-nine size (presumably a result of sexual selection via male-male competition), hamadryas males use their greater size and strength against females as well as against other males. The most striking form of aggression in hamadryas society is the *neck-bite,* in which a male bites the nape of a female's neck. Although such bites rarely break the skin or produce blood, they do gradually wear away the hair on the back of the neck and head, and it is usually quite obvious which females have been victim to frequent neck-biting because the back of their neck and head is hairless and covered with wounds in various stages of healing. In most cases, however, neck-bites do not seem to physically harm females; rather, it is the *anticipation* of a neck-bite that seems to provoke the greatest submissive reaction. Overall, hamadryas females appear to live in constant fear of aggression by males, as they are exceptionally skittish of human observers (and other potential threats in the form of conspecifics and predators) compared to other female baboons.

In this chapter, we attempt to quantify aggression by hamadryas males toward females and elucidate its causes and contexts in light of the sexual coercion model developed by Smuts and Smuts (1993). Possible dyads in which hamadryas male-female aggression can occur are limited for three reasons. First, social interaction usually occurs only within OMUs. Second, virtually all females in hamadryas society are part of an OMU (i.e., there are no "solitary" females). Third, a hamadryas male rarely behaves aggressively toward another male's female unless he is trying to appropriate her. As such, our examination of male aggression toward females pertains only to leader males and the females in their OMUs.

We ask the following questions: (1) In what context(s) does hamadryas male aggression toward females typically occur? (2) Do leader males vary in their rates of aggression? And if so, then (3) what are the primary factors contributing to variation in aggression by hamadryas males?

Several predictions can be derived from what we already know or assume about the hamadryas social system. First, if aggression functions to condition females, as has been previously assumed (Kummer 1968a; Swedell 2006), then it should occur primarily in the context of takeovers and should vary with female proximity maintenance. Females who remain closer to their leader males should receive less aggression from them. Second, if grooming reflects a female's motivation to remain near and interact with her leader male, then females who spend more time grooming their leader males might receive less aggression from them. Third, if estrous females, who have been shown to remain closer to their leader males than anestrous females (Swedell 2006), do so because they are herded more often when they are estrous compared to when they are anestrous, then they should receive more overall aggression. If, however, their closer proximity is simply a function of heightened proceptivity, then we may not find this difference. Fourth, Kummer (1968a) suggested that females with newborn infants are given more latitude in their patterns of proximity and are not threatened or neck-bitten by the leader male as readily as other females. Moreover, Swedell (2006) found that females with newborn infants tended to remain closest to their leader male during the first couple of months after birth. Both authors suggested that this pattern was due to the greater amount of protection given to (and sought by) females with newborn infants. If females with young infants do remain closer to their leader males and/or are given some special status, then they should receive less aggression than other females.

Methods

The data reported here derive from a band of hamadryas baboons at the Filoha field site in the lowlands of central Ethiopia. The band under study ("Group 1" in Swedell 2000, 2002, 2006; "Band 1" in subsequent publications) consists of over 200 individuals and has been under observation intermittently since 1996. Data from two main study periods contribute to these analyses: October 1996 through September 1998 (LS) and March 2005 through February 2006 (AS). Data consist of four types: 30-minute continuous focal samples on individual females (1996–1998), 10-minute point time scan samples on OMUs (1996–1998), 2-minute point time scan samples on OMUs (2005–2006), and *ad libitum* data from both observation periods (which contribute to the general data set but were not used to calculate frequencies of behavior, e.g., patterns of proximity and grooming).

Over 2000 hours of focal, scan, and *ad libitum* observations contribute to the data reported here, though continuous focal behavioral data—that which is most valuable for calculations of frequencies of aggressive events—varies widely across females and OMUs (ranging from 1.5 to 10 hours for individual females and 1.5 to 12 hours for OMUs). Because of the wide variation in observation time across OMUs, we set a minimum of 3 hours of focal data on OMUs (sum total of focal data on all females in that OMU) for some comparisons across OMUs and leader males (e.g., Figures 10.3a, 10.4a). Agonistic behavior between males and females was classified as in Table 10.1. Definitions of behavioral elements and additional information on the study site and subjects can be found in Swedell (2006).

Table 10.1   Agonistic behavior between
hamadryas leader males and females

Male Threat or Aggression	Female Responses
Stare threat	Grimace
Eyebrow raise	Kick
Chase	Present
Grab	Scream
Hit	Crouch
Push	Run away
Neck-bite	Run to
Bite on back	Groom
Possession grip	

All data reported here were originally collected for other purposes rather than with an explicit focus on male aggression. The patterns generated by these data are therefore preliminary and subject to further study and corroboration.

Results

Agonistic Behavioral Elements

Table 10.1 shows the agonistic behavioral elements characterizing the relationship between male and female hamadryas baboons. These elements are listed in approximate order of intensity, from mild threats such as *stare threats* and *eyebrow raises* to more intense forms of aggression such as *chasing, grabbing,* and *neck-biting*. Neck-biting involves a male biting a female on the nape of her neck; this is the most common form of aggression by male hamadryas baboons toward females. (Females do occasionally neck-bite juveniles and other females, but neck-biting is primarily a male behavior.) Individual neck-bites do not usually break the skin, but a female who has received multiple neck-bites within a short period of time will sometimes bleed on the back of her head, neck, or ears, and the hair is often worn off in these areas. *Biting on the back* is similar to a neck-bite but lower on the back. Holding a female in a *possession grip* involves a male standing over a female or sitting behind her, embracing her and physically restraining her from moving away. These latter two behavioral elements are uncommon and—in addition to *pushing*—have been observed *only* in the context of takeovers. Figure 10.1 shows a leader

Figure 10.1 A leader male embraces a female during an attempted takeover in which the female received substantial aggression. The blood on their fur is from injuries to both the male and the female. Photo by Helga Peters.

male and a female embracing, both bleeding from injuries incurred during an aggressive takeover.

Females usually respond to male threats by *grimacing, kecking,* and *presenting,* and they respond to actual aggression by *screaming* and *crouching.* Figure 10.2 shows a female crouching and screaming in response to a neck-bite from her leader male. Hamadryas females also often respond to male aggression by *running to* their leader male—even if he was the source of the aggression—and grooming him intensely.

Baseline Rates of Agonism and Aggression

Table 10.2 shows the specific contexts in which aggression occurs under baseline conditions. We define "baseline" as any time *other than* within two weeks of a takeover. The list on the left shows the most common precursors to male aggression, and the list on the right shows the most common female responses to male aggression. The most frequent behavioral elements are noted with an asterisk.

Figure 10.3a shows the baseline rates of agonism by males toward females for the 15 leader males for whom we have at least one year of *ad libitum* observations

Figure 10.2 A hamadryas female crouches and screams in response to a neck-bite from her leader male. Photo by Helga Peters.

Table 10.2 Most frequent contexts in which threat or aggression by leader males toward females occurs under baseline conditions.[a]

Prior to Threat or Aggression	Following Threat or Aggression
Female moves away from leader male*	Female kicks, screams, presents, and/or crouches*
Female grooms female in another OMU	Female approaches and/or grooms leader male*
Female copulates with nonleader male	Female runs away from leader male
	Leader male grooms female

[a] Baseline conditions: any time other than within two weeks before or after a takeover.
* Most frequent behavioral elements.

plus at least three hours of continuous focal data on female members of that male's OMU. Figure 10.3b shows the same data for all males for whom we have at least one year of *ad libitum* observations. For most males, the values are equal to zero because no aggression toward females by these males was observed under baseline conditions.

Limiting ourselves to data from males for whom we had at least three hours of focal data as well as actual data on physical aggression under baseline

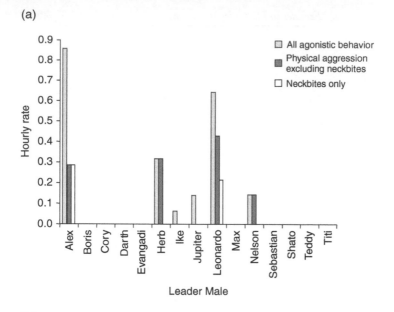

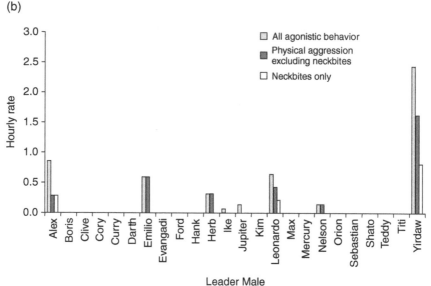

Figure 10.3 Variation among leader males in baseline rates of agonism. "Baseline" is defined as any time *other than* within two weeks before or after a takeover. The first column represents all agonistic behavior listed in Table 10.1 (i.e., both threats and physical aggression), the second column represents all physical aggression (grab, hit, bite on back, chase, push, possession grip), and the third column represents neck-bites only. For males with zero values, no agonism of any kind was observed under baseline conditions (either during focal samples or during *ad libitum* observations). (a) Leader males for whom we have at least one year of *ad libitum* observations and at least three hours of continuous focal data on their females. (b) All males for whom we have at least one year of *ad libitum* observations.

conditions—that is, all males in Figure 10.3a for whom values were not equal to zero (N=6)—we generated the following median rates of agonism by hamadryas leader males toward females under baseline conditions: 0.25 neck-bites per hour, 0.30 instances of non–neck-bite physical aggression per hour, and 0.23 hourly instances of overall agonism (i.e., all agonistic elements listed in Table 10.1). Because we did not include males for whom no aggression was observed under baseline conditions, these values are likely overestimates of rates of aggression. Bearing this in mind, we estimate an overall average rate of about one aggressive event per four hours of focal observation.

Agonism and Aggression during Takeovers

Figure 10.4 compares agonism and aggression by leader males under baseline conditions with rates during takeovers. Our data surrounding takeovers include the day of the takeover itself and within two days post-takeover. (We have too few data from the two-week periods prior to and following takeovers for a proper comparison of behavioral patterns during this time and have thus limited our comparison to the period for which we have data.) Because we had varying amounts of focal data from males during takeovers (N=5) compared to baseline conditions (N=15), a different set of males contributes to each mean value in Figure 10.4a. Using these two sets of males, a Mann-Whitney rank-sum test showed a significant difference between rates of agonism (p=0.001) and neck-biting (p=0.002) across the two conditions, although the difference between rates of non–neck-bite aggression across the two conditions did not reach statistical significance (p=0.065). For five leader males, we had focal data on both conditions (Figure 10.4b). The difference between baseline and takeover agonism (Mann-Whitney rank-sum test; p=0.011) and neck-biting (p=0.025) in these males across the two conditions was statistically significant, though the difference in non–neck-bite aggression was not (p=0.158). In both of the above comparisons, the difference in non–neck-bite aggression across the two conditions was not statistically significant; however, our sample size of these behavioral elements was smaller than our sample sizes for overall agonism and neck-biting.

Our data are clear: males directed aggression toward their females at especially high rates during takeovers. Of the five leader males for whom we have data from both conditions, three of these males showed no aggression toward females at all during observations under baseline conditions. For four of the five

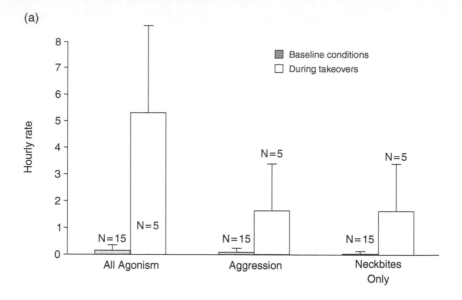

(a)

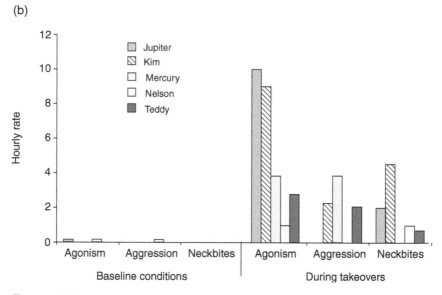

(b)

Figure 10.4 Baseline rates of agonism compared to agonism during takeovers. "Baseline" is defined as any time *other than* within two weeks before or after a takeover; "during takeovers" is defined as the day of or within two days after a takeover. (a) Comparison of means of two samples of males for whom we have data on one or both conditions. Each bar represents the mean of a sample of males (with standard deviations shown). The 15 males represented in the baseline sample are the same as those shown in Figure 10.1a, whereas the five males represented in the takeover sample are the only leader males for whom we have continuous focal data on their newly acquired females on the actual day of the takeover. Bars on the left

males from whom we have data during takeovers, aggression was directed by the new leader male toward the female being taken over (and she was subsequently successfully taken over by that male). In one case the aggression was directed by a defeated leader male (Mercury), who subsequently lost his female.

Correlates of Aggression Rates

As a means of elucidating the factors underlying variation among leader males in their baseline aggression rates, we examined the relationship between rates of agonism by leader males and several other factors: the number of females in a male's OMU, the average distance between a female and her leader male, the amount of time females spent grooming their leader males, the amount of time leader males spent grooming their females, and female reproductive state. For Figures 10.5–10.7, data points represent males (Figure 10.5) or females (Figures 10.6, 10.7) for whom we have at least one year of *ad libitum* observations as well as data on one of the above variables.

Figure 10.5 shows the relationship between rates of agonism by leader males (including all male agonistic behavior listed in Table 10.1) and the number of females in that male's OMU. There is no relationship between these two variables.

Figure 10.6 shows the relationship between rates of agonism by leader males toward individual females (not the males' overall agonism rate as shown in Figure 10.3b, but agonism toward a given female) and the average distance between that female and that male. Data points represent all male-female dyads for whom we have both data on aggression by that male toward that female *and* data on proximity between them (calculated from 10-minute point time samples during continuous samples on focal females). There is a positive correlation between these two variables (Spearman Correlation; $r_s = 0.61$, $p = 0.009$).

Figure 10.7 shows the relationship between the rate of agonism by leader males toward individual females and the percentage of a female's available

---

represent all agonistic behavioral elements listed in Table 10.1; central bars represent physical aggression (grab, hit, bite on back, chase, push, possession grip), and bars on the right represent neck-biting only. (b) Comparison of rates of agonism, physical aggression, and neck-biting in the same males ($N = 5$) across both conditions.

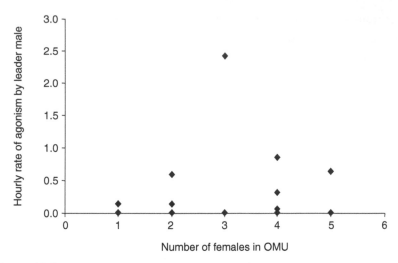

Figure 10.5 Scatter plot showing the relationship between the number of females in a leader male's OMU and his hourly rate of overall agonism. Data points represent leader males.

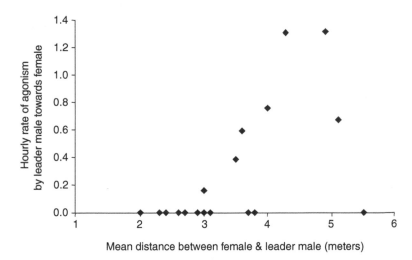

Figure 10.6 Scatter plot showing the relationship between average distance between a female and her leader male and the overall rate of agonism by that male toward that female. Proximity calculated from ten-minute point time samples (Spearman Correlation $r_s = 0.61$, $p = 0.009$). Data points represent individual females.

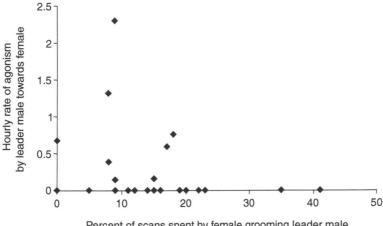

Figure 10.7 Scatter plot showing the relationship between percentage of a female's "available social time" spent grooming her leader male and the overall rate of agonism by that leader male toward that female. "Available social time" is defined as point time samples during which a female is not traveling, foraging, eating, or drinking, but can be viewed as "available" for social interaction (see Swedell 2002, 2006). Percentage of time spent grooming was calculated as a percentage of 10-minute point time samples during which a female was grooming her leader male. Data points represent individual females.

social time spent grooming her leader male. "Available social time" is defined as point time samples during which the female was not traveling, foraging, eating, or drinking, but could be viewed as "available" for social interaction (see Swedell 2006). The percentage of time spent grooming was calculated as the percentage of scan samples representing a female's available social time during which she was grooming her leader male. There is no statistically significant relationship between these two variables (R=−0.166; p=0.379). There is also no relationship between rates of agonism by leader males and the percentage of time (also calculated from point time samples) spent by leader males grooming their females (R=−0.005; p=0.983). Owing to space constraints, this plot is not shown.

Figure 10.8a shows the proportion of females in each reproductive state that received agonism by leader males under baseline conditions. Because many observations contributing to this data set were *ad libitum*, we cannot control for differences in observation time on females in differing reproductive conditions. We can, however, compare variation in agonism across reproductive state to

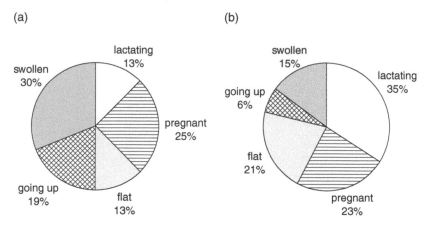

Figure 10.8 (a) Proportion of agonistic events by leader males under baseline conditions that occurred when the recipient female was in each reproductive state. (b) Proportion of females in study group in each reproductive condition at a given point in time. "Going up" refers to the period of tumescence of the sexual skin prior to reaching its full size; "swollen" refers to maximal swelling; "flat" refers to a nonswollen female who is between the swelling phases of her monthly cycle (i.e., not pregnant or lactating); "lactating" refers to a female with a suckling infant; and "pregnant" refers to females showing the pink paracallosal skin characteristic of pregnancy in baboons (Altmann 1970) and/or obvious weight gain and cessation of cycling.

what would be expected by chance. We assume here that all females in the study group were equally likely to be observed receiving aggression, which, over the course of the entire study period, we think is a reasonable assumption. Using the typical distribution of reproductive states across females in the study group at a given point in time as our "expected" values, shown in Figure 10.8b, we compared observed proportions of agonism to these expected proportions using Chi-square analysis and found a statistically significant difference between observed and expected (Chi-square = 58.24, $p < 0.001$). In particular, lactating females received less agonism than expected, whereas swollen (both going up and fully swollen) females received more agonism than expected.

Discussion

Hamadryas baboons have perhaps the most male-dominated society across the primate order, and—with the exception of pinnipeds and some cetaceans—possibly across mammals as a whole (Kummer 1968a; Smuts and Smuts 1993;

Swedell 2006). It is thus striking that aggression is not more prevalent in hamadryas society overall, especially compared to other baboons. In essence, hamadryas males seem to use aggression in a more directed fashion than other primates such as chimpanzees and savanna baboons. Rather than coercing females every time they come into estrus, while at the same time competing aggressively with other males for access to those females (Kitchen et al., Chapter 6, Muller et al., Chapter 8 in this volume), hamadryas males use aggression primarily during takeovers, as a means of conditioning females. Unlike other one-male group-living animals, in which—in theory—males need not coerce females to the same degree that they do in multimale groups, hamadryas one-male units are surrounded by many potential male competitors. That a one-male mating system can even exist within a multimale society illustrates the power of coercion in maintaining the hamadryas OMU as a socially and reproductively exclusive unit within a larger social group.

Aggression as Conditioning after Takeovers

A comparison of rates of agonism during takeovers with baseline rates (Figure 10.4) reveals three main differences. First, rates of aggression are far higher during takeovers than at other times. Compared to about one aggressive event every four hours under baseline conditions, during takeovers aggression occurs *at least once per hour* and often far more than that. In most cases, this aggression is directed by the takeover male toward the female he is taking over (or attempting to take over); in one case in our current data set, the previous leader male also used aggression to attempt to retain his female. Second, there is much wider variation among males on the day of a takeover than at other times. This may be partly due to the small amount of focal data we have for days of takeovers. However, these numbers do represent actual rates of aggression during the short time that the focal animals were observed. (These differences are more likely due to variation in female motivation to stay with the takeover male, as illustrated by the difference between Who and Bea in the opening paragraphs of this chapter.) Third, males exhibit more *kinds* of aggression in the context of takeovers than at other times. Of the list of aggressive elements shown in Table 10.1, *pushing, biting on the back,* and *possession grip* have been observed *only* in the context of takeovers.

The difference between baseline rates of agonism and rates on days of takeovers is consistent across all males for whom we have data on both conditions.

For most leader males, no aggression of any kind was observed under baseline conditions, either during focal samples or during *ad libitum* observations (Figure 10.3). Unfortunately, we do not have enough data to track aggression rates for the subsequent days and weeks following a takeover, but our subjective impression is that aggression rates diminish rapidly from day one and by the end of the first week have nearly reached baseline levels.

Essentially, then, hamadryas females undergo a behavioral conditioning process after a takeover. A female's new leader male trains her through aggression to follow him, remain near him, and cease interacting with her previous leader male. After this conditioning process is over, a female automatically maintains spatial proximity to her leader male with little further need for prompting.

This association between takeovers and aggression adds a new dimension to comparisons between "male dispersing" and "female dispersing" taxa. Unlike the typical pattern in other baboons in which females are philopatric and males disperse, hamadryas baboons are characterized by male philopatry and some amount of female dispersal (Hammond et al. 2006). However, unlike most other primates in which female dispersal occurs—such as mountain gorillas, chimpanzees, atelines, and several colobines—hamadryas baboon females do not disperse voluntarily, nor do they appear to have any inherent motivation to emigrate (Pusey 1979; Crockett 1984; Struhsaker and Leland 1987; Watts 1990; Sterck 1997; Printes and Strier 1999). Rather, they "are transferred" among OMUs by leader males through the use of physical aggression (Swedell 2005, 2006).

Other Functions and Correlates of Male Aggression

It thus seems clear that the main function of hamadryas male aggression toward females is to condition females during and immediately after takeovers. The vast majority of male aggression toward females occurs during takeovers and attempted takeovers, and many leader males in the Filoha population have never been observed to neck-bite a female outside of this context. Aggression does occur at other times, though, so it presumably has other, less important, functions as well.

One way to address other possible functions is to look at the specific contexts in which aggression occurs, shown in Table 10.2. Although we do not have a large enough sample size to analyze these data quantitatively, we can at least point to two general patterns consistent with previous research (e.g.,

Kummer 1968a). First, other than during takeovers, the main contexts in which males are aggressive toward females fall into two general categories: females moving away from their leader male (even only a meter or two) and females interacting with individuals outside the OMU. There appears to be little consistency among males in the actual distances that provoke aggression, as some leader males respond with threats or aggression when their females move only 1 m further away than they were previously (e.g., 4 m instead of 3), whereas others regularly tolerate distances of 20 m or more. Second, the outcome of male aggression appears to be to bring the female closer to the male. A female hamadryas baboon is far more likely to respond to male aggression by moving toward him and grooming him than by running away from him. Female hamadryas appear to be especially motivated to groom their leader male after a social conflict or other situation involving a real or perceived threat, even if that threat is the leader male himself (Swedell 2006). Following such a conflict or stressful situation, females often groom their leader male intensely, and this intense grooming appears to have a calming effect. Females thus appear to seek comfort in—and give comfort to—the very individual that is most aggressive toward them.

Figure 10.3 reveals a substantial amount of variation in aggression by leader males. If we assume that this variation among males represents real differences and is not simply an effect of small sample size, what factors might underlie it? Although we do not have sufficient data on male aggression outside of the context of takeovers to investigate this question in depth, we performed some simple regression analyses to investigate possible correlates of male aggression. One possibility is the number of females in a one-male unit. Might males with more females to keep track of show higher rates of aggression than males who have fewer females? This does not seem to be the case, as Figure 10.5 reveals no relationship between these two variables.

Rather than being a feature of a one-male unit as a whole, however, perhaps variation in male aggression is related to individual dyadic relationships between females and their leader males. This can be explored by looking at variation in female behavior. For example, within an OMU, females vary widely in their proximity maintenance to the leader male (Swedell 2006). If male aggression toward females functions to herd females and bring them closer, then perhaps variation among females in average distance to their leader male influences the amount of aggression they receive. Our data support this idea: the rate of agonism by leader males toward individual females does increase with the average distance between them. Because hamadryas females typically follow their leader males and maintain proximity to them rather than the reverse, this variable can be viewed as a measure of proximity maintenance on the part of the female. It thus appears that females that stay closer to their leader male receive less aggression from him. So why do all females not simply stay as close as possible to their leader male so as to avoid aggression from him? The answer to this question likely lies in individual variation in optimal strategies, as well as additional factors explored later in this chapter. For some females, the costs of aggression are probably lower than the benefits of pursuing social, and sometimes sexual, relationships with other individuals, including both female relatives and nonleader males.

In addition to proximity, aggression by males toward females may also be mediated by grooming. Might females who spend more time grooming their leader males incur less aggression than females who spend less time grooming them? The answer to this question is unclear, but Figure 10.7 suggests that it might be "yes." There is no statistically significant relationship between the hourly rate of agonism by leader males toward individual females and the percentage of a female's available social time spent grooming her leader male. However, the lack of a significant relationship between female-male grooming and male aggression is probably a result of the cluster of points along the X-axis that represent dyads with both low grooming rates *and* aggression rates of zero. It light of this pattern, it is noteworthy that there are no dyads in which both the aggression *and* the grooming rate are high. This suggests, perhaps, that the amount of aggression received by females who do not spend much time grooming their leader males is based on other factors, but that females who spend a substantial amount of time grooming their leader males are indeed "rewarded" by receiving less aggression from them. It should be noted, of course, that proximity and grooming are themselves correlated: females who stay closer to their leader males also groom them more. We cannot disentangle the effects of these two variables without a larger data set.

Female reproductive condition also appears to play a role in the amount of aggression a female receives, with interesting similarities and differences to both a closely related primate with different patterns of bonding, the chacma baboon, and a more distantly related primate with similar patterns of bonding, the mountain gorilla. First, we found support for the idea that females with young infants receive less aggression than other females (Figure 10.8). Our results contrast with the pattern seen in chacma baboons (Kitchen et al., Chapter 6 in this volume), in which lactating females receive just as much aggression as others. Our results are, however, consistent with the pattern seen in mountain gorillas, in which lactating females receive the least aggression overall (Robbins, Chapter 5 in this volume). Both mountain gorillas and hamadryas baboons, compared to other baboons and chimpanzees, are characterized by a high degree of paternity certainty and male protection of females and their offspring (Robbins, Chapter 5).

Second, consistent with evidence from other baboons, macaques, and chimpanzees (Smuts and Smuts 1993), as well as captive hamadryas baboons (Zinner et al. 1994), we found that fully half of the instances of nontakeover-related agonism by leader males occurred when the female recipient had a sexual swelling (Figure 10.8). Such a high frequency is far above what we would expect given the low number of swollen females in the band at any one time. Thus, despite the close proximity of females to their leader males throughout their reproductive cycle, hamadryas leader males are nevertheless more likely to respond with aggression when a swollen female moves away from him or interacts with another individual than when a nonswollen female does so. This pattern supports the notion that hamadryas male aggression functions to control female sexuality, as such a model predicts that females should receive more aggression when they are most fecund (Muller et al., Chapter 1 in this volume).

Additional factors may also influence male aggression toward females in hamadryas society. One of these factors is female tenure in a one-male unit. Although we do not yet have enough data from various points during each female's tenure with a male to test this idea, we suspect that females receive the most aggression from their leader males shortly after being taken over and then receive less aggression over time, especially after additional females have joined their OMU (Swedell 2006).

Finally, there is the possibility that males simply differ in their temperament (Clarke and Boinski 1995), such that some males react more strongly than other males to females who wander far away or interact with outsiders. One indication of such variation is that some males tolerate grooming among females of different OMUs (Swedell 2006). Other males, by contrast, do not allow their females to do more than inspect an infant from a different OMU and will react to such behavior with threats or aggression. For example, Boris, who shows a baseline aggression rate of zero (Figures. 10.3a, 10.3b), twice tolerated grooming between one of his females and a female in another OMU without reacting agonistically. Alex, by contrast, who shows one of the highest baseline aggression rates (Figures 10.3a, 10.3b), chased one of his females over 40 m for grooming a female in another OMU and would likely have neck-bit her had he caught her (which he did not, as she escaped into the terminal branches of a tree).

Summary and Conclusions

In general, hamadryas male aggression toward females does not occur with the same regularity as it does in other baboons. Rather, it is concentrated at critical times, when a male is attempting to integrate a new female into his OMU. During and after takeovers, aggression functions to condition a female to switch her following response to a different leader male. The contrast between hamadryas and other baboons (cf. Kitchen et al., Chapter 6 in this volume) in their patterns of male aggression illustrates the differences among baboons in the relationship between aggression and social organization. In chacmas and other "savanna" baboons, male competition over access to females is a regular state of affairs, and males use aggression against females to both compete with other males and coerce females into mating with them (Kitchen et al., Chapter 6 in this volume). Aggression in most male baboons is an emergent property of— rather than a determining factor of—social organization. By contrast, in the hamadryas system males use aggression in a much more limited but directed fashion, primarily to establish and maintain the cohesion of their OMUs by controlling female behavior. Without this particular use of male aggression, hamadryas one-male units would likely not even exist. Unlike other baboons, therefore, hamadryas male aggression against females is inextricably tied to social organization in that it is the cohesive force behind the one-male unit social structure and one of the proximate mechanisms behind the hamadryas social system as a whole (Kummer 1971).

Aggression occurring outside the context of takeovers is far less consistent in its patterns. In general, a leader male is aggressive toward his females when they move too far away or leave the social boundaries of the one-male unit. However, leader males do not appear to be uniform in their perception of "allowable" distances and extra-OMU interactions. Whereas all hamadryas males are aggressive toward females after takeovers, males appear to be far more variable in the amount of aggression they show toward females at other times.

Our results suggest that three factors in particular may underlie variation in hamadryas male aggression: female proximity maintenance to males, time spent by females grooming males, and female reproductive state. Although our sample size is still too small to state this with great certainty, our preliminary data suggest that females who stay closer to their leader male and spend more time grooming him will, in turn, incur less aggression from him. Our results also suggest that females are most likely to receive aggression from their leader males when they are sexually swollen and least likely to receive aggression when they are lactating. These results are consistent with our understanding of the hamadryas social system, in which males coerce females into semipermanent one-male units, females respond to male aggression by staying close to them and grooming them, and hamadryas leader males provide protection for females and their offspring.

In conclusion, although the fitness costs to females of male aggression have yet to be determined, it is clear so far that hamadryas male aggression toward females falls within the definition of sexual coercion provided by Smuts and Smuts (1993) and, more specifically, the definition of coercive mate guarding (via herding, punishment, and sequestration) described by Muller and colleagues in Chapter 1 of this volume. From this preliminary analysis, it appears that the vast majority of aggression by hamadryas males toward females serves to bring a female closer to her leader male, punish a female for interacting with nonleader males, and condition a female to remain near her leader male over time—all of which increase the chance of future copulation and conception with the leader male, while decreasing the female's chance of copulation and conception with other males. The coercive tactics used by hamadryas males, combined with the likely benefits to females of male protection (Swedell and Saunders 2006), appear to result in a system in which males control female behavior so absolutely that actual rates of aggression are quite low (compared to chacma baboons, for example; see Kitchen et al., Chapter 6 in this volume). However, females are very quick to respond to male threats and are exceptionally responsive to external threats compared to other female baboons. To the extent that infanticide falls within the parameters of sexual coercion as well (Smuts and Smuts 1993; Muller et al., Chapter 1 in this volume), it is an additional form of sexual coercion used by hamadryas males and, unlike female-directed aggression, has a more quantifiable cost to female recipients (Swedell and Tesfaye 2003; Swedell 2006). Future study of the correlates of male aggression outside the context of takeovers, as well as variation in behavioral and hormonal indicators of aggression-induced stress in females, should help to further elucidate the proximate determinants of male aggression toward females, the costs to females of such aggression, and the relationship between male aggression and hamadryas baboon social organization.

Acknowledgments

We are most grateful to the Wildlife Conservation Department of Ethiopia for permission to conduct this research. We also thank the Awash National Park Baboon Research Project for the use of their field vehicle. Funding for L. Swedell's research in 1996–1998 was provided by the National Science Foundation (#9629658), the Wenner-Gren Foundation (#6034), the L.S.B. Leakey Foundation, and the National Geographic Society (#6468-99). Funding for A. Schreier's research in 2005–2006 was provided by the City University of New York PSC-CUNY Research Award Program (award #66588-0035 to L. Swedell), the New York Consortium in Evolutionary Primatology, and the City University of New York Ph.D. Program in Anthropology. We are also grateful to Shahrina Chowdhury for her assistance with data entry and Marina Cords for delivering an earlier version of this chapter at the XXIth Congress of the International Primatological Society in Entebbe, Uganda. Finally, we thank Martin Muller and Richard Wrangham for the invitation to contribute to this volume and for their many helpful comments.

References

Abegglen, J. J. *On Socialization in Hamadryas Baboons.* London: Associated University Presses, 1984.

Altmann, S. A. "The Pregnancy Sign in Savannah Baboons." *Laboratory Animal Digest* 6 (1970): 7–10.

Bergman, T. J. "Hybrid Baboons and the Origins of the Hamadryas Male Reproductive Strategy." In *Reproduction and Fitness in Baboons: Behavioral, Ecological, and Life History Perspectives,* eds. L. Swedell and S. R. Leigh, pp. 81–103. New York: Springer, 2006.

Clarke, A. S., and S. Boinski. "Temperament in Nonhuman Primates." *American Journal of Primatology* 37 (1995): 103–125.

Colmenares, F. "Clans and Harems in a Colony of Hamadryas and Hybrid Baboons: Male Kinship, Familiarity and the Formation of Brother-Teams." *Behaviour* 121 (1992): 61–94.

———. "Greeting Behaviour between Male Baboons: Oestrus Females, Rivalry and Negotiation." *Animal Behaviour* 41 (1991): 49–60.

Colmenares, F., M. G. Lozano, and P. Torres. "Harem Social Structure in a Multiharem Colony of Baboons (*Papio* spp.): A Test of the Hypothesis of the 'Star-Shaped' Sociogram." In *Current Primatology, Volume II: Social Development, Learning and Behavior,* ed. J. J. Roeder, pp. 93–101. Strasbourg: Université Louis Pasteur, 1994.

Colmenares, F., F. Zaragoza, and M. V. Hernández-Lloreda. "Grooming and Coercion in One-Male Units of Hamadryas Baboons: Market Forces or Relationship Constraints?" *Behaviour* 139 (2002): 1525–1553.

Crockett, C. M. "Emigration by Female Red Howler Monkeys and the Case for Female Competition." In *Female Primates: Studies by Women Primatologists,* ed. M. F. Small, pp. 159–173. New York: Alan R. Liss, 1984.

DeVore, I. "Male Dominance and Mating Behavior in Baboons." In *Sex and Behavior,* ed. F. A. Beach, pp. 266–289. New York: John Wiley and Sons, 1965.

Hammond, R. L., L. J. Lawson Handley, B. J. Winney, M. W. Bruford, and N. Perrin. "Genetic Evidence for Female-Biased Dispersal and Gene Flow in a Polygynous Primate." *Proceedings of the Royal Society London B* 273 (2006): 479–484.

Huffman, M. A. "Consort Intrusion and Female Mate Choice in Japanese Macaques." *Ethology* 75 (1987): 221–234.

Kummer, H. *Social Organization of Hamadryas Baboons: A Field Study.* Chicago: University of Chicago Press, 1968a.

———. "Two Variations in the Social Organization of Baboons." In *Primates: Studies in Adaptation and Variability,* ed. P. C. Jay, pp. 293–312. New York: Holt, Rinehart and Winston, 1968b.

———. "Immediate Causes of Primate Social Structures." Proceedings of the 3rd International Congress in Primatology, Zurich 1970. 1971.

———. "Dominance versus Possession: An Experiment on Hamadryas Baboons." In *Symposia of the IVth International Congress of Primatology, Volume I: Precultural Primate Behavior,* ed. E. W. Menzel Jr., pp. 226–231. Basel: Karger, 1973.

———. *In Quest of the Sacred Baboon.* Princeton, N.J.: Princeton University Press, 1995.

Kummer, H., W. Götz, and W. Angst. "Triadic Differentiation: An Inhibitory Process Protecting Pair Bonds in Baboons." *Behaviour* 49 (1974): 62–87.

Kummer, H., and F. Kurt. "Social Units of a Free-Living Population of Hamadryas Baboons." *Folia Primatologica* 1 (1963): 4–19.

Morin, P. A. "Reproductive Strategies in Chimpanzees." *Yearbook of Physical Anthropology* 36 (1993): 179–212.

Printes, R. C., and K. B. Strier. "Behavioral Correlates of Dispersal in Female Muriquis *(Brachyteles arachnoides)*." *International Journal of Primatology* 20 (1999): 941–960.

Pusey, A. E. "Intercommunity Transfer of Chimpanzees in Gombe National Park." In *The Great Apes,* eds. D. A. Hamburg and E. R. McCown, pp. 465–479. Menlo Park, Calif.: Benjamin/Cummings, 1979.

Schreier, A., and L. Swedell. "Evidence for Clans in a Population of Wild Hamadryas Baboons." *American Journal of Physical Anthropology Supplement* 44 (2007): 209–210.

Sigg, H., A. Stolba, J. J. Abegglen, and V. Dasser. "Life History of Hamadryas Baboons: Physical Development, Infant Mortality, Reproductive Parameters and Family Relationships." *Primates* 23 (1982): 473–487.

Smuts, B. B. and R. W. Smuts. "Male Aggression and Sexual Coercion of Females in Nonhuman Primates and Other Mammals: Evidence and Theoretical Implications." *Advances in the Study of Behavior* 22 (1993): 1–63.

Sterck, E. H. M. "Determinants of Female Dispersal in Thomas Langurs." *American Journal of Primatology* 42 (1997): 179–198.

Struhsaker, T. T., and Lysa Leland. "Colobines: Infanticide by Adult Males." In *Primate Societies,* eds. B. B. Smuts, D. L. Cheney, R. M. Seyfarth, R. W. Wrangham, and T. T. Struhsaker, pp. 83–97. Chicago: University of Chicago Press, 1987.

Swedell, L. "Two Takeovers in Wild Hamadryas Baboons." *Folia Primatologica* 71 (2000): 169–172.

———. "Affiliation among Females in Wild Hamadryas Baboons *(Papio hamadryas hamadryas)*." *International Journal of Primatology* 23 (2002): 1205–1226.

———. "Dispersal by Force: Residence Patterns of Wild Female Hamadryas Baboons." *American Journal of Physical Anthropology Supplement* 40 (2005): 202.

———. *Strategies of Sex and Survival in Hamadryas Baboons: Through a Female Lens.* Upper Saddle River, N.J.: Pearson Prentice Hall, 2006.

Swedell, L., and J. Saunders. "Infant Mortality, Paternity Certainty, and Female Reproductive Strategies in Hamadryas Baboons." In *Reproduction and Fitness in Baboons: Behavioral, Ecological, and Life History Perspectives,* eds. L. Swedell and S. R. Leigh, pp. 19–51. New York: Springer, 2006.

Swedell, L., J. Saunders, M. Pines, A. Schreier, and B. Davis. "Alternative Reproductive Strategies in Male Hamadryas Baboons: Leaders, Followers, and Solitary Males." *American Journal of Physical Anthropology Supplement* 46 (2008): 203.

Swedell, L., and T. Tesfaye. "Infant Mortality after Takeovers in Wild Ethiopian Hamadryas Baboons." *American Journal of Primatology* 60 (2003): 113–118.

Watts, D. P. "Ecology of Gorillas and Its Relation to Female Transfer in Mountain Gorillas." *International Journal of Primatology* 11 (1990): 21–45.

Zinner, D., M. Schwibbe, and W. Kaumanns. "Cycle Synchrony and Probability of Conception in Female Hamadryas Baboons *Papio hamadryas*." *Behavioral Ecology and Sociobiology* 35 (1994): 175–183.

III

SEXUAL COERCION AND MATE GUARDING IN HUMANS

11

Coercive Violence by Human Males against Their Female Partners

Margo Wilson and Martin Daly

Studying conflict and violence in human relationships sounds like it should be easy. We begin, after all, with an intuitive understanding of the species of interest and an affinity for its preferred habitats, and it is a great convenience that a seemingly limitless supply of people have relevant experience that they are often willing—even eager—to share. However, the ability of human beings to recount their past actions and their subjective experiences to others is a two-edged sword, tempting researchers to rely excessively and credulously on accounts that are often self-serving or otherwise biased. The questionable validity of self-report data has engendered considerable controversy in the specific research domain of interest here, namely, men's violence against women, and we will revisit this issue.

The human animal's speechifying is by no means its only challenging peculiarity. The backbone of primatological research is observation of behavior in the natural habitat, but this cannot be made a foundational component of human research for reasons of privacy. Of course, other primate species may also behave somewhat differently when they are readily visible than when they are not, and this can distort or at least delay an observer's apprehension of social realities, but this is a uniquely pervasive issue in human research. Moreover, whereas few would object on ethical grounds to unobtrusive observation of the secret lives of chimpanzees by such means as hidden cameras, spying on the private behavior of our fellow human beings for scientific purposes is simply unacceptable.

In many animals, including most primates, who mates with whom is of great interest to conspecific third parties. But here, too, *Homo sapiens* is unique in the extent to which mating relationships are politicized. Marriage engenders alliance among extensive kin groups, and family members exert considerable influence on both mate choice and the stability and quality of marriages. Coercive violence against women is seldom completely private, and anticipated reactions of third parties can both encourage and deter it, as is discussed in greater detail by Rodseth and Novak (Chapter 12 in this volume). And then, in addition to all these complications, there is cultural diversity: According to anthropologists, there are societies in which wife-beating is normative and a large majority of women report having experienced it, and there are others in which such abuse is rare and is condemned (Counts et al. 1999). This cross-cultural diversity complicates the task of describing and explaining the reality of human male violence against women, but it also offers opportunities to test hypotheses about the factors that influence its prevalence and severity.

Why Are Wives Assaulted?

Why do men commit violence against their female partners? The substantial social scientific literature on this topic offers a variety of partial answers. One of the most popular has been that the culprit is societal legitimation of male entitlement. In an oft-repeated phrase, Straus (1980) maintained that "a marriage license is a hitting license." But although there can be little doubt that wife-beating is indeed more prevalent where it is condoned than where it is condemned, to maintain that men assault their wives because they are entitled to do so begs the crucial questions of motive and function. Men are entitled to eat sawdust if they so desire, or to amputate their own extremities, but they have no inclination to do these things. Why should a man be any more inclined to assault the woman who is his partner in life and the (actual or potential) mother of his children?

Almost as popular as the entitlement account, and often appended to it as a fuller explanation, has been the notion that men do *not* in fact harbor specific inclinations to assault their partners, but simply use wives as convenient punching bags when they feel the urge to assault someone else (e.g., Farrington 1980). Grounded in a discredited Freudian/Lorenzian view of behavior as the release of pent-up energy, the idea here is that men whose aspirations are thwarted outside the home "take out their frustrations" on a safe, legitimate target. Evidence that setbacks such as loss of employment are associated with elevated rates of wife assault is frequently cited as support for this interpretation, but this argument is weak, because the relevance of such risk factors is equally to be expected if wife assault is an instrumental means of intimidating women who might otherwise desert their abusers. An additional reason for doubting the hypothesis that wife assault is to be understood as redirected, frustration-induced aggression is that there are other equally safe and legitimate targets who are even more defenseless, namely, children and pets, and yet wives are often preferentially targeted and even effortfully pursued when they try to escape their attackers.

The feminist thesis that wife abuse is a product of "patriarchy" comes closer to the heart of the matter. Of course, simply pointing an accusing finger at evil patriarchal ideology again begs the crucial questions of motive and function, but at least since Dobash and Dobash (1979), most feminists writing on this topic have recognized that men assault their wives not merely because it is legitimate for them to do so, but in order to *maintain control.*

In 1993, *Statistics Canada* conducted a national survey on *Violence against Women* (Johnson and Sacco 1995). The 12,300 respondents included 8385 women who were co-residing with male partners and who reported their experiences of threats and sexual and physical violence by those partners, among other things. Before each woman was asked about violence by her partner, she was asked whether certain statements applied to him:

1. "He is jealous and doesn't want you to talk to other men."
2. "He tries to limit your contact with family or friends."
3. "He insists on knowing who you are with and where you are at all times."
4. "He calls you names to put you down or make you feel bad."
5. "He prevents you from knowing about or having access to the family income, even if you ask."

An affirmative response to each of these questions was significantly associated with affirmations of partner violence, and as the severity and chronicity of partner violence increased, so too did the number of these controlling behaviors that were alleged (Wilson et al. 1995). This Canadian survey provided the prototype for a subsequent national survey in the United States, where the association between partner violence and these same (and additional) autonomy-limiting actions was replicated (Miller 2006).

Although one must acknowledge that these correlations could just reflect variability in women's willingness to engage in broad-brush "badmouthing" of their partners, they do suggest that the same men who are relatively inclined to monopolize, control, and sequester their wives are the ones who are also relatively inclined to assault them. We therefore interpret these survey results as support for the hypothesis of psychological links between proprietary inclinations and violence against wives.

Feminists and other social scientists have come to the same conclusion. Indeed, students of such violence popularly describe it as domestic or intimate "terrorism" (e.g., Johnson and Ferraro 2000). Their point is that the chronic threat and occasional exercise of violence keep women fearful and compliant, and also deter them from leaving. What these feminist analyses lack, however, is a persuasive account of why the male mind should be prone to construing wives as alienable property, why certain categories of women—especially those who are young and fertile (Figure 11.1)—are more frequently assaulted than others, and why some men are more controlling than others. To make sense of these things, an evolution-minded understanding of sexual politics is essential.

Sexual Proprietariness Is a Psychological Adaptation of the Human Male

Since Trivers (1972), animal behaviorists have understood that the sex that provides more parental investment (usually females) is a limiting resource for the fitness of the less investing sex (usually males) and is therefore an object of

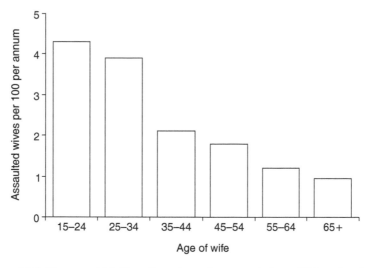

Figure 11.1 Rates at which Canadian wives who responded to a 1993 national *Violence against Women Survey* professed to have been physically assaulted by their husbands in the preceding year, as a function of the respondent's age. After Wilson, Johnson, and Daly (1995).

competition. Although human males make substantial paternal investments in young, at least when compared to most other primates, they seldom if ever match women in this regard. Ethnographic and genetic evidence is consistent in indicating that *Homo sapiens* is like other mammals in the fact that male fitness variance exceeds female fitness variance (e.g., Wilder et al. 2004). Because men have both a higher ceiling on their potential progeny than women and a higher probability of dying childless, intrasexual competition is more severe among men than among women. It is no surprise that men should construe that competition as a sort of "zero-sum game" in which one's gain is another's loss, especially in light of the small but genuine risk of misattributed paternity (Anderson 2006) and the fact that women sometimes prefer already mated men over available bachelors (e.g., Borgerhoff Mulder 1992).

Because human male fitness is and always has been limited primarily by access to the reproductive efforts of fertile women, we have proposed that male sexual proprietariness is an evolved motivational/cognitive subsystem of the human brain/mind (Wilson and Daly 1992a). Sexual jealousy is a sexually differentiated mental state, with that of men being relatively focused on specifically sexual fidelity and that of women on threats to emotional commitment and investment (e.g., Buss et al. 1992; Daly et al. 1982; Teisman and Mosher 1978). This discovery has inspired many attempted counter-demonstrations, perhaps driven by distaste for the idea that women and men are in any way psychologically dissimilar, but the evidence is consistent and abundant (Buss and Haselton 2005). It is also unsurprising when we consider that cuckoldry risk is sexually asymmetrical.

In proposing that men take a proprietary view of women's sexuality and reproductive capacity, we mean that men are motivated to lay claim to particular women as songbirds lay claim to territories, as leopards lay claim to a kill, or as people of both sexes lay claim to valuables. But proprietariness has the further implication of a sense of entitlement, which, if not unique to our species, is at least uniquely highly developed. Trespass against one's property provokes not only a hostile response to the trespasser, but moralistic feelings of aggrievedness and indignation as well. These sentiments have a more broadly social function: whereas hostile feelings motivate aggressive action against rivals, in our species as in many others, the human sentiment of moral indignation motivates appeals to third parties to recognize the trespass as a wrong against the property holder and hence as *justifying* retaliation or more collective sanctions. Exactly what form these sanctions take is, of course, culturally variable, but moral indignation against those who violate one's property rights is, arguably, a human universal and an aspect of our evolved psychology that would be pointless without the social complexity entailed by coalitions, reputations, and politics. However much we may deplore men's coercive violence against their female partners, we will misunderstand it if we fail to recognize that the perpetrators are moral agents engaged in acts that they perceive as righteous.

Because claims of proprietary entitlement are responses to rivalry over limited resources, they necessarily exist in an arena of actual or potential conflicts of interest. When the resources in question are inanimate, the conflicts are relatively simple matters of rivalry, although even here there can be extensive involvement of third parties and hence much moral positioning. It is when the "property" itself is another person, however, that things become really complicated, for in addition to the conflicting ambitions of rival aspirant owners or users, this kind of property has its own preferences and goals, and may have relatives or other allies. Women's variably effective resistance to men's coercion and violence thus reflects, in part, the relative power of the two parties as affected by their respective material and social resources (e.g., Smuts 1992).

The extent to which a woman's relatives afford her protection against misuse by her husband and in-laws is cross-culturally variable, largely as a result of variability in the extent to which the transfer of a woman at marriage from her natal family to her husband's family is institutionalized as a contractual transfer of property. For this and other reasons, the behavioral manifestations of male sexual proprietariness are culturally and historically variable, and one can easily become absorbed by the details: bridewealth and refunds for infertility, guarantees of virginity, compensations for adultery, chaperoning, purdah, and so forth. But we should not let this fascinating diversity obscure the common thread of intense concern with proprietary entitlements in female sexuality and reproductive capacity across the gamut of human material and social conditions. In traditional societies, for example, adultery is universally defined as illegitimate sexual contact involving a married woman and is universally deemed an offense against her husband; the marital status of the male adulterer is universally irrelevant (Daly et al. 1982). Such cross-cultural consistency can only be understood as a reflection of universal human motives and emotions.

Anthropological and historical evidence suggests that wherever there is significant variance in resources, high-status men have parlayed their wealth and power into polygynous monopolization of wives and/or concubines and have been concerned to "protect" their women from potential rivals (Betzig 1986).

Only in relatively recent years, with the inequities engendered by agricultural surpluses and the rise of complex, role-differentiated societies, could extreme polygyny and extreme sequestering of women have become possible. The most despotic harem holders then began to confine women in cells guarded by eunuchs, to maintain records of their menstrual cycles, to farm them out to the harems of underlings when they got too old, and even to kill and replace them *en masse* in the event of security failures and possible cuckoldry (Dickemann 1981; Betzig 1986).

Harem acquisition is a novelty on an evolutionary timescale. It is neither an evolved adaptation in its own right nor a selection pressure that has had any significant effect on human evolution. But the phenomenon is of interest to evolutionists nonetheless. Harems provide evidence about our species' evolved psychology in the same way that refined sugar does: they provide testimony about evolved appetites. In the first place, the harem phenomenon falsifies the hypothesis that the reason men aspire to be polygynists is that wives are economic assets. Proponents of this view have insisted that wives are sought as a means to wealth, power, and status rather than as perquisites thereof, but harem occupants were typically maintained at great expense and prevented from being productive. Second, this oddly recurring institution also demonstrates that men who collect women are not simply pursuing sexual variety, for in every independent invention of the harem phenomenon, exclusivity of sexual access appears to have been one of the despot's main preoccupations.

Only the richest and most powerful men could collect harems, but millions of men have guarded and constrained "their" women by practices that seem to depart from those of despots only in degree. Veiling of women, chaperoning, purdah, and other forms of "claustration" up to the extreme of incarceration are common social institutions of patrilineal, patrilocal societies, and the significance of these practices becomes apparent when one notes that it is typically only women of reproductive age who are restricted in these ways; prepubertal girls and, especially, postmenopausal women often enjoy much more freedom. Coercive control is also achieved by genital mutilation (WHO Study Group 2006), whose evident function is to thwart women's sexual appetites and even their penetrability until surgical reopening. "Chastity belts" provide less violent solutions to the same problem.

Claustration practices are status-graded (Dickemann 1981): the higher her social status, the more restricted the woman. This is a seemingly paradoxical correlation until one considers that high-status men have both an intense concern about the paternity of their heirs and the wherewithal to confine women even when doing so requires foregoing the women's productive labor or is otherwise costly. Chinese foot-binding was a status-graded practice that not only displayed the male owner's capacity to dispense with the woman's labor but also conveniently rendered her incapable of flight.

Although economic inequality obviously persists in modern societies, the variance in men's access to women has been much reduced. Most notable in this regard is the spread of legislated monogamy. We would expect that coercive control of women should decrease in prevalence and intensity when the variance in men's access to women is reduced. However, in the absence of legitimate patriarchal entitlement to confine and control women, the prevalence of individual resort to "domestic terrorism" could actually be exacerbated. So is this problem shrinking, or is it getting worse in an age of women's liberation? The question cannot be answered with confidence. One might expect that recent trends in the prevalence of intimate partner violence would be well documented, but good victimization surveys with a focus on the problem are relatively recent and have not yet been replicated at intervals, with the result that available data are simply inadequate to say with certainty whether wife abuse is increasing, declining, or stable.

Canada's 1993 *Violence against Women* survey indicated that about 3% of those who had a current intimate partner had been physically attacked by him within the preceding year, and 15% at some time in the relationship. One reason to fear that the problem may be worsening is that younger respondents in this survey reported not just higher recent incidences of such violence than did older respondents, but slightly higher lifetime incidences as well, despite fewer years at risk. Such contrasts could, however, reflect biases in recall or reporting of past experiences rather than genuine cohort differences. At least one piece of evidence suggests that wife assault is becoming less prevalent: rates of lethal assaults on wives, which are of course more reliably documented than nonlethal assaults, have declined in recent decades in Canada and in other Western countries as well.

A case could be made that most claustration practices represent the responses of men to the threats presented by other men and that the agency of the women is beside the point. However, practices such as clitoridectomy and infibulation suggest that women are being "protected" not only from predatory males but from their own inclinations as well. If men are indeed concerned that their wives may harbor adulterous inclinations, are these fears realistic? At least since Trivers (1972), evolutionists have understood that in effectively polygynous species such as ours, male fitness is much more directly a function of the number of mates than is female fitness, and that polygamous inclinations are therefore likely to be sexually differentiated. This view gibes with some popular thinking, too. According to a possibly apocryphal story, for example, the American writer Dorothy Parker awoke from a dream with a feeling of profound insight, wrote herself a note, and discovered in the morning that she had written "Hogamus, higamus, man is polygamous. Higamus, hogamus, woman, monogamous." In accordance with this stereotype, many writers (of both sexes) have claimed that men who are fearful or suspicious of female infidelity are "projecting" their own lusty motives onto women. However, that view is no longer tenable.

In the 1980s, animal behaviorists were forced to reconsider the stereotype of the monogamous female when genetic methods began to reveal a surprising, and at least occasionally very high, incidence of "extra-pair paternity" among socially monogamous, biparental birds (Westneat and Stewart 2003). The ensuing research has demonstrated that there are many potential benefits of extra-pair mating for females, including gaining access to material resources, fertility assurance, bet-hedging by diversifying paternity, recruiting superior genes for offspring, and deterring infanticide or other harms that might subsequently be inflicted by males who had not been the female's extra-pair partners. But even before the genetic data on avian paternity called into question the meaning of social monogamy and biparental care, many primatologists already suspected that strict female monogamy did not prevail over the course of hominid evolution.

When females mate with multiple males during a single fertile period, the different males' ejaculates must "compete" for the paternity of her offspring. Where such sperm competition has been chronic, there has been selection for increases in sperm count and ejaculate volume, and thus in testis size. An association between male testis size and female polygamy has now been demonstrated in several mammalian orders, but it was first noted in primates (Harcourt et al. 1988), and its discoverers did not fail to remark that, although human testes are smaller than those of the most promiscuous species, they are nevertheless larger than would be expected under monogamy. This observation has substantially bolstered the notion that ancestral women were not strictly monogamous in their sexual behavior and hence that selection may have equipped the human female with facultative inclinations to cuckold their primary partners by clandestine adultery, or to maintain simultaneous mating relationships, or both. These ideas sit well with a variety of testimonial and survey evidence that the inclination to mate polygamously is not restricted to males, and they have had substantial impact on recent research into women's sexuality (e.g., Gangestad et al. 2005; Haselton and Gangestad 2006; Pillsworth and Haselton 2006). They also help make sense of men's precoccupations with female fidelity. Although the incidence of clandestine extra-pair paternity in human beings has probably seldom or never been so high as suggested in some sensational treatments of the topic, it has also not been negligible (Anderson 2006).

Homicide Data as a Window on Coercive Violence

Self-report data must always be interpreted with caution, both because of faulty memory and because of biased self-presentation and other social agendas. An additional problem of particular relevance to research on violent victimization is the variability in subjective responses to similar acts. Different women may describe essentially identical spousal behavior very differently, and what intimidates one may infuriate another and be shrugged off by a third. Some family violence researchers have tried to get around these problems of subjectivity by steering clear of ill-defined terms such as "violence" and restricting their questions to "behavioral" items: "Has person X ever pushed you? Ever slapped you? Have you slapped him?", and so forth. The hope that this approach will yield more "objective" data is naive, however, and it may even make matters worse by conflating serious terrorism with relatively trivial, even playful, behavior.

The most widely used survey instrument in wife assault research is just such a behavioristic compendium: the Conflict Tactics Scale (CTS). This instrument (or minor modifications thereof) is the basis of scores of published studies, and yet there is little reason to think that it yields a valid reflection of past events. Its test-retest reliability—whether respondents tell the same tale if asked again after an interval—has apparently never been assessed, and the evidence on interpartner reliability is not encouraging. In at least two published studies in which the CTS was separately administered to male and female partners, concordance between their responses was astonishingly low. In other words, whether X professed to have hit Y was virtually unrelated to whether Y professed to have been hit by X (see Dobash et al. 1992).

Unfortunately, inferences based on other sorts of data on nonlethal violence must be even shakier than those based on such survey data. Arrests or court proceedings, for example, will provide a researcher with a sample of assaults, but we can be certain that other, comparable acts escaped inclusion in the data, and we can be equally certain that the included cases will be a biased sample with respect to variables of possible interest. No one would credit a claim that rich men are less likely to assault their wives than poor men if it were based solely on arrest data, and no one should.

It is for these reasons that many years ago, we decided to make use of homicide data in order to test hypotheses about variables that exacerbate or mitigate interpersonal conflict (Daly and Wilson 1988). We believe that this strategy has been successful, but we grant that it depends on an assumption that could, in some cases, be importantly wrong, namely, that homicides are for the most part similar in motivation to nonlethal assaults. In the case of uxoricides (killings of wives), this means that most cases reflect the same coercive, proprietary inclinations that apparently motivate domestic terrorism, rather than being disposals of unwanted wives.

Disposals of unwanted wives undeniably occur, but examination of case files indicates that they constitute a small minority of uxoricides. Paradoxically, many more wife-killers appear motivated to hold on to their wives than to be rid of them: a wife's taking action to leave her husband is the single most frequent eliciting circumstance for uxoricide. A recurrent feature of such cases is the killer's declaration that "If I can't have her, no one can." In fact, despite reduced access or "opportunity," women with violent, controlling husbands are more likely to be slain in the immediate aftermath of separation than while co-residing (Figure 11.2).

Women of high reproductive value are especially likely to be killed: the rate at which women are slain by their husbands is a monotonically declining function of their age (Figure 11.3), which closely matches the age-related patterning of nonlethal assaults by male partners according to survey data (Figure 11.1) and is furthermore just the opposite of what we should expect if uxoricides were predominantly to be understood as disposals of women whom men no longer value.

Most homicide researchers seeking a typology of the killers' "motives" have classified uxoricides precipitated by the wife's adultery and those precipitated by her efforts to desert her husband together, in a single category of cases attributed to "jealousy." The adaptive problems of adultery and desertion are in a

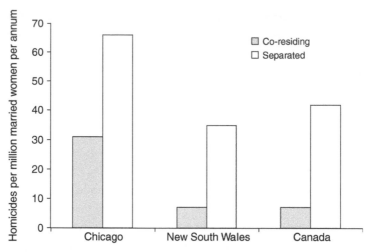

Figure 11.2 Rates at which married women were killed by their husbands in three samples (Chicago 1965–1990; New South Wales, Australia 1968–1986; and Canada 1974–1990), according to whether the couple were co-residing or separated at the time of the homicide. This is a conservative portrayal of the risk associated with separation, because the denominators for the "separated" rates include all married women living apart from their husbands, whereas most of the homicides occurred within a few weeks of the separation. After Wilson and Daly (1993a).

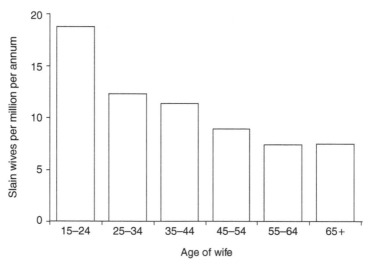

Figure 11.3 Uxoricide rate in registered marriages in Canada (1974–1992) as a function of the victim's age. After Wilson, Johnson, and Daly (1995).

sense distinct, since only adultery puts the man at risk of investing in a rival's child, but the two types of cases appear similar in the indignant, proprietary aggressiveness of the killers, who seem to react to adultery and desertion as more or less the same affront. And of course they are, in another sense, very much alike: the important commonality from an evolution-minded perspective is that the man is losing control of female reproductive capacity and hence losing ground in the reproductive competition between men. So if we accept this broad category of "jealous" uxoricides—that is, those motivated by actual or suspected infidelity and/or by actual or suspected desertion of the marriage—then what proportion of the cases does this broad motivational category capture? The answer is a large majority. Daly and Wilson (1988) reviewed the information on ostensible motives in every sample of well-described spousal homicides that they could find, representing a wide range of cultures. "Jealousy," as defined above, was apparently the dominant motivational factor in more than half the uxoricides in *every* such sample and usually in more than 80% of the described cases.

At first glance, interpreting violence against wives as a product of motives that evolved to deter wifely infidelity or desertion sounds like it could only apply to violence that stops short of lethality. Why would a man wishing to control and retain his wife ever kill her? However, this objection presupposes that killers are more foresightfully rational than is warranted by the evidence or than we mean to imply. The proposition that motives and emotions have evolved to serve the actor's interests is not a claim that they engender rational, purposive action. If, for example, we propose that men (and other male animals) have evolved to resent and resist sexual infidelity by their partners "in order to" protect their paternity and promote their reproductive success, the phrase "in order to" refers to the adaptive functions that sexual proprietariness served for ancestral males, not to conscious purpose. It is for this reason that we have referred to spousal homicides as "slips in a dangerous game" of brinksmanship between coercive, threatening men and women attempting to retain their autonomy (Daly and Wilson 1988).

In several writings, David Buss has defended an alternative interpretation (e.g., Buss 2005). He argues that the human mind likely contains an evolved uxoricidal "module" (among other relationship-specific "homicide modules"), whose adaptive function resides partly in disposing of someone who is thwarting the killer's pursuit of fitness and partly in making an intimidatory social display to third parties. Buss challenges what he calls our "slip-up hypothesis" by noting that wife-killings are commonly deliberate rather than accidental, but as discussed above, the proposition that uxoricide is a maladaptive by-product of motives whose evolved function is nonlethal coercion is not a proposition about conscious intent. Buss also cites a high incidence of detailed homicidal fantasies as evidence for specific homicidal adaptations, but in our view, this is another *non sequitur*. Consider the fact that, although a large proportion of undergraduate men indeed report that they have fantasized killing someone, an even larger proportion report that they have fantasized that they were playing video games (Kai 2003); no one would infer from this that men's minds contain video-game-playing adaptations. Most importantly, Buss's proposal that the human mind contains evolved machinery dedicated to uxoricide requires that ancestral men who killed wives thereby promoted their fitness more effectively than those who merely assaulted them, and did so with sufficient frequency and regularity for selection to have fashioned the proposed adaptation(s). This proposition seems to us utterly implausible. Men who kill their wives antagonize their wives' kinsmen, deprive their own children of maternal care, and inspire collective sanctions. Our reading of the anthropological literature convinces us that these consequences were no less costly in premodern societies than they are today.

Thus, it is our view that uxoricide is in general nonadaptive and "spiteful" in the sense that it is likely to impose a net cost on the killer as well as the victim. And if the functionality of the underlying motives resides in keeping and controlling wives, then killing them seems all the more dysfunctional. But the men who pursue and kill the women who have left them are clearly not engaged in the instrumental pursuit of what others would consider their self-interest. The most striking evidence in this regard is the fact that such men are much more likely than other killers to commit suicide as part of the same violent event (Wilson et al. 1997). Many such men leave notes or other evidence that the uxoricide and the suicide were, from the outset, elements of a unitary project whose completion will be a source of defiant pride.

A satisfactory explanation of these puzzling facts is most likely to come from an improved theoretical understanding of the "economics" of threat. A threat is an effective social tool, and more often than not an inexpensive one. But it loses its effectiveness if the threatening party is seen to be bluffing—that is, to be unwilling to pay the occasional cost of following through when the threat is ignored or defied. Such follow-through often appears spiteful—a risky or an expensive response that is too late to be useful—and yet an effective threat must convince its target that such follow-through will nevertheless occur. Thus, although killing an estranged wife seems clearly to be counterproductive, threatening one who might leave can be self-interested, and so can pursuing her with further threats, as can displays of anger and of ostensible obliviousness to costs. Effective threatening behavior must not "leak" signs of bluff, and the best way to appear sincere may be to be sincere (Frank 1988). Even so, most men who coerce, pursue, and threaten women do not go so far as to kill them, and those who do may be considered dysfunctional overreactors in a game of brinksmanship.

Sources of Variability in Intimate Partner Violence

An evolutionary perspective provides more than an explanation for what appear to be privileged, potent psychological links between sexual rivalry, infidelity, and partner loss, on the one hand, and violent inclinations, on the other. It also provides a theoretical framework from which further predictions about the patterned incidence of such violence can be derived (Wilson and Daly 1993b; Wilson et al. 1997; Shackelford et al. 2005). If male sexual proprietariness represents complex evolved psychological adaptation, designed by selection to promote male fitness in ancestral environments, and if it entails costs, then sexually proprietary attention and action should have evolved to be allocated in response to cues of expected utility. Many male birds, for example, guard their mates closely during their fertile phase, but cease once the last egg is laid. In human beings, we hypothesize that such variable cuing of sexual proprietariness affects behavioral variability both within and between societies. Some phenomena, such as age-related changes in fertility, are cross-culturally general and likely to account for within-society variability in more or less similar ways. Others, such as the risk imposed by desperate, disenfranchised male rivals, vary across societies and may therefore be expected to account for some of the between-society variability.

As noted earlier, wife-beating (unlike uxoricide) can often be interpreted as coercive and hence self-interested. However, violence is a risky way to manipulate others, for rather than making its victims *wish* to comply, it sometimes inspires them to defiance whenever the opportunity arises. In the case of violence against wives, assaults can elicit severe and even lethal self-defense or retaliation (Wilson and Daly 1992b), as well as vengeance by the victims' relatives. It follows that such violence is often the recourse of desperate men lacking access to the positive incentives that might inspire more voluntary compliance. Moreover, a seemingly inexpensive and effective initial use of violence can embroil one in escalating hostilities, closing off any possibility of negotiation. In the case of wife-beating, one implication would seem to be that a man is especially likely to resort to beating his wife when he perceives his control of his spouse to be tenuous and he lacks material or other incentives to retain her. In other words, a man may be especially likely to engage in coercive violence if he perceives his wife's current mate value as higher than his own (although attempting to raise his value in his wife's eyes by giving more is an alternative, nonviolent response to the same problem; see Gangestad and Simpson 2000).

Within societies, much of the variability in male jealousy, and hence coercion, must be attributable to variable attributes of women. A major determinant of a woman's mate value is of course her age. Her "reproductive value" (expected future fertility) is maximal at puberty and begins to decline steeply in her 30s; the strong relevance of age to sexual attractiveness and marriage market value parallels these changes (Kenrick and Keefe 1992). Thus, young wives have more and better options for both extramarital affairs and mate-switching than older wives, and their partners have more reason to be jealous of them, which evidently explains their higher rates of violent victimization by those partners. It must be acknowledged, however, that the direct relevance of wife's youth to husband's violence is hard to prove and remains debatable, because other possibly relevant variables are confounded with a woman's age, including parity and childlessness, duration of the union, the age of her male partner, and perhaps her social and material resources, too. Since young men are the most violent age-sex class generally, an obvious hypothesis is that the man's age is actually the more relevant factor behind the risk pattern seen in Figures 11.2 and 11.3. This does not seem to be the case, however, or at least not the whole story, since young women married to much older husbands are actually substantially more likely to be slain than those married to men of their own age (Wilson et al. 1997).

Of course, the risk that a wife's infidelity will produce a cuckoo in the nest also varies in relation to state-dependent fecundability; one might therefore hypothesize that jealous monitoring and violence will vary accordingly. In a pioneering study of human mate guarding, Flinn (1988) recorded the identity, whereabouts, and activities of everyone he saw during standardized walks through a Caribbean village, and found that men both spent more time with potentially fertile partners (those who were currently undergoing menstrual cycles) than with those who were pregnant or postmenopausal, and also displayed more agonism toward those partners (and toward other men) when they were cycling than if in other reproductive conditions. Recent research suggests that similar patterning of mate guarding across the menstrual cycle occurs in urban America, too (Gangestad et al. 2002, 2005; Pillsworth and Haselton 2006).

The degree of coercive constraint of wives, including violence, can also be expected to reflect cues of the local, contemporaneous intensity of male sexual competition and poaching. Relevant cues may include encounter rates with potential male rivals; whether they are alone or in all-male groups *versus* being accompanied by women (cues of "bachelor pressure"); whether local bachelors have reason (e.g., poor life prospects) to be reckless; cues of the status, attractiveness, and resources (hence, mate value) of rivals relative to oneself, and of other social groups or categories (lineages, castes, etc.) relative to one's own social group or category; and cues of local marital (in)stability. But many theoretically relevant parameters such as relative cohort size, expected life span, local marital stability, and local prevalence of adultery cannot be cued simply by stimuli present at the time of behavioral decisions. They must instead be apprehended cumulatively over large parts of the life span. In ancestral environments, no one collated and communicated the sort of statistical information that would permit optimal social decision making. People had to induce the magnitude of opportunities and risks from personal experience and from the credited testimony of trusted interactants, especially kin. This implies that people gradually develop mental models of their social universe that cannot be abruptly modified or discarded. The "inertial" aspects of such a developmental process help explain why enculturation "sticks" and why some men, like the boxer Mike Tyson, persist in coercive violence even after acquiring sufficient means to achieve their social goals by using lower-risk, positive incentives.

When individual men monopolize multiple women, others are consigned to bachelorhood, and male-male competition is exacerbated (Betzig 1986). On this basis, we might expect marital coercion and violence to be more extreme in polygynous than in monogamous populations. In a cross-cultural study, Levinson (1989) indeed found a significant correlation across nonstate societies between wife-beating and a dimension ranging from polyandrous through monogamous to polygynous marriage. We suspect that the real relationship is even stronger than this analysis has indicated, both because Levinson rank-ordered polygyny in a manner unrelated to the crucial consideration of the variance in men's access to women, and because marital polygyny is an imperfect indicator of the breeding system and hence of the intensity of intrasexual competition. A related reason for a link between polygyny and wife abuse may be mediated by patrilocality: in polygynous societies, a wife is typically transferred at marriage from her natal family's home to her husband's, and may lose the protective benefit of having kin nearby. According to Chagnon (1992: 149), for example, Yanomamö women "dread the possibility of being married to men in distant villages, because they know that their brothers will not be able to protect them".

We might also predict that men would be more sexually demanding, threatening, and coercive when circumstances dictate that their wives are relatively unmonitored, such as when the men are sea-going foragers. However, an alternative response to constraints on wife-monitoring may be to abandon patrilineage and engage in a combination of mating effort and investment in maternal kin instead (e.g., Hartung 1985). A consequence of prioritizing matrilineal links is then likely to be matrilocality and the wife's greater access to genetic relatives who may deter abuse by her husband. That said, jealous violence against wives is conspicuous even in matrilineal societies with avuncular inheritance (Daly et al. 1982).

Sadly, female choice may itself be a force that has helped to maintain partner violence in our species. Certainly, it is easy to envision circumstances in which a man's capacity to use violence effectively might enhance his attractiveness to women. For example, where women are at risk of being abducted, as among the Yanomamö, or even where sexual harassment and assault are prevalent, a husband with a fierce reputation can be a social asset. Dare we hope that in modern societies, selection is now acting against male violence?

References

Anderson, K. G. "How Well Does Paternity Confidence Match Actual Paternity? Evidence from Worldwide Nonpaternity Rates." *Current Anthropology* 47 (2006): 513–520.

Betzig, L. L. *Despotism and Differential Reproduction: A Darwinian View of History.* Hawthorne, N.Y.: Aldine de Gruyter, 1986.

Borgerhoff Mulder, M. "Women's Strategies in Polygynous Marriage: Kipsigis, Datoga and Other East African Cases." *Human Nature* 3 (1992): 45–70.

Buss, D. M. *The Murderer Next Door: Why the Mind Is Designed to Kill.* New York: Penguin, 2005.

Buss, D. M., and M. Haselton. "The Evolution of Jealousy." *Trends in Cognitive Science* 9 (2005): 506–507.

Buss, D. M., R. J. Larsen, D. Westen, and J. Semmelroth. "Sex Differences in Jealousy: Evolution, Physiology and Psychology." *Psychological Science* 3 (1992): 251–255.

Chagnon, N. A. *Yanomamö.* New York: Harcourt Brace, 1992.

Counts, D. A., J. K. Brown, and J. C. Campbell. eds. *To Have and to Hit: Cultural Perspectives on Wife Beating.* Champaign: University of Illinois Press, 1999.

Daly, M., and M. I. Wilson. *Homicide.* New York: Aldine de Gruyter, 1988.

Daly, M., M. I. Wilson, and S. J. Weghorst. "Male Sexual Jealousy." *Ethology and Sociobiology* 3 (1982): 11–27.

Dickemann, M. "Paternal Confidence and Dowry Competition: A Biocultural Analysis of Purdah." In *Natural Selection and Human Social Behavior*, eds. R. D. Alexander and D. W. Tinkle, pp. 439–475. New York: Chiron, 1981.

Dobash, R. E., and R. P. Dobash. *ViolenceAgainst Wives: A Case against the Patriarchy.* New York: Free Press, 1979.

Dobash, R. P., R. E. Dobash, M. I. Wilson, and M. Daly. "The Myth of Sexual Symmetry in Marital Violence." *Social Problems* 39 (1992): 71–91.

Farrington, K. M. "Stress and Family Violence." In *The Social Causes of Husband-Wife Violence*, eds. M. A. Straus and G. T. Hotaling, pp. 94–114. Minneapolis: University of Minnesota Press, 1980.

Flinn, M. V. "Mate-Guarding in a Caribbean Village." *Ethology and Sociobiology* 9 (1988): 1–28.

Frank, R. H. *Passions within Reason: The Strategic Role of the Emotions.* New York: W. W. Norton, 1988.

Gangestad, S. W., and J. A. Simpson. "The Evolution of Human Mating: Trade-Offs and Strategic Pluralism." *Behavioral and Brain Sciences* 23 (2000): 573–587.

Gangestad, S. W., R. Thornhill, and C. E. Garver. "Changes in Women's Sexual Interests and Their Partners' Mate Retention Tactics across the Menstrual Cycle: Evidence for Shifting Conflicts of Interest." *Proceedings of the Royal Society of London Series B* 269 (2002): 975–982.

Gangestad, S. W., R. Thornhill, and C. E. Garver-Apgar. "Women's Sexual Interests across the Ovulatory Cycle Depend on Primary Partner Developmental Instability." *Proceedings of the Royal Society of London Series B* 272 (2005): 2023–2027.

Harcourt, A. H., P. H. Harvey, S. G. Larson, and R. V. Short. "Testis Weight, Body Weight and Breeding System in Primates." *Nature* 293 (1981): 55–57.

Hartung, J. "Matrilineal Inheritance: New Theory and Analysis." *Behavioral and Brain Sciences* 8 (1985): 661–688.

Haselton M. G., and S. W. Gangestad. "Conditional Expression of Women's Desires and Men's Mate Guarding Across the Ovulatory Cycle." *Hormones and Behavior* 49 (2006): 509–518.

Johnson, H., and V. Sacco. "Researching Violence against Women: Statistics Canada's National Survey." *Canadian Journal of Criminology* 37 (1995): 281–304.

Johnson, M. P., and K. J. Ferraro. "Research on Domestic Violence in the 1990s: Making Distinctions." *Journal of Marriage and the Family* 62 (2000): 948–963.

Kai, M. "Content and Prevalence of Fantasies." Unpublished BSc thesis, McMaster University, 2003.

Kenrick, D. T., and R. C. Keefe. "Age Preferences in Mates Reflect Sex Differences in Human Reproductive Strategies." *Behavioral and Brain Sciences* 15 (1992): 75–133.

Levinson, D. *Family Violence in Cross-Cultural Perspective.* Newbury Park CA: Sage, 1989.

Miller, J. "A Specification of the Types of Intimate Partner Violence Experienced by Women in the General Population." *Violence against Women* 12 (2006): 1105–1131.

Pillsworth E. G., and M. G. Haselton. "Male Sexual Attractiveness Predicts Differential Ovulatory Shifts in Female Extra-Pair Attraction and Male Mate Retention." *Evolution and Human Behavior* 27 (2006): 247–258.

Shackelford, T. K., A. T. Goetz, D. M. Buss, H. A. Euler, and S. Hoier. "When We Hurt the Ones We Love: Predicting Violence against Women from Men's Mate Retention." *Personal Relationships* 12 (2005): 447–463.

Smuts, B. "Male Aggression against Women: An Evolutionary Perspective." *Human Nature* 3 (1992): 1–44.

Straus, M. A. "The Marriage License as Hitting License: Evidence from Popular Culture, Law and Social Science." In *The Social Causes of Husband-Wife Violence,* eds. M. A. Straus and G. T. Hotaling. Minneapolis: University of Minnesota Press, 1980.

Teisman, M. W., and D. L. Mosher. "Jealous Conflict in Dating Couples." *Psychological Reports* 42 (1978): 1211–1216.

Trivers, R. L. "Parental Investment and Sexual Selection." In *Sexual Selection and the Descent of Man, 1971–1971,* ed. B. Campbell, pp. 136–150. Chicago: Aldine, 1972

Westneat, D. F., and I. R. K. Stewart. "Extra-Pair Paternity in Birds: Causes, Correlates, and Conflict." *Annual Reviews of Ecology, Systematics and Evolution* 34 (2003): 365–396.

WHO Study Group on Female Genital Mutilation and Obstetric Outcome. "Female Genital Mutilation and Obstetric Outcome: WHO Collaborative Prospective Study in Six African Countries." *Lancet* 367 (2006): 1835–1841.

Wilder, J. A., Z. Mobasher, and M. F. Hammer. "Genetic Evidence for Unequal Population Sizes of Human Females and Males." *Molecular Biology and Evolution* 21 (2004): 2047–2057.

Wilson, M. I., and M. Daly. "The Man Who Mistook His Wife for a Chattel." In *The Adapted Mind: Evolutionary Psychology and the Generation of Culture,* eds. J. H. Barkow, L. Cosmides, and J. Tooby, pp. 289–322. New York: Oxford University Press,1992a.

———. "Who Kills Whom in Spouse Killings? On the Exceptional Sex Ratio of Spousal Homicides in the United States." *Criminology* 30 (1992b): 189–215.

———. "Spousal Homicide Risk and Estrangement." *Violence and Victims* 8 (1993a): 3–16.

————. "An Evolutionary Psychological Perspective on Male Sexual Proprietariness and Violence against Wives." *Violence and Victims* 8 (1993b): 271–294.

Wilson, M. I., M. Daly, and J. Scheib. "Femicide: An Evolutionary Psychological Perspective." In *Feminism and Evolutionary Biology*, ed. P. A. Gowaty, pp. 431–465. New York: Chapman Hall, 1997.

Wilson, M. I., H. Johnson, and M. Daly. "Lethal and Nonlethal Violence against Wives." *Canadian Journal of Criminology* 37 (1995): 331–361.

12

The Political Significance of Gender Violence

Lars Rodseth and Shannon A. Novak

Violence against women is a worldwide phenomenon, but it varies in both frequency and severity from one population to another (Counts et al. 1999; Garcia-Moreno et al. 2006). In a few small-scale societies, such as the Wape of Papua New Guinea (Mitchell 1999), gender violence of any kind is reported to be extremely rare. In many other settings, however, women are frequently subjected to physical harm, including abuse by their in-laws or their own natal kin (Brown 1997; Gruenbaum 2001; Sen 2001). Intimate partner violence, especially wife-beating, may be socially condoned and even commended (e.g., Abraham 2000; McClusky 2001).

Despite this variation, gender violence seems to be quite uniform in at least one respect. Most of its victims are fertile females—not girls and not grandmothers, but women of childbearing age. The pattern is evident in every region of the world, but a clinical study in Bradford, England provides one vivid illustration (Novak 1999, 2006). In a sample of 673 women who had been admitted to the Bradford Royal Infirmary over a 12-month period, victims of domestic assault were compared with patients who had suffered other kinds of injuries. The mean age of the assault victims was 30, and almost 90% of them were younger than 40. Accident patients, by contrast, had a mean age of 36 and many of them (36%) were 40 or older. Taking the sample as a whole, we see that women younger than 20 and those older than 40 were similar in that their greatest risk of injury came from accidents rather than assaults. Between 20 and 40, however, accident cases were far outnumbered by cases of domestic violence. These data suggest that a woman's risk of being assaulted increases as she enters her childbearing years and decreases as she approaches menopause.

In evolutionary psychology and related fields, the prevailing interpretation of such a pattern is that gender violence is a form of sexual coercion and mate guarding (Symons 1979; Daly and Wilson 1988, 1994, Chapter 11 in this volume; Smuts 1992, 1995; Wilson and Daly 1992, 1998; Smuts and Smuts 1993; Buss 1994, 2000). Most assaults on women, according to this view, are motivated by male jealousy as a result of real or imagined sexual infidelity. In short, men batter women to control their sexuality. Such abuse declines precipitously when a woman reaches menopause because her reproductive value has similarly declined. At some point around age 40, she is no longer worth bullying.

While this kind of argument has become the conventional wisdom among evolutionary scientists, an alternative explanation should be carefully considered. In many social settings, as a woman approaches 40, her political capital begins to soar (Brown 1982; Kerns and Brown 1992). No longer the isolated wife, tethered to infants and toddlers, she is able to take on a more public persona. With much social experience in relatively small circles, she can now cultivate allies beyond the household and the extended family. At the same time, she is likely to be protected by her grown offspring, including married sons who depend on her for their own political reasons. One cost of mate guarding, after all, is that the guard himself must withdraw to some extent from public life. A vigilant husband is often keeping tabs on his wife when he might have been investing in wider, political alliances. A vigilant *mother*, it may be argued, allows him to do both (Brown 1997).

Modeling the Competition

There may be something to either or both of these explanations. The point of the example is not to decide the matter one way or the other, but to question the assumptions that guide us directly to a *sexual* interpretation of gender violence, before we have seriously considered the wider political context. What we tend to see in a case of wife-beating, for example, is a private imbroglio that has popped into public view. At its root, the phenomenon is intimate and sexual, but special circumstances have turned it into something collective and political. Thus, whatever its public ramifications, wife-beating can always be traced to the psychology of the conjugal bond—and specifically to what Margo Wilson and Martin Daly call male sexual "proprietariness" (1992:289):

By "proprietary," we mean first that men lay claim to particular women as songbirds lay claim to territories, as lions lay claim to a kill, or as people of both sexes lay claim to valuables, . . . Proprietariness has the further implication, possibly peculiar to the human case, of a sense of *right* or *entitlement*. Trespass by rivals provokes not only hostility but a feeling of grievance, a state of mind that apparently serves a more broadly social function. Whereas hostile feelings motivate action against one's rivals, grievance motivates appeals to other interested persons to recognize the trespass as a wrong against the property holder and hence as a justification for individual retaliation or a grounds for more collective sanctions.

A fertile female is a prize, "an individually recognizable and potentially defensible resource packet," as Wilson and Daly put it, and the rest of their argument follows from this premise. The wider community comes into play in one respect only—as the final guarantor of personal property rights. "Other interested persons" may be recruited to punish a trespasser, for example, or to defend the right of the property holder to do so. Whatever punishment is meted out to the "property" herself directly reflects the holder's interest in monopolizing her as a sexual resource.

The idea that wife-beating might be an expression of something more than sexual proprietariness—that it might serve, for example, to appease, impress, manipulate, or orchestrate "other interested persons" for long-term strategic gains—does not enter into the analysis. What is missing from this model is, in a word, politics (cf. Ortner and Whitehead 1981; Rodseth 1990).

As a model, of course, it cannot and should not include everything. Models are useful precisely because they oversimplify the world and thus draw our attention to what is analytically important (Weber 1949). Evolutionists have good theoretical reasons to focus on sexual proprietariness and to develop tools that help us grasp its role in social behavior. When a model has served its purpose, however, and we become interested in some dimension of reality that has been, until now, simply left out of the analysis, it is time to try a new model.

Here we argue that human gender violence takes at least three distinct forms, only one of which follows the proprietary pattern (cf. Johnson 1995; Johnson and Ferraro 2000). Thus, when a woman is *battered* by her husband or intimate partner, the violence is likely to be private, personal, and driven by sexual jealousy, just as Wilson and Daly describe it. At the other extreme is violence that manifests a communitywide *sexual antagonism*—the aloofness or hostility of all the men in a political network toward all things feminine, including their wives.

In contrast to battering, this kind of violence is likely to be public, political, and driven by sexual aversion. Between these two extremes is a hybrid pattern we call *domestic discipline*. Although it usually takes place in private, the violence is designed in this case to send a political message and not merely to control a sexual resource. In all three forms, gender violence is geared to the surrounding political system.

A String of Situations

Our argument is based on a specific conception of human social organization in comparison with the patterns of other primates (Wrangham 1987; Rodseth et al. 1991; Rodseth and Wrangham 2004; Rodseth and Novak 2006). One critical aspect of human grouping is perhaps the most obvious of all: people carry out their lives, not in herds, troops, or isolated families, but in fission-fusion societies. During waking hours, people move about in temporary and variable parties, yet retain membership in a more or less well-defined social network. Members of the same network often travel independently, and only occasionally does an entire village, for example, assemble face-to-face. Our social life in this perspective is nothing more (and nothing less) than "a string of situations, that is, encounters when persons are physically copresent" (Collins 2004:18). Only a few other primates, including chimpanzees, are known to lead such episodic lives, tacking between intense social interaction and moments or hours of solitude.

In a fission-fusion society, lone individuals are vulnerable to assault (Wrangham 1986, 1999), and lone females in particular are vulnerable to sexual assault (Smuts 1992; Smuts and Smuts 1993). Among chimpanzees at Gombe, a male will sometimes use violence to isolate an estrous female from others who might intervene on her behalf (Tutin 1979; Goodall 1986:457). This tactic carries risks, especially if the consorts withdraw to a "border zone" where they are likely to encounter members of neighboring communities (Goodall 1986:464–465). Nevertheless, a female subjected to persistent bullying may be forced into a liaison lasting days or weeks (Wrangham and Peterson 1996:145):

> Signs of her having been coerced normally soon fade, and the pair may move together peacefully. But he may yet turn and attack her brutally, and sometimes she shows her unhappiness about the situation by escaping when she hears other males calling.

This pattern of violence, as Wrangham and Peterson (1996:146) note, is comparable to human battering—yet the differences may be as interesting as the similarities. Certainly, like a human batterer, the male chimpanzee seeks to isolate and control a female who is *known* to him. Assailant and victim, in other words, are not strangers but members of the same community. In the human case, however, the nature of the relationship is often much closer than this. Unlike chimpanzees, humans form exclusive and relatively stable sexual bonds. As a result, the kind of battering inflicted by a male chimpanzee for hours or days may be carried on by a man for months or years.

Thus a battered wife confronts a situation that would be alien to a chimpanzee. Her assailant is not a temporary but a continuing menace, and she cannot escape him by simply running away. Indeed, she is most likely to be assaulted while *facing*, rather than fleeing, her assailant. A comparison of trauma in female chimpanzees with the same kind of trauma in battered women presents a striking contrast: while the chimps are injured in the back of the skull, women are consistently targeted in the region of the eyes, nose, and mouth (Novak and Hatch, Chapter 13 in this volume). Women, it would seem, are usually not chased down but *faced* down. One reason for this is obvious. A woman can fight back with both weapons and words in ways that a chimpanzee cannot. Words also allow her to plead or negotiate with her assailant— taking the risk, in the process, of being hit in the face. In many cases, however, the most important difference between human and simian patterns of gender violence comes down to *architecture*. Most women, after all, live in a built environment that both protects and confines them.

Around the House

Any domestic enclosure tends to keep women under the watch of intimate others. Young wives in particular may be quite constrained in where they can go and who they can see: "In some societies, all younger women are restricted to the household because they are believed to be sexually voracious and in need of constant supervision" (Brown 1992:19). One example comes from the Bakgalagadi, an agropastoralist people of the Kalahari: "A young wife is under the constant scrutiny of her affines, particularly her mother-in-law, and . . . is expected to work hard and to produce children" (Solway 1992:51–52). Similarly, among Garífuna farmers in Belize, "a young woman is supposed to work in the vicinity of her house and to leave her yard only with a specific destination in mind"

(Kerns 1992:105). Even in industrial and postindustrial settings, women tend to move in smaller orbits than those of men in the same community (Pain 1991; Hanson and Pratt 1995:102–103; Madriz 1997). In a study of dual-income families in Columbus, Ohio, women who were employed full-time commuted farther than their male counterparts, yet still experienced a higher level of "fixity constraint" in the sense that their daily activities had to be organized around certain recurring, "household-serving" tasks (Kwan 2000).

Humans in general, of course, are tied to "home bases" in a way that chimpanzees and most other primates are not (Isaac 1989; Wrangham et al. 1999). Indeed, only humans have *domestic* lives, it may be argued, for the simple reason that only humans maintain households (as manifested in regular, intimate reunions at a sleeping site) within communities (as manifested in occasional, larger reunions for cooperative activities). This is not to suggest that all human societies have sharply defined "private" and "public" spheres. The private/public distinction, as it is usually understood in Western contexts, should not be projected onto human societies in general (Leacock 1981; Collier and Yanagisako 1987). Yet *some* distinction between household and community life is a basic part of the human condition: "Although varying in structure, function, and societal significance, 'domestic groups' which incorporate women and infant children, aspects of child care, commensality, and the preparation of food can always be identified as segments of a larger overarching social whole" (Rosaldo 1980:398). This statement, supported by more than a century of ethnographic and archaeological research, encapsulates a pattern of social organization that is apparently both culturally universal and uniquely human (Rodseth and Novak 2000, 2006).

The critical issue for present purposes is the degree of domestic *privacy*—the extent to which household activities are shielded from the public gaze. Privacy has been a surprisingly neglected topic in evolutionary circles, especially given the intriguing remarks by Donald Symons in one of the founding texts of evolutionary psychology (1979:67): "Human beings almost always seek whatever privacy is available to engage in sexual activities." Why people, of all primates, should be so shy of copulating in the open is a basic anthropological question to which there is no clear answer.

In chimpanzees, private copulation is possible only in consortships. These relationships require separation from the rest of the community and a period of social isolation, often initiated by violent attacks on the female partner. In this sense, privacy in chimpanzees is a fleeting arrangement, imposed by violence and maintained by physical distance. The potential costs of such an arrangement are twofold. In venturing to the margins of the community range, consorts are vulnerable to attacks by members of a neighboring community. At the same time, prolonged absence from the core area imposes a "political" cost on the male in particular, whose alliances with other males must be tested and reinforced through regular interaction.

Could the "household," in its earliest manifestation, have been a way for males to isolate and control their consorts without actually removing them from the midst of the community? To pursue this question would take us far afield, but it does point in an interesting direction. If the household is a consortship *inside* the community, the problem of privacy becomes acute: how can the isolation of the wandering pair be reconstituted within a camp or village, where the community is not a widely dispersed social network but an immediate and tangible *audience?*

In the spirit of Erving Goffman (1959, 1967), a few anthropologists have looked at the domestic realm as a "backstage" where household members negotiate their parts and coordinate their performances (Gregor 1977; Wilson 1988). Such an approach highlights the *flow of information,* and the nature of barriers to that flow, between a household and the wider world. Does information pass easily, we may ask, from the domestic to the public and back again, or is it selectively filtered and sometimes blocked?

The most conspicuous barrier is a wall, which not only obstructs the view of outsiders but often limits what women and children in a family know about the public realm. The effectiveness of architecture as a means of information control is suggested by the North Indian case of *pirzade,* Muslim women of saintly descent who describe themselves as "frogs in a well"—"people with intellectual and physical horizons limited to the tiny patch of sky above their heads" (Jeffery 1979:11). While this is an extreme example, it is part of a widespread pattern among both Muslims and Hindus from Afghanistan to Bangladesh. Throughout this region, we encounter what Sylvia Vatuk (1992:156) calls a "plethora of social institutions designed to restrict young women to the 'private,' domestic arena; limit their physical mobility, social contacts, and activities; and make them generally dependent on and subordinate to the authority of the male members of their families."

If the "well" of the *pirzade* represents the maximum of domestic privacy, the minimum is approached in cases of mobile foragers who do not (or did not, until recently) build permanent structures at all. Thus the !Kung San consider it "bad manners" to look directly at a husband and wife who are engaged in sexual intercourse, though the pair may be concealed by nothing more than a blanket (Lee 1979:461). Domestic life in general is often carried out in full view of the camp, suggesting little distinction between private and public space. The benefits of this arrangement were apparent to one !Kung woman who was familiar with the sedentary lives of her pastoralist neighbors. As she explained to Patricia Draper (1993:47),

> The !Kung women are better off. Among the Herero, if a man is angry with his wife he can put her in their house, bolt the door and beat her. No one can get in to separate them. They only hear her screams. When we !Kung fight, other people get in between.

Around the world, most women are beaten behind walls. In private quarters, there is often nowhere for a victim to run and no one to turn to for help. Thus the built environment, whatever safety it may offer from outside dangers, poses dangers of its own to women, especially when they are cut off from both information and their potential allies in the wider world.

Conjugal Coercion

Perhaps the most important deterrent to domestic violence is the likelihood of others intervening. This simple but powerful principle was first distilled from the ethnographic record in the 1970s. Based on a comparative analysis of gender relations under various rules of postmarital residence, Louise Lamphere (1974) suggested that wives benefit by maintaining social ties with parents, siblings, and other natal kin who may be called upon for aid. Similarly, John and Beatrice Whiting (1975) argued that women find safety in numbers and that wives who are isolated on farmsteads, for example, are likely to suffer especially high levels of abuse.

Although this argument cannot explain all the ethnographic variation, there is clearly a connection between a woman's isolation and her vulnerability to assault (Smuts 1992; Counts et al. 1999). In rural North Carolina, for example, 95% of a sample of battered women reported that "leaving the house for any reason invariably resulted in accusations of infidelity which culminated in assault" (Hilberman and Munson 1978:461). Similarly, among the Maya of Belize, "Jealous husbands often see their wives' travel as 'looking for another man' and retaliate with physical abuse" (McClusky 2001:260).

When a woman has no available allies to intervene, she is especially vulnerable to intimidation and manipulation by a violent husband. If the abuse becomes chronic and severe, it may be described as *battering*—a distinct mode of human aggression that

- thrives on domestic privacy and conjugal propinquity,
- is intended to preserve an exclusive pair-bond, and
- runs the risk of serious or fatal injury over the long term.

This kind of violence has been described in the sociological literature as "intimate terrorism" (Johnson 1995; Johnson and Ferraro 2000). To be *intimate*, of course, the violence must be part of a close, personal, and fundamentally private relationship. This is precisely what distinguishes battering from more "political" forms of gender violence.

Battering in this sense clearly fits the model of sexual proprietariness developed by Wilson and Daly (1992). A batterer seeks to consolidate his power over a sexual and domestic partner. He employs various "autonomy-limiting" tactics, including assault, to control his partner's movements and social interactions. Debilitating injury or homicide may be explained as an unintended consequence of violence that was designed to manipulate and intimidate (Daly and Wilson 1988; Wilson and Daly 1998:204–207).

Despite its destructive effects, battering does not necessarily imply a generalized antagonism toward women. On the contrary, it seems to reflect a powerful emphasis—indeed, an *over*emphasis—on conjugal intimacy and exclusivity. A batterer tends to be obsessed with his partner and their domestic relationship, which comes to eclipse all other social commitments. Although the victim is deliberately cut off from family and friends, the batterer's own relationships outside the home are often attenuated as well. In North America, for example, one of the best predictors of marital violence is the prolonged unemployment of the husband (Heise 1998:274; cf. Totten 2003). Having suffered a loss of face as well as income, an unemployed husband is likely to spend more time at home. His sheer presence in the household makes it easier for him to monitor, antagonize, and fight with his wife (cf. Felson et al. 2003). Meanwhile, the withering of relationships that were rooted in his work environment tends to shift most of his attention and social energy to the marriage bond.

Let us be clear about it: our argument is *not* that wives are being used in this situation as convenient "punching bags" by frustrated husbands who would rather be assaulting their co-workers or bosses. The principal emotion behind this kind of abuse is not frustration but anxiety or fear. The collapse of a man's social network provokes desperate efforts to salvage what is left to him, which is often his wife and not much else (Daly and Wilson 1988, 1994; Wilson and Daly 1998). The fact that his efforts may fail or that his violence may destroy the marriage merely confirms what we already know—that desperate and "downwardly mobile" persons are often unable to achieve their goals, or even to act effectively in pursuit of them.

Because battering seems to be associated with the erosion of male alliances, it presents a puzzle for theories of gender violence that focus on the sexual aggression and misogyny of "men in groups." In a 1992 article that is still the best available analysis of gender violence from an evolutionary perspective, Barbara Smuts argued that "Men face a trade-off between the development of bonds with wives and the development of bonds with other men; in other words, the elaboration of strong marital bonds interferes with the development of effective male alliances, and vice-versa" (1992:17; see also Smuts 1995; Rodseth and Novak 2000). Given this trade-off, Smuts hypothesized that "men will be less likely to beat their wives in societies in which marital bonds are emphasized and will be more likely to beat their wives in societies in which these bonds are sacrificed in favor of male alliances" (1992:18). Yet we find this hypothesis to be implausible, because wife-beating in its most destructive form—battering—seems to stem from a husband's fixation on the marital bond.

Communicating with Women

In some social contexts, however, strong male alliances do increase the likelihood of violence against women. Many societies in the ethnographic record are marked by *sexual antagonism* (Herdt and Poole 1982)—a pattern of aloofness and hostility between the sexes, reflected in residential separation, rituals that exclude women, fear of sexual intercourse or menstrual pollution, and pervasive physical aggression against women (Whiting and Whiting 1975; Paige and Paige 1981; Draper and Harpending 1982). The most extreme cases are concentrated in two widely separate regions, Melanesia and the Amazon rainforest (Gelber 1986; Gregor and Tuzin 2001). In village-based, hunter-gardener societies throughout "Melazonia," men often live together in a communal house, visiting their wives to take meals or provide game, but striving to keep themselves pure of feminine influences.

Another classic example of sexual antagonism comes from central Australia. In the late nineteenth century, the Aranda were reported to practice group sexual assault as a rite of female initiation. Based on the original ethnography of Spencer and Gillen (1899), Peggy Sanday provides this summary account (1981:154):

> At age 14 or 15 the girl is taken out into the bush by a group of men for what is known as the vulva-cutting ceremony. A designated man cuts the girl's vulva, after which she is gang raped by a group of men that does not include her future husband. When the ceremony is concluded, the girl is taken to her husband and from then on no one else has the right of access to her.

A case such as this, which is hardly unique in the ethnographic record (e.g., Marshall 1971:152; Gregor 1985, 1990), suggests a pattern of gender violence that is the mirror image of wife-battering. This assault takes place not in the "closet" of domestic life but outside the community altogether; the act is not private and immoral but public and ceremonial; and the victim is not monopolized by one man but shared by the entire group. Indeed, the only man specifically excluded from the process is the husband-to-be.

Thus the vulva-cutting ceremony may be seen as a ritual enactment of the trade-off between male alliances and the marriage bond. The network of males is first removed from the village, sharply distinguishing them from the domestic sphere, and then their fraternal solidarity is dramatically reinforced through a group sexual assault. In other cases, women are threatened with gang rape as a punishment for serious offenses against the masculine order—crimes such as adultery, menstrual pollution, or merely viewing the sacred objects that are safeguarded in the men's house. The political logic of such a moral system has been explored by Jane Collier and Michelle Rosaldo (1981:297, emphasis added):

> Gang rape is a peculiarly appropriate sanction for women who wander beyond male control . . . because it simultaneously reveals . . . women's inability to forge social relationships through the use of their sexuality, asserts the controlling and even creative power of male prowess, and *affirms male solidarity in the face of women who disrupt their bonds.*

The actual occurrence of gang rape, as Collier and Rosaldo note, is probably "extremely rare." Just the threat of collective male aggression seems to be enough, in many cases, not only to intimidate women (Murphy and Murphy 1974:136; Gelber 1986:25) but to reinforce the bonds of men in a fraternal interest group (Murphy 1957; Otterbein 1979).

Sexual antagonism takes many forms, and violence against women is often an ideological symbol rather than a customary practice. Yet the pattern of everyday life reflects people's fantasies and fears as well as their experiences and observations. In highland Ecuador, for example, "the brutality of some episodes of domestic violence guarantee its salience in women's minds, and gossip underscores it as a continuous threat. Accounts of wives maimed or beaten to death assure that women remain painfully aware that violence is a potential course of action open to husbands in any marital altercation" (McKee 1999:168–169). Husbands and wives may quarrel only rarely and reside together in a physically intimate setting, based on a kind of "antagonistic tolerance" (to adapt a phrase coined by Robert M. Hayden [2002] in a quite different context). Their ordinary interactions may be governed by tacit understandings that were crystallized in a few extraordinary moments of violence, perhaps in their own household, perhaps in another.

Aloofness and hostility between the sexes, we would argue, are ultimately rooted in men's anxiety about intrasexual competition as a threat to their political solidarity. By abusing or terrorizing women, including his wife, a man is often saying something about his social priorities. He is denying that his marriage or any sexual relationship takes precedence over his "fraternal" bonds—his political alliances with other men. In contrast to battering, then, this kind of violence

- thrives on sexual antagonism and conjugal aloofness,
- is intended to preserve a fraternal interest group, and
- results in the victim's humiliation and occasional serious injury.

Although its chief objective is fraternal solidarity, such violence also allows men to assert themselves individually without directly challenging their rivals. In Papua New Guinea, for example, "women provide a background against which men can highlight their own achievements, a foil for their display of 'strength.' The contrast itself seems to stimulate men and enhance their sense of self-esteem" (Berndt 1962:403). This argument has been extended and elaborated by Marilyn Gelber in her survey of *Gender and Society in the New Guinea Highlands* (1986:85–86):

The treatment of women seems to be an important means of publicly demonstrating irascibility, the potential for violence, and the threshold for tolerance of others' behavior. Brutality toward women may be a kind of implied threat toward other men. Women are particularly opportune subjects for display of this kind because, being unimportant politically, they are also politically neutral; that is, the way they are treated has few adverse repercussions on the relationships among men of the local group.

Thus, in traditional Melanesia, where women in general are often defined by men as moral "rubbish," a particular woman can be treated with impunity in a way that a man could never be. The irony is that their political insignificance is precisely what lends women their *utility* as a means of communication among men. Under this system of collective patriarchy, assaulting a woman can serve as a political *signal*—not to the victim herself (whose subordinate status can be taken for granted) but to the assailant's potential challengers within the men's house. The point of the violence may be to "shock and awe" a masculine audience, without directly challenging any particular political rival (cf. Rodseth and Wrangham 2004:399).

Chimpanzee Semiotics

Humans are probably not the only primates in which male aggression against females carries political significance. As noted in other chapters in this volume, male baboons, gorillas, and chimpanzees have all been observed to include females in their status displays. An especially vivid example is provided by Kitchen et al. (Chapter 6 in this volume), who suggest that female baboons are sometimes attacked to intensify the effect of a male's "loud call," which serves in part to challenge and intimidate other males.

Such a pattern is brought to light, it should be noted, only by preserving the wider social context in which events take place; an approach that focuses tightly on the dyad of assailant and victim or that pools data from many different social contexts would be unlikely to perceive the kind of "audience effects" that are so well documented by Kitchen et al. When a contextual approach has been taken in the analysis of other highly competitive multimale, multifemale groups, audience effects of various kinds have been detected (e.g., Smuts 1985; Goodall 1986; Cheney and Seyfarth 2007; de Waal 2007).

Chimpanzees are of special relevance to the present argument because their social organization resembles the "Melazonian" pattern of sexual antagonism. At East African sites such as Gombe and Kanyawara, all the adult males are dominant over all the females. Males are related by birth and establish something comparable to a "fraternal interest group" (Boehm 1992;

Rodseth and Wrangham 2004). The status of adult males, furthermore, is never seriously threatened by females, who are subjected to a kind of collective patriarchy (or "fratriarchy," given the absence of anything like human fatherhood).

Most aggression against female chimpanzees occurs not in consortships, which are relatively rare, but in congregations of adult males. Group copulation in particular often leads to violence. When a chimpanzee is in estrus, "she typically becomes the nucleus of a party comprising many or all of the community males" (Goodall 1986:450). Before she reaches maximum tumescence, every member of the party has an opportunity to mate. Eventually, however, she is monopolized by the highest ranking male, who stakes his claim by herding, threatening, or attacking her, while aggressively excluding his rivals. At Kanyawara, the average rate of aggression in large parties of males roughly doubled when a maximally swollen, parous female was present (Muller 2002:117).

Mating is not the only social context in which rates of aggression tend to increase. Even anestrous females are often subjected to threats or attacks by males, and much of the violence is inflicted while other males are present. At Gombe, according to Goodall (1986:349), at least one-quarter of male attacks on females occurred during reunions, when traveling or foraging parties met up after a lengthy separation. Indeed, "Males were aggressive to females considerably more often during reunion than in any other contexts" (Goodall 1986:344). Conceivably, such aggression is a form of sexual coercion: the male attacks the female to reestablish his dominance and ensure future mating opportunities.

Yet another interpretation is available. Dominance relationships in general are tested during reunions, and male aggression may be directed toward individuals of either sex (Bygott 1979; Muller 2002). As Goodall (1986:342) points out: "That females were victims most often is not surprising, partly because they are a less risky proposition." As a result, she suggests, "females are more likely to become scapegoats in instances of redirected aggression." Extending this line of argument, we hypothesize that a male sometimes targets a female to make a political point—not because she represents any serious challenge to him, but precisely because she is politically insignificant and can thus serve as a safe (yet believable) proxy for a real political rival. Branches, empty kerosene cans, and sometimes primatologists have been used in the same way by male chimpanzees whose foremost intention, presumably, was to have an effect on observers rather than the object at hand (cf. Muller et al., Chapter 8 in this volume). Every violent display conveys information, not only to the principal target, but to any bystander who might be inclined to challenge the aggressor in the future.

Our approach to chimpanzee behavior is influenced, no doubt, by our general conception of human social organization. Based on many ethnographic examples, we have described human society as a bifurcated habitat, with more or less distinct "niches" of domestic and public life (Rodseth and Novak 2000, 2006). The somewhat surprising conclusion we have drawn here is that chimpanzees confront a social landscape that is similarly (though much less sharply) bifurcated, with consortship providing a quasi-domestic context and reunion a quasi-public one. What may seem to be rather far-fetched "reasoning by analogy" gains credence, we would argue, when it is realized that violence in consortships clearly resembles human battering, while gender violence in reunions is not unlike the kind of public brutality that springs from a general masculine antagonism toward women.

Community Dues

Somewhere between the extremes of private coercion and public display fall many or most cases of violence against women. It is unusual, in other words, for such violence to be "purely" coercive or "purely" semiotic, with no mixing of motives at all. To say this much is just to acknowledge the complex etiology of all human behavior, as Daly and Wilson did long ago with this fitting illustration (1983:343):

> Why is that man over there bullying his wife? Because of his testosterone? His upbringing? His Y chromosome? His culture? His social status? His religion? In order that he not be cuckolded? Because he no longer loves her? Because of a frustrating day at the office? Sadism? Poverty? Sunspots?

All these explanations, the authors noted, may have some merit, and to invoke one factor is not necessarily to deny or reject the others. At the same time, "in order that he not be cuckolded" is an explanation of general interest to evolutionary theorists, while sadism and sunspots can (usually) be left out of the analysis.

Cuckoldry, however, is itself a complex phenomenon, and its prevention serves multiple adaptive ends. The beating of a wife for her (suspected) infidelity may seem to be a straightforward matter of sexual "proprietariness," yet even here the violence has an important political dimension that is only dimly illuminated, if at all, by models of mate guarding and sexual jealousy. What is really at stake when a woman "cheats" on her husband? Based on sexual selection theory (Bateman 1948; Trivers 1972), the all too ready answer has been the husband's investment in another man's offspring. And indeed, if just these three adults—cuckolded husband, unfaithful wife, and sneaky paramour—were involved in the "transaction," it might be enough to calculate the cost in terms of misdirected investment and leave it at that. If humans behaved like gibbons, in other words, and formed independent breeding groups with just one pair-bond apiece, such a simple analysis would be appropriate (cf. Palombit 1994, 1999; Sommer and Reichard 2000). But humans do not behave like this, even in the most thinly dispersed or socially atomized populations. The fact that conjugal families are always embedded within a wider political community suggests that sexual transactions, however privately conducted, are likely to have public implications.

How can we do justice to these implications without losing sight of the sexual proprietariness that undoubtedly fuels a great deal of gender violence? At minimum, we must recognize and take account of the fact that "anticipated reactions of third parties can both encourage and deter" what goes on in private (Daly and Wilson, Chapter 11 in this volume). Next, we might try to ascertain who these "third parties" might be. Certainly in-laws and other relations are likely to get involved: "Marriage engenders alliance among extensive kin groups, and family members exert considerable influence on both mate choice and the stability and quality of marriages" (Daly and Wilson, Chapter 11). Still missing from the analysis, however, is what we ordinarily call the *community*—an immediate, enveloping, and politically significant social network that includes some who are not necessarily family members or even personal friends.

The community has seldom been given its due in evolutionary analysis. Most of the classic work in sociobiology focused on either kinship or sexuality— topics of fundamental importance in the study of any sexual species, from humans to hyacinths (Hamilton 1964; Williams 1966; Trivers 1972). Such ecumenical subject matter drew social and natural scientists into one vast and stimulating conversation, and this in itself is a major breakthrough in the history of ideas (Rodseth and Novak 2006). Yet the sex-and-kinship paradigm of the 1970s is ill-equipped to deal with certain features of human societies that are quite unusual in the natural world but little theorized in the social sciences. One of these features—fission-fusion sociality—provided the starting point for our argument.

We turn now to the political community—a more obvious feature, no doubt, but one that continues to slip our conceptual net.

From Domestic Discipline to Public Power

In many instances, wife-beating is not just tolerated by the community or sanctioned by the men's house, but commended by both sexes as a sign of traditional authority and domestic discipline. This is often the case in tribal and peasant societies that maintain a "moderate" sexual antagonism—not as extreme as the men's-house complex in Melanesia and Amazonia, but still far from a European or North American model of spousal companionship and complementarity. Among the Bedouin of western Egypt, for example, both men and women extol the virtues of a domineering husband (Abu-Lughod 1986:89):

> Admired men were often described as difficult *(wa'r)* or tough *(jabbār)*. Large in stature and physically strong, such men assert their will. Women claim, for instance, that "real men" control all their dependents and beat their wives when the wives do stupid things.

Similarly, in a mestizo village south of Mexico City, a wife who dabbled in witchcraft was seen as fully deserving of what she received (Romanucci-Ross 1973:111–112): "Leticia's husband beat her soundly, for (as he explained to me) a man cannot allow his wife to make a *tarugo* ('stupid fool') of him and that if she continued these witching games he would have to beat her again (though this was primarily an assertion of his manliness)." In this kind of context, a wife-beater has no reason to hide his aggression and may in fact seek to display it.

Examples such as these suggest that wife-*battering*, which is seen as deplorable in virtually all societies, should be distinguished from "ordinary" beatings, which are often condoned or even lauded by the local community (Counts 1990, 1992; Campbell 1999). The difference has been elaborated by Judith K. Brown (1999:4):

> In many nonindustrial societies, husbands beat their wives in what Beatrice Whiting (personal communication 1986) refers to as a "physical reprimand." . . . The distinction between wife beating and wife battering is necessary to accommodate data from these societies, where men who beat their wives are not "abnormal" or "deviant" but merely behaving in a culturally expected manner.

Indeed, if a woman has somehow damaged her husband's reputation, a physical reprimand may be the only legitimate way for him to recover his social standing. An especially revealing example is provided by Lauris McKee (1999:170), based on her fieldwork in a mestizo community of highland Ecuador:

> If a man feels his domestic authority is compromised, his marital situation soon becomes a subject of village talk. If the "problem" persists, ridicule or insults from a coterie of male villagers who consider themselves hypermasculine, as well as a man's own reading of his gender role, may prod him into beating his wife.

Social expectations of this kind may be more formally developed in codes of conduct and even enshrined in law. Thus the Lusi-Kaliai, horticultural villagers of West New Britain, justify wife-beating by appealing to the same general principles that have been discussed on the national level by the Law Reform Commission of Papua New Guinea (Counts 1999:76):

> A husband has the right (even the duty) to strike his wife if she flirts with other men or commits adultery; if she draws blood in punishing their children; if she fails to meet her domestic obligations, such as preparing meals, caring for their children, keeping their house and its grounds clean and tidy, or working in the gardens; if she behaves in a way that publicly shames him or his kin; if she fails to assist her husband in meeting his ceremonial obligations; if she fights with her co-wives or abuses his children by another woman.

In this case as in many others, adultery is considered grounds for corporal punishment, but so are lesser offenses that reflect on the husband's status as both a householder and a political actor (cf. Wilson and Daly 1992:310–313). During fieldwork among the Lusi-Kaliai in 1985, Dorothy Ayers Counts heard women remark that "a male relative who failed to punish his wife for carelessness in her domestic chores was himself responsible for the disheveled state of their household" (1999:76). In a quite different context, among Mayan villagers, wife-beating is most commonly justified as punishment for "laziness, infidelity, or disrespect for the husband's parents" (McClusky 2001:253). In other societies around the world, though a somewhat different range of offenses may be listed, masculine reputation is similarly buffered from the "stupid" or "shameful" things that wives are "known" to do.

Because patriarchal authority and political ambition tend to run together in such societies, the domestic sphere becomes an important proving ground for leadership in general. An extreme example of this pattern has been described by Mary Elaine Hegland (1991, 1999) in a rapidly changing cultural and political context. Between 1966 and 1979, Hegland studied gender relations in a large village in southwestern Iran. Here the fluctuation and eventual collapse of state control required local factions to build and maintain their own militias. As a result, according to Hegland (1999:235),

> young men were valued as fighters while women were valued as reproducers and as the means of creating or solidifying relations with other groups through marriage. . . . Women were also valuable as a means of demonstrating control and authority. The protection of dependents of both sexes and their submission and obedience effectively signaled to other political actors the dominance and political strength of a political leader and his faction.

"Wife abuse," as Hegland puts it, "became one more arena for men to demonstrate or jockey for political power" (1999:239). Even when it is carried out in private, aggression against women seems once again to assume—if not require—a masculine audience in the public sphere. Yet the "message" that is being sent to other interested parties is hardly a matter of property rights ("help me defend my property, or at least do not interfere with my own legitimate efforts to do so"), for such rights are not at issue. The point of the abuse, according to Hegland's interpretation, is not to warn other men away from a sexual resource but to impose discipline on a kind of political performance— the public behavior of the patriarch's dependents. Only an impressive demonstration of their submission and obedience is likely to keep his political (as opposed to sexual) rivals at bay, and in this way sustain a small zone of security among the ruins of the state.

Leveling the Field

In humans, we have argued, male violence against women runs the gamut from shameful behavior that is intentionally hidden to virtuous behavior that is intentionally displayed. Although we acknowledge the role of sexual coercion and mate guarding, especially in the case of battering, we have tried to tease out certain political factors as well. Our suggestion, in short, is that male violence against women is not a unitary phenomenon but takes at least three distinct forms. In the form of battering, it is promoted by the social isolation of the conjugal pair. As an expression of sexual antagonism, it is fueled by the extension and elaboration of male alliances. An emphasis on such alliances is likely to *reduce* gender violence as a form of private coercion, but *increase* it as a form of public spectacle. Finally, as a "physical reprimand," wife-beating may well take place in private, yet send a public message nonetheless. Such a reprimand is part of a general pattern of domestic discipline that

- takes advantage of domestic privacy to improve a public performance,
- aims to protect the reputation of the assailant and his household, and
- results in occasional rather than chronic or cumulative injuries.

In sharp contrast to battering, which also takes place in private, this kind of violence is consistent with the social values of other community members, and even the victim herself. It takes an unusual person of either sex to adopt and sustain a moral position outside the mainstream of a traditional political community.

Many are accused of doing so, however, just by acting unconventionally or in their own self-interest. If a woman commits adultery, for example, it is often construed as a threat to the community as a whole. Under these circumstances, wife-beating may be a special case of *social leveling*—the moralistic repudiation of selfish, deceitful, or excessively ambitious behavior. The egalitarianism of small-scale societies, according to Christopher Boehm (1993, 1999), depends on ridicule, ostracism, and other leveling mechanisms to neutralize what he calls "bullies" and "upstarts." Because his analysis focuses on the "main political actors," Boehm has little to say about socially problematic women (1999:7–8):

> When I surveyed hundreds of band-level and tribal societies that were egalitarian (Boehm 1993) to see what was done about upstarts who were hungry for power, the problem personalities were males—group leaders, shamans, proficient hunters, homicidal psychotics, or other men with unusual powers or strong tendencies toward political ambition. This group included bullies intent on aggrandizing power—or on taking more than their share of women. It almost always seems to be males who try to dominate their peers.

Boehm adds an important qualification: "Were contemporary world ethnography more complete, it surely would include some reports of female upstarts" (1999:8). As it stands, however, the ethnographic record suggests to Boehm that women are generally not "the problem personalities" who are "hungry for power."

The irony is that the "natives" themselves tend to see things differently. In the same bands and tribes that Boehm analyzed, women are widely regarded as "selfish" and "greedy," with a tendency to "cause trouble among otherwise peaceful men" (Collier and Rosaldo 1981:296). In Amazonia, for example, social strife is routinely traced to feminine sources: "To put it bluntly, women are the archenemies of the social order" (Roe 1982:230; see also Gregor and Tuzin 2001:321–322). In this ideological light, according to Collier and Rosaldo (1981:296),

> women, as well as men, turn against women when men fight with one another. Ethnographers report cases of women telling men of other women's transgressions, approving of wife beating, and even beating other women themselves. . . . Finally, cultural conceptions that make women ready scapegoats for unwanted trouble simultaneously suggest to people that the most effective strategy for achieving peaceful relations among men is to control troublemaking women.

Nor are such attitudes confined to small-scale societies. In Fiji, for example, among Muslims and Hindus of various class backgrounds, "An abused woman who 'talks too much' or who aggravates her husband or his kin is given little sympathy, since such behavior is considered a form of provocation" (Lateef 1999:226). Whatever the actual sources of social strife, people in many societies tend to blame women. As a result, we would argue, a lot of women get beaten.

Conclusion

The role of "society" in gender violence can be easily trivialized as a matter of what is culturally permitted or enjoined. This is not the point of our argument. We do not see the "system" as an autonomous or ultimate source of human behavior (nor, by the way, do most cultural anthropologists these days—a point that has been slow to dawn on many evolutionists still tilting at cultural determinism). We are trying instead, as good Darwinists must, to place the phenomenon in a wide ecological and historical context. In particular, we are trying to take in more of the social landscape than is usually considered in evolutionary accounts of gender violence in order to see what difference this might make in our scientific perceptions.

As long as cuckoldry is observed through the keyhole, so to speak, it will never appear as anything but a transaction among three parties in an isolated domestic space. Only if we follow these parties out of the boudoir and into the plaza (or the courthouse, or the club) do we begin to appreciate the community, right alongside the conjugal unit, as a vehicle of reproductive competition.

The cuckold is one and the same person with his wife and with his "mates"; his dilemma is to act accordingly. This is not a choice, it should be noted, between siring offspring on the one hand and seeking prestige, wealth, or power on the other. The causal chain leading from copulation to reproduction is short and direct compared to the one that proceeds from the men's house or the stock market. Yet this tells us nothing about the relative importance of mating effort and political effort to a man's genetic survival. If sexual intimacy spells the doom of the political order, placing the entire village at risk, then what is usually described as a "mating opportunity" is actually a dangerous trap—which is exactly the way men see it in sexually antagonistic societies. Only in a world that has been effectively pacified and bureaucratized is the burden of self-defense and political security lifted from the local community—freeing its able-bodied men, it would seem, to focus on the sexual "poachers" in their midst.

A politically sophisticated Darwinism is available to us. It was conceived by Darwin himself in *The Descent of Man* (1871). While laying out the principles of sexual selection theory, he took time to speculate on the political psychology of "primeval men" trapped in cycles of intertribal warfare, yet bound to compete with their comrades for reproductive opportunities within the tribe. A century later, when sexual selection was "rediscovered" by theoretical biologists, only one of them—Richard D. Alexander—followed Darwin's lead in arguing for arms races within and between human groups as the engine of our social evolution (Alexander 1971, 1979, 1987, 1990). Most Darwinists, with an occasional nod to what Robert Trivers (1971) had called moralistic aggression, went on to investigate sexuality and kinship in a kind of political vacuum. More recently, political violence and war have resurfaced in the work of evolutionary anthropologists (e.g., Keeley 1996; Wrangham 1999; Otterbein 2004; Wiessner 2006). Darwin's theme of within-group morality as the complement of between-group competition can again be heard, in quite different registers, in the work of Christopher Boehm (1999) and Raymond C. Kelly (2000). Yet all these contributions have been ancillary, in a sense, to the sex-and-kinship paradigm that has been so successful in the study of nonhuman species.

This is not another call for the recognition of human uniqueness on the basis of complex culture (or language, or big brains, or an opposable thumb). We do believe that Aristotle was on to something when he called man a political animal, and we acknowledge that human intelligence really *is* "Machiavellian"

(Byrne and Whiten 1988; Harcourt and de Waal 1992; Byrne 2001). But it is not our aim to substitute for a one-sided sexual model an equally one-sided political model of human psychology and social evolution. Each of these models is equally possible, as Max Weber might have told us, but only if it serves as the starting point rather than the conclusion of an investigation.

References

Abraham, Margaret. *Speaking the Unspeakable: Marital Violence among South Asian Immigrants in the United States.* New Brunswick, N.J.: Rutgers University Press, 2000.

Abu-Lughod, Lila. *Veiled Sentiments: Honor and Poetry in a Bedouin Society.* Berkeley: University of California Press, 1999.

Alexander, Richard D. "The Search for an Evolutionary Philosophy of Man." *Proceedings of the Royal Society of Victoria* 84 (1971):99–120.

———. *Darwinism and Human Affairs.* Seattle: University of Washington Press, 1979.

———. *The Biology of Moral Systems.* New York: Aldine de Gruyter, 1987.

———. "How Did Humans Evolve? Reflections on the Uniquely Unique Species." University of Michigan Museum of Zoology Special Publication 1 (1990):1–38.

Bateman, A. J. "Intra-Sexual Selection in *Drosophila.*" *Heredity* 2 (1948): 349–368.

Berndt, Ronald M. *Excess and Restraint.* Chicago: University of Chicago Press, 1962.

Boehm, Christopher. "Segmentary 'Warfare' and the Management of Conflict: Comparison of East African Chimpanzees and Patrilineal-Patrilocal Humans." In *Coalitions and Alliances in Humans and Other Animals,* eds. A. H. Harcourt and F. B. M. de Waal, pp. 137–173. Oxford: Oxford University Press, 1992.

———. "Egalitarian Society and Reverse Dominance Hierarchy." *Current Anthropology* 34 (1993): 227–254.

———. *Hierarchy in the Forest: The Evolution of Egalitarian Behavior.* Cambridge, Mass.: Harvard University Press, 1999.

Brown, Judith K. "Cross-Cultural Perspectives on Middle-Aged Women." *Current Anthropology* 23 (1982): 143–156.

———. "Lives of Middle-Aged Women." In *In Her Prime: New Views of Middle-Aged Women,* eds. Virginia Kerns and Judith K. Brown, pp. 17–30. Urbana and Chicago: University of Illinois Press, 1992.

———. "Agitators and Peace-Makers: Cross-cultural Perspectives on Older Women and the Abuse of Young Wives." In *A Cross-Cultural Exploration of Wife Abuse: Problems and Prospects,* ed. Aysan Sev'er, pp. 79–99. Queenston, Ontario: Edwin Mellen Press, 1997.

———. "Introduction: Definitions, Assumptions, Themes, and Issues." In *To Have and to Hit: Cultural Perspectives on Wife Beating,* eds. Dorothy Ayers Counts, Judith K. Brown, and Jacquelyn C. Campbell, pp. 3–26. Urbana: University of Illinois Press, 1999.

Buss, David M. *The Evolution of Desire.* New York: Basic Books, 1994.

———. *The Dangerous Passion: Why Jealousy Is as Necessary as Love and Sex.* New York: Free Press, 2000.

Bygott, J. D. "Agonistic Behaviour, Dominance and Social Structure in Wild Chimpanzees of the Gombe National Park." In *The Great Apes,* eds. D. A. Hamburg and E. R. McCown, pp. 404–427. Menlo Park, Calif.: Benjamin Cummings, 1979.

Byrne, Richard W. "Social and Technical Forms of Primate Intelligence." In *Tree of Origin: What Primate Behavior Can Tell Us about Human Evolution,* ed. Frans B. M. de Waal, pp. 145–172. Cambridge, Mass.: Harvard University Press, 2001.

Byrne, Richard W., and Andrew Whiten, eds. *Machiavellian Intelligence: Social Expertise and the Evolution of Intellect in Monkeys, Apes, and Humans.* New York: Oxford University Press, 1988.

Campbell, Jacquelyn C. "Sanctions and Sanctuary: Wife Battering within Cultural Contexts." In *To Have and to Hit: Cultural Perspectives on Wife Beating,* eds. Dorothy Ayers Counts, Judith K. Brown, and Jacquelyn C. Campbell, pp. 261–285. Urbana: University of Illinois Press, 1999.

Cheney, Dorothy L., and Robert M. Seyfarth. *Baboon Metaphysics: The Evolution of a Social Mind.* Chicago: University of Chicago Press, 2007.

Collier, Jane F., and Michelle Z. Rosaldo. "Politics and Gender in Simple Societies." In *Sexual Meanings: The Cultural Construction of Gender and Sexuality,* eds. S. B. Ortner and H. Whitehead, pp. 275–329. Cambridge: Cambridge University Press, 1981.

Collier, Jane F. and Sylvia J. Yanagisako. "Introduction." In *Gender and Kinship: Essays toward a Unified Analysis,* eds. J. F. Collier and S. J. Yanagisako, pp. 1–13. Stanford, Calif.: Stanford University Press, 1987.

Collins, Randall. *Interaction Ritual Chains.* Princeton, N.J.: Princeton University Press, 2004.

Counts, Dorothy Ayers. "Introduction." In *Domestic Violence in Oceania,* ed. Dorothy Ayers Counts, pp. 1–6. Special issue of *Pacific Studies* 13(1990).

———. "The Fist, the Stick, and the Bottle of Bleach: Wife Bashing and Female Suicide in a Papua New Guinea Society." In *Contemporary Pacific Societies: Studies in Development and Change,* eds. V. Lockwood, T. Hardy, and B. Wallace, pp. 249–259. New York: Prentice Hall, 1992.

———. "'All Men Do It': Wife Beating in Kaliai, Papua New Guinea." In *To Have and to Hit: Cultural Perspectives on Wife Beating,* eds. Dorothy Ayers Counts, Judith K. Brown, and Jacquelyn C. Campbell, pp. 73–386. Urbana: University of Illinois Press, 1999.

Counts, Dorothy Ayers, Judith K. Brown, and Jacquelyn C. Campbell, eds. *To Have and to Hit: Cultural Perspectives on Wife Beating,* second edition. Chicago: University of Illinois Press, 1999.

Daly, Martin and Margo Wilson. *Sex, Evolution, and Behavior,* second edition. Boston, Mass.: Wadsworth, 1983.

————. *Homicide.* New York: Aldine de Gruyter, 1988.

————. "Evolutionary Psychology of Male Violence." In *Male Violence,* ed. John Archer, pp. 253–288. London: Routledge, 1994.

Darwin, Charles. *The Descent of Man, and Selection in Relation to Sex.* London: John Murray, 1871.

Draper, Patricia. "Contrasts in Sexual Egalitarianism in Foraging and Sedentary Context." In *Family Patterns, Gender Relations,* ed. Bonnie J. Fox, pp. 33–48. Ontario: Oxford University Press, 1993.

Draper, Patricia, and Henry Harpending. "Father Absence and Reproductive Strategies: An Evolutionary Perspective." *Journal of Anthropological Research* 38 (1982): 255–273.

Felson, Richard B., J. Ackerman, and S. J. Yeon. "The Infrequency of Family Violence." *Journal of Marriage and Family* 65 (2003): 622–634.

Gruenbaum, Ellen. *The Female Circumcision Controversy: An Anthropological Perspective.* Philadelphia: University of Pennsylvania Press, 2001.

Garcia-Moreno, Claudia, Henrica A. F. M. Jansen, Mary Ellsberg, Lori Heise, and Charlotte H. Watts. "Prevalence of Intimate Partner Violence: Findings from the WHO Multi-Country Study on Women's Health and Domestic Violence." *The Lancet* 368 (2006): 1260–1269.

Gelber, Marilyn G. *Gender and Society in the New Guinea Highlands: An Anthropological Perspective on Antagonism toward Women.* Boulder, Colo. : Westview Press, 1986.

Goffman, Erving. *The Presentation of Self in Everyday Life.* Garden City, N.Y.: Doubleday, 1959.

————. *Interaction Ritual: Essays on Face-to-Face Behavior.* Garden City, N.Y.: Doubleday, 1967.

Goodall, Jane. *The Chimpanzees of Gombe: Patterns of Behavior.* Cambridge, Mass.: Belknap Press of Harvard University Press, 1986.

Gregor, Thomas A. *Mehinaku: The Drama of Daily Life in a Brazilian Indian Village.* Chicago: University of Chicago Press, 1977.

————. *Anxious Pleasures: The Sexual Lives of an Amazonian People.* Chicago: University of Chicago Press, 1985.

————. "Male Dominance and Sexual Coercion." In *Cultural Psychology,* eds. J. Stiegler, R. Shweder, and G. H. Herdt, pp. 477–495. Chicago: University of Chicago Press, 1990.

Gregor, Thomas A. and Donald Tuzin, eds. *Gender in Amazonia and Melanesia: An Exploration of the Comparative Method.* Berkeley: University of California Press, 2001.

Hamilton, William D. "The Evolution of Social Behavior." *Journal of Theoretical Biology* 7 (1964): 1–52.

Hanson, Susan, and Geraldine Pratt. *Gender, Work and Space.* London and New York: Routledge, 1995.

Harcourt, Alexander H., and Frans B. M. de Waal, eds. *Coalitions and Alliances in Humans and Other Animals.* Oxford: Oxford University Press, 1992.

Hayden, Robert M. "Antagonistic Tolerance: Competitive Sharing of Religious Sites in South Asia and the Balkans." *Current Anthropology* 43 (2002): 205–231.

Hegland, Mary Elaine. "Political Roles of Aliabad Women: The Public/Private Dichotomy Transcended." In *Women in Middle Eastern History: Shifting Boundaries in Sex and Gender,* eds. Nikki R. Keddie and Beth Baron, pp. 215–230. New Haven, Conn.: Yale University Press, 1991.

————. "Wife Abuse and the Political System: A Middle Eastern Case Study." In *To Have and to Hit: Cultural Perspectives on Wife Beating,* eds. Dorothy Ayers Counts, Judith K. Brown, and Jacquelyn C. Campbell, pp. 234–251. Urbana: University of Illinois Press, 1999.

Heise, Lori L. "Violence against Women: An Integrated, Ecological Framework." *Violence against Women* 4 (1998): 262–290.

Herdt, Gilbert ,and Fitz John P. Poole. "'Sexual Antagonism': The Intellectual History of a Concept in New Guinea Anthropology." *Social Analysis* 12 (1982): 3–28.

Hilberman, E., and K. Munson. "Sixty Battered Women." *Victimology* 2 (1978): 460–470.Isaac, Glynn. *The Archaeology of Human Origins: Papers by Glynn Isaac,* ed. Barbara Isaac. Cambridge: Cambridge University Press, 1989.

Jeffery, Patricia. *Frogs in a Well: Indian Women in Purdah.* London: Zed Press, 1979.

Johnson, M. P. "Patriarchal Terrorism and Common Couple Violence: Two Forms of Violence against Women." *Journal of Marriage and the Family* 57 (1995): 283–294.

Johnson, M. P., and K. J. Ferraro. "Research on Domestic Violence in the 1990s: Making Distinctions." *Journal of Marriage and the Family* 62 (2000): 948–963.

Keeley, Lawrence H. *War Before Civilization.* Oxford: Oxford University Press, 1996.

Kelly, Raymond C. *Warless Societies and the Origin of War.* Ann Arbor: University of Michigan Press, 2000.

Kerns, Virginia. "Female Control of Sexuality: Garífuna Women at Middle Age." In *In Her Prime: New Views of Middle-Aged Women,* eds. Virginia Kerns and Judith K. Brown, pp. 95–111. Urbana and Chicago: University of Illinois Press, 1992.

Kerns, Virginia, and Judith K. Brown, eds. *In Her Prime: New Views of Middle-Aged Women,* second edition. Urbana and Chicago: University of Illinois Press, 1992.

Kwan, Mei-Po. "Gender Differences in Space-Time Constraints." *Area* (2000) 32: 145–156.

Lamphere, Louise. "Strategies, Cooperation, and Conflict Among Women in Domestic Groups." In *Woman, Culture, and Society,* eds. M. Z. Rosaldo and L. Lamphere, pp. 97–112. Stanford, Calif.: Stanford University Press, 1974.

Lateef, Shareen. "Wife Abuse among Indo-Fijians." In *To Have and to Hit: Cultural Perspectives on Wife Beating,* eds. Dorothy Ayers Counts, Judith K. Brown, and Jacquelyn C. Campbell, pp. 216–233. Urbana: University of Illinois Press, 1999.

Leacock, Eleanor B. *Myths of Male Dominance.* London: Monthly Review Press, 1981.

Lee, Richard B. *The !Kung San: Men, Women and Work in a Foraging Society.* Cambridge: Cambridge University Press, 1979.

Madriz, Esther. *Nothing Bad Happens to Good Girls: Fear of Crime in Women's Lives.* Berkeley: University of California Press, 1997.

Marshall, Donald S. "Sexual Behavior on Mangaia." In *Human Sexual Behavior*, eds. Donald S. Marshall and R. C. Suggs, pp. 103–162. New York: Basic Books, 1971.

McClusky, L. J. *"Here, Our Culture is Hard": Stories of Domestic Violence from a Mayan Community in Belize.* Austin: University of Texas Press, 2001.

McKee, Lauris. "Men's Rights/Women's Wrongs: Domestic Violence in Ecuador." In *To Have and to Hit: Cultural Perspectives on Wife Beating*, eds. Dorothy Ayers Counts, Judith K. Brown, and Jacquelyn C. Campbell, pp. 168–186. Urbana: University of Illinois Press, 1999.

Mitchell, William E. "Why Wape Men Don't Beat Their Wives: Constraints toward Domestic Tranquility in a New Guinea Society." In *To Have and to Hit: Cultural Perspectives on Wife Beating*, eds. Dorothy Ayers Counts, Judith K. Brown, and Jacquelyn C. Campbell, pp. 100–109. Urbana: University of Illinois Press, 1999.

Muller, Martin N. "Agonistic Relations among Kanyawara Chimpanzees." In *Behavioural Diversity in Chimpanzees and Bonobos*, eds. Christophe Boesch, Gottfried Hohmann, and Linda F. Marchant, pp. 112–124. Cambridge: Cambridge University Press, 2002.

Murphy, Robert F. "Intergroup Hostility and Social Cohesion." *American Anthropologist* 59 (1957): 1018–1035.

Murphy, Yolanda, and Robert F. Murphy. *Women of the Forest.* New York: Columbia University Press, 1974.

Novak, Shannon A. "Skeletal Manifestations of Domestic Assault: A Predictive Model for Investigating Gender Violence in Prehistory." Ph.D. dissertation, Department of Anthropology, University of Utah. Salt Lake City, 1999.

———. "Beneath the Façade: A Skeletal Model of Domestic Violence." In *Funerary Archaeology of Human Remains*, eds. Rebecca Gowland and Christopher Knüsel, pp. 238–252. Oxford: Oxbow Press, 2006.

Ortner, Sherry B., and Harriet Whitehead. "Introduction: Accounting for Sexual Meanings." In *Sexual Meanings: The Cultural Construction of Gender and Sexuality*, eds. Sherry B. Ortner and Harriet Whitehead, pp. 1–27. Cambridge: Cambridge University Press, 1981.

Otterbein, Keith F. "A Cross-Cultural Study of Rape." *Aggressive Behavior* 5 (1979): 425–435.

———. *How War Began.* College Station: Texas A&M University Press, 2004.

Paige, Karen Ericksen, and Jeffery M. Paige. *The Politics of Reproductive Ritual.* Berkeley: University of California Press, 1981.

Pain, Rachel. "Space, Sexual Violence, and Social Control: Integrating Geographical and Feminist Analysis of Women's Fear of Crime." *Progress in Human Geography* 15 (1991): 415–431.

Palombit, R. A. "Extra-Pair Copulations in a Monogamous Ape." *Animal Behaviour* 47 (1994): 721–723.

———. "Infanticide and the Evolution of Pair Bonds in Nonhuman Primates." *Evolutionary Anthropology* 7 (1999): 117–129.

Rodseth, Lars. "Prestige, Hunger, and Love: Plumbing the Psychology of Sexual Meanings." *Michigan Discussions in Anthropology* 9 (1990): 107–127.

Rodseth, Lars, and Shannon A. Novak. "The Social Modes of Men: Toward an Ecological Model of Human Male Relationships. *Human Nature* 11 (2000): 335–366.

———. "The Impact of Primatology on the Study of Human Society." In *Missing the Revolution: Darwinism for Social Scientists,* ed. Jerome H. Barkow, pp. 187–220. New York: Oxford University Press, 2006.

Rodseth, Lars, and Richard W. Wrangham. "Human Kinship: A Continuation of Politics by Other Means?" In *Kinship and Behavior in Primates,* eds. Bernard Chapais and Carol M. Berman, pp. 389–419. New York: Oxford University Press, 2004.

Rodseth, Lars, Richard W. Wrangham, Alisa M. Harrigan, and Barbara B. Smuts. "The Human Community as a Primate Society." *Current Anthropology* 32 (1991): 221–254.

Roe, Peter G. *The Cosmic Zygote: Cosmology in the Amazonian Basin.* New Brunswick, N.J.: Rutgers University Press, 1982.

Romanucci-Ross, Lola. *Conflict, Violence, and Morality in a Mexican Village.* Palo Alto, Calif.: National Press Books, 1973.

Rosaldo, Michelle Z. "The Use and Abuse of Anthropology: Reflections on Feminism and Cross-Cultural Understanding." *Signs* 5 (1980): 389–417.

Sanday, Peggy Reeves. *Female Power and Male Dominance: On the Origins of Sexual Inequality.* Cambridge: Cambridge University Press, 1981.

Sen, Mala. *Death by Fire: Sati, Dowry Death, and Female Infanticide in Modern India.* New Brunswick, N.J.: Rutgers University Press, 2001.

Smuts, Barbara B. *Sex and Friendship in Baboons.* Cambridge, Mass.: A: Harvard University Press, 1985.

———. "Male Aggression against Women: An Evolutionary Perspective." *Human Nature* 3 (1992): 1–44.

———. "The Evolutionary Origins of Patriarchy." *Human Nature* 6 (1995): 1–32.

Smuts, Barbara B., and Robert W. Smuts. "Male Aggression and Sexual Coercion of Females in Nonhuman Primates and Other Mammals: Evidence and Theoretical Implications." *Advances in the Study of Behavior* 22 (1993): 1–63.

Solway, Jacqueline S. "Middle-Aged Women in Bakgalagadi Society (Botswana)." In *In Her Prime: New Views of Middle-Aged Women,* eds. Virginia Kerns and Judith K. Brown, pp. 49–58. Urbana and Chicago: University of Illinois Press, 1992.

Sommer, V., and U. Reichard. "Rethinking Monogamy: The Gibbon Case." In *Primate Males: Causes and Consequences of Variation in Group Composition,* ed. P. M. Kappeler, pp. 159–168. Cambridge: Cambridge University Press, 2000.

Spencer, W. Baldwin and Frank J. Gillen. *The Native Tribes of Central Australia.* New York: Dover, 1968 [1899].

Symons, Donald. *The Evolution of Human Sexuality.* New York: Oxford University Press, 1979.

Totten, M. "Girlfriend Abuse as a Form of Masculinity Construction among Violent, Marginal Male Youth." *Men and Masculinities* 6 (2003): 70–92.

Trivers, Robert L. "The Evolution of Reciprocal Altruism." *Quarterly Review of Biology* 46 (1971): 35–57.

———. "Parental Investment and Sexual Selection." In *Sexual Selection and the Descent of Man,* ed. B. Campbell, pp. 136–179. Hawthorne, N.Y.: Aldine de Gruyter, 1972.

Tutin, C. E. G. "Mating Patterns and Reproductive Strategies in a Community of Wild Chimpanzees *(Pan troglodytes schweinfurthii)." Behavioral Ecology and Sociobiology* 6 (1979): 29–38.

Vatuk, Sylvia. "Sexuality and the Middle-Aged Woman in South Asia." In *In Her Prime: New Views of Middle-Aged Women,* eds. Virginia Kerns and Judith K. Brown, pp. 155–170. Urbana and Chicago: University of Illinois Press, 1992.

de Waal, Franz. *Chimpanzee Politics: Power and Sex among the Apes,* 25th anniversary edition. Baltimore, Md.: Johns Hopkins University Press, 2007.

Weber, Max. *The Methodology of the Social Sciences,* eds. Edward A. Shils and Henry A. Finch. Glencoe, IL: Free Press, 1949.

Whiting, John W. M., and Beatrice B. Whiting. "Aloofness and Intimacy of Husbands and Wives: A Cross-Cultural Study." *Ethos* 3 (1975): 183–207.

Wiessner, Polly. "From Spears to M16s: Testing the Imbalance of Power Hypothesis among the Enga." *Journal of Anthropological Research* 62 (2006): 165–191.

Williams, George C. *Adaptation and Natural Selection.* Princeton, N.J.: Princeton University Press, 1966.

Wilson, Peter J. *The Domestication of the Human Species.* New Haven, Conn.: Yale University Press, 1988.

Wilson, Margo, and Martin Daly. "The Man Who Mistook His Wife for a Chattel." In *The Adapted Mind,* eds. Jerome H. Barkow, Leda Cosmides, and John Tooby, pp. 289–322. New York: Oxford University Press, 1992.

———. "Lethal and Nonlethal Violence against Wives and the Evolutionary Psychology of Male Sexual Proprietariness." In *Rethinking Violence Against Women,* eds. R. Emerson Dobash and Russell P. Dobash, pp. 199–230. Thousand Oaks, Calif.: A: Sage, 1998.

Wrangham, Richard W. "Ecology and Social Relationships in Two Species of Chimpanzee." In *Ecological Aspects of Social Evolution: Birds and Mammals,* eds. Daniel I. Rubenstein and Richard W. Wrangham, pp. 352–378. Princeton, N.J.: Princeton University Press, 1986.

———. "The Significance of African Apes for Reconstructing Human Social Evolution." In *Primate Models of Hominid Evolution,* ed. W. G. Kinzey, pp. 51–71. Albany, N.Y.: SUNY Press, 1987.

———. "The Evolution of Coalitionary Killing." *Yearbook of Physical Anthropology* 42 (1999): 13–30.

Wrangham, Richard W., J. H. Jones, G. Laden, D. Pilbeam, and N. Conklin-Brittain. "The Raw and the Stolen: Cooking and the Ecology of Human Origins." *Current Anthropology* 40 (1999): 567–594.

Wrangham, Richard W., and Dale Peterson. *Demonic Males: Apes and the Origins of Human Violence.* Boston: Houghton Mifflin, 1996.

13

Intimate Wounds: Craniofacial Trauma in Women and Female Chimpanzees

Shannon A. Novak and Mallorie A. Hatch

Because physical assaults can leave distinctive traces in hard tissue, it is often possible to recognize cases of serious violence based on physical evidence alone. In humans, the skeletal effects of interpersonal violence have been described in terms of wound location, fracture morphology, and state of healing (e.g., Brickley and Smith 2006; Judd 2006). Most of the research along these lines has focused on injuries suffered in warfare (e.g., Walker 1989; Owsley and Jantz 1994; Smith 2003). When war is not in evidence, however, and adult female skeletons exhibit healed lesions of traumatic origin, the injuries are sometimes attributable to gender violence—especially male assaults on their sexual partners (Shermis 1982; Martin 1997; Novak 2006).

If similar inferences could be drawn in nonhuman primates, the osteological record might be combined with direct observations of behavior to shed light on male aggression against females. Yet skeletal trauma has received little attention in the primatological literature, in part because of doubts about museum collections as representative samples of natural populations (cf. Randall 1944; Buikstra 1975). Much of the research that has been done focuses on postcranial fractures to assess the risks of arboreal living (reviewed in Lovell 1991). Cranial injuries have received only modest attention, and sex differences in trauma patterns remain virtually unknown (Jurmain and Kilgore 1998).

In this chapter, we analyze a series of skulls from Liberian chimpanzees *(Pan troglodytes verus)* to compare patterns of trauma by age and sex. Our findings indicate a very high frequency of violent injuries in both males and females. These injuries vary by sex, however, in terms of their anatomical distribution, suggesting that females, unlike males, were commonly assaulted from behind. These results are compared with the findings of other skeletal studies of chimpanzee populations from East Africa (Jurmain 1989, 1997; Carter et al. 2008), and Central and West Africa (Lovell 1987, 1990; Jurmain 1997; Jurmain and Kilgore 1998). Finally, we compare the cranial trauma identified in the Liberian chimpanzees to the soft-tissue injuries suffered by 194 women who had been assaulted by their husbands or sexual partners (Novak 2006).

Violence in Chimpanzees

To interpret the skeletal evidence, we must first consider the general patterns of social interaction and violence in chimpanzees. In East African populations, males are regularly aggressive toward females (Muller et al., Chapter 8 in this volume). The health consequences of such aggression are largely unknown, but violence does contribute to chimpanzee mortality. At Mahale, for example, the most frequent cause of death is illness, followed by senescence, and then by intraspecific aggression (Nishida et al. 2003). Moreover, Wrangham et al. (2006:22) emphasize that the greatest risk of both lethal and nonlethal attacks are by members of one's own community rather than by neighboring chimpanzees. Nonlethal attacks are common, with males experiencing a median rate of 2301 attacks per 100,000 hours and females at a rate of 911 attacks per 100,000 hours (Wrangham et al. 2006:20). Although such attacks may be quite frequent, Goodall (1986:97) notes that most of these intracommunity fights are not severe: "They usually last less than a minute and seldom result in serious injury."

During an attack a chimpanzee may hit, kick, scratch, and grapple with his target, or inflict more series injuries by biting, stamping on, dragging, and throwing his victim to the ground (Goodall 1986:317). At Gombe, the majority of attack-related injuries recorded during a seven-year period were minor cuts or scratches, though some 17% of the wounds were deemed serious. Although not all of these injuries are detailed, those described as the most severe appear to be primarily soft-tissue wounds (Goodall 1986:99–100, Table 5.6). It is important to note that observers are likely to underestimate such wounds, as they can be difficult to see under a thick coat of hair.

Goodall does, however, report two cases that would have undoubtedly left a bony signature: an infant with a broken femur or pelvis, and an adolescent with a broken foot. A particularly interesting case for our purposes here is that of

Fifi. As a late adolescent of 14 years, Fifi was attacked by an adult male of her own community. Fifi had a deep puncture wound on her head that became infected, and it was suspected that the attacker's canine tooth remained embedded in her skull (Goodall 1986:99). If we could have examined Fifi's lesion, we would likely have seen a blunt-force injury from the canine surrounded by concentric and radiating fractures. Variations in these characteristics help to distinguish blunt force from other types of trauma, such as sharp force or projectile (Galloway 1999). As the latter two types of wounds result from uniquely human weapons, we should expect blunt-force injuries to be the most common skeletal trauma in chimpanzees.

Male chimpanzees, overall, consistently display more aggression than females. Muller (2002:114), for example, found that male chimpanzees at Kanyawara were aggressive 14 times more often than females and that other males were the most likely targets of their displays. By contrast, Goodall (1986) found that most aggression was directed toward adult females and that females were the most likely targets of *physical* assaults (Goodall 1986:347–349). While sexual coercion has been proposed for much of the physical aggression directed at females (Smuts and Smuts 1993; Muller 2002), the assaults at Gombe illustrate the range of contexts—reunion, meat eating, sexual behavior, and no obvious context—in which females found themselves as targets of male violence.

Although females appear to risk injury primarily from male assaults, it is important to consider other contexts that might produce trauma. Females do indeed attack other females, especially in feeding contests or when protecting their offspring (e.g., Goodall 1986; Boesch and Boesch-Achermann 2000; Muller 2002). These attacks, however, occur substantially less often than attacks by males. At Gombe, 33% of the female aggression was targeted at other adult females, while males targeted adult females in nearly half of their agonistic acts. For Kanyawara females, "Such aggression was rarely severe," argues Muller (2002:119), "it consisted primarily of mild threats or brief charging displays used to supplant feeding competitors."

The most aggressive targeting by resident females is against young females who are attempting to immigrate. In such cases, males sometimes intervene to protect these immigrants, and the aggression subsides as female dominance relationships become established (Nishida 1989). In some communities, however, low-ranking females may face a different kind of assault by higher ranking females. At Gombe, in particular, infanticide and attempted infanticide place the mother at risk of injury as she defends her offspring (Pusey et al. 1997). Such attacks, however, appear to be unusual in most chimpanzee communities and seem to be exacerbated by resource competition at Gombe.

Although we have focused on assaults as a cause of injury, accidents must also be considered. Falls, in particular, would be expected to produce some trauma, though only the worst cases are likely to result in skeletal injuries. At Gombe over a two-year period (1978–1979), Goodall recorded 51 cases of falls in which the chimpanzees "actually hit the ground; often they managed to save themselves by grabbing branches" (1986:98). In a study of skeletal remains from Gombe, Jurmain (1989:236) reported a high frequency of healed fractures, and suggested that "many of these fractures, especially of the distal appendages, probably resulted from falls." Significantly, Jurmain did not attribute any *cranial* injuries to a fall. Similarly, a recent study of chimpanzee skeletons from Kibale National Park, Uganda, identified extensive skeletal trauma, the most severe of which was attributed to falls (Carter et al. 2008). Again, these falls appear to have produced only postcranial injuries.

Based on behavioral observations, a typical fall occurs during an arboreal pursuit, when a panicked victim misses a jump or slips off a branch. At Gombe, some 80% of the fall victims were males, though most of these were juveniles or infants (Goodall 1986:100, Table 5.7). "Adult males fell most often during aggressive incidents, often when two were fighting each other" (Goodall 1986:98). There were four cases of falls by adult females, three of which were the result of aggressive encounters with other females. Although most of the falls did not result in serious wounds, one young adult male "injured his hand or wrist, both of which swelled" (Goodall 1986:98), and one middle-aged male died instantly with a broken neck (1986:111, Table 5.10; see also Carter et al. 2008:400).

To our knowledge, cranial trauma as a result of a fall has not been observed in wild chimpanzees. If such trauma were to occur, the likely effect would be a diffuse, blunt-force lesion. No such lesion was found in the Liberian chimpanzee crania. All the healed wounds in this series were localized depression fractures or punctures that appear to be the result of powerful blows or bites.

Study Sample

The Liberian Chimpanzee Osteological Collection (LCOC) is housed at the Peabody Museum of Archaeology and Ethnology at Harvard University. The collection includes some 294 wild-shot chimpanzee skulls, crania, and mandibles that were sent to Professor Earnest Hooten by a Methodist missionary, George Harley. Based at the town of Ganta in northern Liberia, Harley obtained the majority of the crania during the 1940s and 1950s from various sources in the surrounding region. According to annotation on the skulls, a number of the specimens were acquired from nine villages (Kpein, Zuluyi, Dingamo, Flumpa, Gipo, Lepula, Kawi, Bialatuo, and Zoutuo) in Nimba County. Some were collected from Bongo and Grand Gedeh counties, and five came from across the border in Côte d'Ivoire.

The ecology of the region in the 1950s was quite different from what it is today (Harley 1939; Peters 1956). Large expanses of forest were present throughout the region, although tracts of land near roads had been used for agriculture. Harley, in a 1943 letter to Hooten, reported that significant populations of chimpanzees inhabited the forests of Nimba: "The woods are literally full of chimps, as soon as you get ten miles away from inhabited areas" (Harley, Peabody Museum archives).

Today, however, only small enclaves of forest remain in the region, with chimpanzee populations concentrated in the Nimba Mountains. Most of what is known about these populations is based not on direct observation but on surveys of nest sites, tools, and feces (e.g., Matsuzawa and Yamakoshi 1996; Humle and Matsuzawa 2004). Since 1976, chimpanzees have been studied at the adjacent site of Bossou, Guinea, but the Bossou population of some 20 individuals is confined to a patch of deciduous forest surrounded by agricultural fields (Sugiyama and Koman 1979a, 1979b; Sugiyama 2004). Apparently, the community is not entirely cut off from the populations in the mountains, as it was recently reported that some Bossou chimps had crossed into Liberian forests (Ohashi 2006). Ohashi's recent survey of Nimba County identifies chimpanzee populations in three areas, all located northeast of the Ganta Mission. The relationship between the existing groups and the chimpanzees that were collected by Harley remains to be determined.

Because there was no primatologist to study Liberian chimpanzees before they were exterminated or pushed into isolated pockets of forest, skeletal analysis may be the only way to shed light on their social behavior in large, relatively undisturbed populations. Our sample consists of 246 crania in the LCOC. Sex was determined for each cranium by dimorphic characteristics in the bone and teeth (Schultz 1969; Schwartz and Dean 2001; Whitehead et al. 2004). Most of the crania were male; females represented just over a third of the sample, and less than 10% were indeterminate for sex[1] (Table 13.1).

Each individual's age was estimated on the basis of morphological changes associated with growth, development, and degeneration in the bone and teeth. Assessment of these changes allowed each cranium to be assigned to one of four categories: juvenile, adolescent, young adult, and old adult (Table 13.2).[2] The majority of the Liberian crania were from young adults and old adults, whereas less than 20% of the sample were juveniles and adolescents.

Undoubtedly, the LCOC is a sample of more than one chimpanzee population. The question arises as to how well the sample represents living populations whose demographics are known. Although these demographics vary widely, the Liberian sample resembles some of these populations in age and sex composition (Table 13.3). For example, the frequency of adult females and adolescent males in the LCOC closely approximates that of the Mitumba community at

Table 13.1   Chimpanzee crania frequencies by sex and age.

	Age				
Sex	Juvenile	Adolescent	Young Adult	Old Adult	Total
Female	0	10	47	38	95 (38.6%)
Male	0	12	71	47	130 (52.8%)
Indeterminate	20	1	0	0	21 (8.5%)
Total	20 (8.1%)	23 (9.3%)	118 (48.0%)	85 (34.6%)	246 (100%)

Table 13.2   Age categories, morphological traits, and chronological age.

Age Category	Cranial and Dental Features	Approximate Chronological Age
Juvenile	Birth to eruption of the permanent second molar	Birth to 8 years[a]
Adolescent	Second molar fully erupted to partial or complete eruption of permanent canines and incomplete eruption of the permanent third molar	8 years to 10 years[a]
Young Adult	Complete eruption of the dentition with an open basilar suture—mild dental wear	10 years to 20 years[b]
Old Adult	Closed basilar suture and obliteration of cranial sutures—moderate to severe dental wear may be exhibited	20 years and above[b]

[a] Age based on dental eruption from Zihlman et al. (2004).
[b] Age based on cranial suture closure from Krogman (1930, 1969).

Table 13.3 Average community composition (adapted from Wrangham et al. 2006, Appendix 9).

Site	Community	Adult Male	Adult Female	Adolescent Male	Adolescent Female	Infant/Juvenile	Total
Budongo	Sonso	12 (24.0%)	11 (22.0%)	4 (8.0%)	6 (12.0%)	17 (34.0%)	50
Gombe	Kahama	4.2 (42.0%)	1.8 (18.0%)	0.6 (6.0%)	1 (10.0%)	2.4 (24.0%)	10
Gombe	Kasekela	8.7 (18.8%)	12.2 (26.3%)	1.8 (3.9%)	3.5 (7.6%)	20.1 (43.4%)	46.3
Gombe	Mitumba	3.1 (14.1%)	7.4 (33.6%)	1.6 (7.3%)	1.9 (8.6%)	8 (36.4%)	22
Kibale	Kanyawara	10 (22.7%)	13.2 (29.9%)	2.7 (6.1%)	3.2 (7.3%)	15 (34.0%)	44.1
Kibale	Ngogo	24 (16.7%)	47 (32.6%)	15 (10.4%)	9 (6.3%)	49 (34.0%)	144
Mahale	K	3.8 (14.8%)	10.7 (41.6%)	0.8 (3.1%)	2.6 (10.1%)	7.8 (30.4%)	25.7
Mahale	M	8.5 (10.8%)	28.6 (36.2%)	7.6 (9.6%)	7.7 (9.7%)	26.6 (33.7%)	79
Tai	Northern	6.7 (10.9%)	21.5 (35.1%)	3 (4.9%)	4.2 (6.9%)	25.8 (42.2%)	61.2
Liberia		118 (48.2%)	85 (34.7%)	20 (8.2%)	10 (4.1%)	20 (8.2%)	245*

*The single adolescent that was indeterminate for sex was removed from this calculation.

Gombe. By contrast, there is an overrepresentation of adult males in the Liberian sample, even when compared to the high of 42% within the Kahama community at Gombe. Adolescent females are somewhat underrepresented in the Liberian sample, but the greatest bias seems to be the low frequency of juveniles. In extant populations, juveniles and infants comprise between 24% (Kahama community, Gombe) and 42.2% (Northern community, Taï) of the community's members.

Trauma Patterns

The Liberian chimpanzee collection exhibited very high rates of healed cranial trauma (Hatch 2006). Nearly 60% of the 246 skulls had at least one healed lesion whose characteristics were consistent with an attack. Here we examine the frequency of such wounds by sex and age, as well as the distribution of these wounds on the cranium. Again, it is important to note that this study focuses on skeletal wounds that had healed before death. In principle, this focus should exclude injuries that were produced when the animal was captured or killed (e.g., machete, trap, or gunshot wounds). Perimortem wounds were seen in this skeletal series but will not be considered here.

Sex and Age

Males and females were equally likely to have at least one healed wound (56.2% and 56.8%, respectively). There was no statistical difference in trauma by sex, indicating that both males and females experienced similar rates of assault that were severe enough to break bone (Table 13.4). Although the sex distribution of wounds was nearly equal, there were distinct differences in age groups.

Old adult males (72.3%) and old adult females (73.4%) had the highest frequencies of individuals with healed trauma (Table 13.5). Because we expect to see cumulative trauma events over an individual's life, this finding is not surprising. Skeletal injuries that occur in infancy and childhood will heal—providing the victim survives—and will be preserved in the bone as an adult. In the Liberian series, this linear relationship between frequency of healed wounds and increasing age holds for both males and females, but only from adolescence on.

Juveniles are the exception to this age-related trend. These individuals had rates of trauma nearly as high (66.7%) as the old adults. This finding is surprising, and there are no published studies about aggression toward juveniles that allow these frequencies to be readily interpreted. Because of individual

Table 13.4   Presence of at least one wound by sex.

Sex	N	Trauma Present	%
Female	95	54	56.8%
Male	130	73	56.2%
Indeterminate	21	14	66.7%
Total	246	141	57.3%

$$\chi^2 = 0.011, \, df = 1, \, P = 0.918^*$$

*Indeterminates were not included in Chi-square statistics. This category includes 20 juveniles and 1 adolescent.

Table 13.5   Presence of at least one wound by sex and age.

Sex	Age	N	Trauma Presence	%
Indeterminate	Juvenile	21	14	66.7%
Female	Adolescent	10	4	40.0%
	Young Adult	47	22	46.8%
	Old Adult	38	28	73.7%
Male	Adolescent	12	5	41.7%
	Young Adult	71	34	47.9%
	Old Adult	47	34	72.3%

variability in rates of healing, we cannot determine a more specific age at which these juveniles were wounded (Roberts and Manchester 2005:91–92). Although these individuals recovered from their injuries, the fact that they did not survive to adolescence may indicate that their condition, or that of their mothers, was somehow compromised.

Location of Wounds

To assess the location of wounds, each cranium was divided into two zones— the face and the vault. The zones were divided by a line running from ear to ear.[3] Wounds located forward of this line were assigned to the face, and those behind the line to the vault. Facial wounds were observed more often in males than in females (Figure 13.1, Table 13.6). A majority of the male skulls (59%)

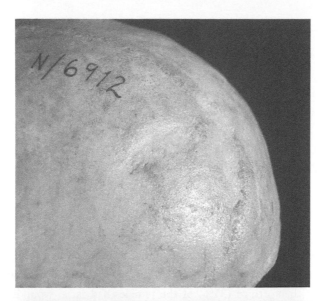

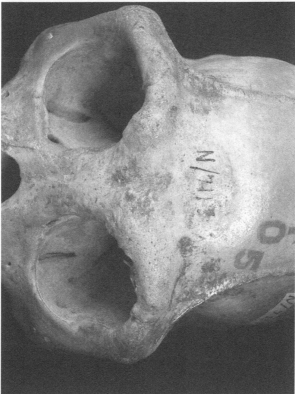

Figure 13.1 Healed blunt-force wound in posterior vault of an old adult female (top) and frontal of a young adult male (bottom). Copyright 2006 the President and Fellows of Harvard College.

Table 13.6   Sex differences in presence of at least one wound in zones of the
face or vault.

Sex	Face			Vault		
	Total N	Trauma Present	%	Total N	Trauma Present	%
Female	95	29	30.5%	95	39	41.1%
Male	130	77	59.2%	130	29	22.3%
Total	225	106	47.1%	225	68	30.2%
	$\chi^2=18.150$, $df=1$, $P=0.001$			$\chi^2=9.146$, $df=1$, $P=0.002$		

had healed wounds in the face, whereas only 22% of them had similar injuries
in the back of the head. Among the females, however, this pattern was nearly
reversed: less than a third (31%) of the females had facial wounds, whereas
41% had healed trauma in the back of the cranium. Putting this another way,
the males had nearly twice as many facial wounds as the females did, but the
females had nearly twice as many injuries to the back of the head. This suggests different responses by males and females during assaults.

In humans, lesions to the anterior cranium are indicative of face-to-face assaults (Knight 1991; Roberts and Manchester 2005). Similar lesions in chimpanzees are likely to have a similar etiology. This suggests that, in the LCOC,
males were more likely than females to have been injured in face-to-face altercations. Females, by comparison, were more likely to have been hit (or bitten) from
behind, suggesting that they had turned away from their attackers (Figure 13.2).

Behavioral observations document cases of females fleeing from males, being
pinned face-down and beaten, or crouching to deflect the blows. Chimpanzee
mothers, in particular, have been observed huddling over their offspring in an
apparent effort to protect them from an attacking male. Such an assault was documented by Goodall (1986:343):

> Severe attack on Melissa by Evered. There was no obvious reason for this at-
> tack, which lasted for about forty seconds. Melissa was in the flat phase of
> her cycle. The infant Gremlin clung ventrally throughout.

Although Goodall does not report injuries to Melissa or Gremlin in this attack,
there were other assaults in which the infant was injured. Michaelmas, a 4-year-
old infant, had his femur or pelvis broken when his mother was attacked (see
Jurmain 1989:234). She was "cycling at the time and subjected to frequent severe attacks" (Goodall 1986:99). Although the sexual selection literature has

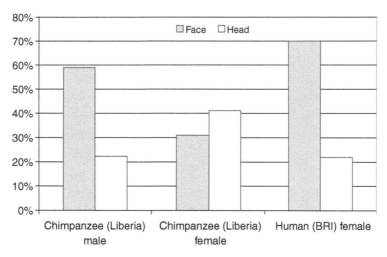

Figure 13.2 Distribution of cranial trauma in Liberian chimpanzee collection and human domestic assault victims.

emphasized the costs incurred by females during assaults, less attention has been given to the impact of secondary wounds on their offspring.

In summary, three trends were observed in the trauma of Liberian chimpanzees. First, females and males had very similar frequencies of antemortem cranial trauma. Second, juveniles and old adults had high levels of trauma, whereas adolescents and young adults had lower levels of trauma. Finally, there was a significant sex difference in the distribution of wounds on the crania. Females were significantly more likely to have lesions to the back of the head, whereas males had significantly more trauma to the facial region. These patterns make sense in terms of aggression documented in other parts of Africa and suggest that the populations from which the Liberian chimpanzees came experienced high levels of physical aggression directed by males toward females.

Variation across Populations and Species

Chimpanzee Trauma Patterns

Only a few studies have analyzed skeletal injuries in chimpanzees (Lovell 1987, 1990, 1991; Jurmain 1989, 1997; Jurmain and Kilgore 1998; Carter et al. 2008). Like the LCOC, the collections on which these studies are based are fraught with sampling problems. The collected skeletons are, for the most part, anonymous. They were acquired over years or decades by hunters and other local residents drawing from an extensive region. Using such a fragmentary record, we must be cautious in characterizing social behavior and even more circumspect in drawing regional comparisons. What follows, then, is a preliminary analysis intended to suggest directions for further research.

In the 1980s, Nancy Lovell (1987, 1990, 1991) examined a series of wild-shot apes, including orangutans, gorillas, and chimpanzees, housed at the Smithsonian Institution in Washington, D.C. (Table 13.7). Most of her sample of 40 chimpanzees belonged to the central subspecies *(P. t. troglodytes)*, although the western *(P. t. verus)* and eastern subspecies *(P. t. schweinfurthii)* were represented in small numbers. Healed trauma was seen in 20% of the crania.[4] Although females displayed more wounds than males, this difference was not statistically significant.

Jurmain (1989) evaluated pathology in a small sample of nearly complete chimpanzee skeletons *(P. t. schweinfurthii)* that researchers had collected and preserved at Gombe National Park. This study was unique in that all the remains belonged to known individuals with documented life histories. Of the 11 crania in the series, 4 exhibited signs of trauma: one old female and one old male had cranial fractures, while two young males had puncture wounds, probably from bites (Table 13.7). Jurmain concluded that "the most common cause of trauma was from interpersonal violence" (1989:229). In a follow-up study, the Gombe sample[5] was compared to 127 crania of wild-shot chimpanzees *(P. t. troglodytes)* collected in Southern Cameroons in the 1930s (Jurmain

Table 13.7   Cranial trauma in samples analyzed by Hatch (2006), Lovell (1987, 1990), Jurmain (1989, 1997), and Carter et al. (2008).

Species	Hatch		Lovell		Jurmain		Carter et al.	
	No.	%	No.	%	No.	%	No.	%
P.t. verus	141/246	57.3%	0/2	0.0%	—	—	—	—
P.t. troglodytes	—	—	5/33	15.2%	7/127	5.5%	—	—
P.t. schweinfurthii	—	—	1/2	50.0%	4/14*	28.6%	8/19	42.1%
P.t. unknown sub-species	—	—	2/3	66.7%	—	—	—	—
Total	141/246	57.3%	8/40	20.0%	11/141	7.8%	8/19	42.1%

*Includes three individuals that were included in the 1997 sample but not in the 1989 sample.

1997; Jurmain and Kilgore 1998). Fractures were found in only 5.5% of the Cameroon series. In both samples, cranial trauma was more common in males than in females, but the difference was not statistically significant.

A recent study by Carter et al. (2008) documented 20 chimpanzee skeletons *(P. t. schweinfurthii)* from Kibale National Park, Uganda. The remains were collected over a three-year period from various locations and communities. Of the 19 crania from Kibale, 8 (42.1%) had fractures or puncture wounds that had healed or partially healed (Table 13.7).[6]

The findings of these studies vary widely. Overall rates of trauma, for example, range from a low of 5.5% in Southern Cameroons to a high of 42.1% in Kibale, Uganda. By comparison, the Liberian series has an extremely high frequency of trauma—ten times higher, in fact, than what was found in the Cameroon sample. At the same time, the cases are similar in that males and females within a sample are about equally likely to exhibit cranial lesions.[7]

The location of such lesions, however, tended to vary by sex. Thus, in the Cameroon series, Jurmain and Kilgore (1998:18) found a statistically significant difference in the location of wounds on male and female skulls. Interestingly, the same sex-biased distribution of wounds was documented in their sample of lowland gorillas *(G. g. gorilla)* from Southern Cameroons. Jurmain and Kilgore (1998:21) found it "puzzling" that cranial vault injuries occurred only in females, especially given that these wounds were "largely independent of facial injuries." Yet this is not so puzzling if behavioral observations are taken into account. When attacked by a male, a female chimpanzee is likely either to flee or to huddle over her infant, leaving the back of her head exposed to her assailant.

Trauma Patterns in Human Victims of Domestic Violence

If we compare head trauma in female chimpanzees with the same kind of trauma in human victims of domestic violence, the human pattern appears to be quite distinct. Thus, in a study conducted at the Bradford Royal Infirmary (BRI) in northern England, 194 victims of domestic assault were compared to 479 women who had been injured in nonvehicular accidents (Novak 2006:242; Allen et al. 2007). A large majority (70%) of the women who had been assaulted by their domestic partners had injuries to the face, whereas only 9% of the accident cases involved similar injuries. At the same time, less than a quarter (21%) of the assault victims had injuries to the vault (defined by the same morphological criteria used for the Liberian chimpanzees). In a comparable series collected in Salt Lake City, Utah, a similar pattern was observed: 86% of the 188 domestic assault victims had injuries to the face, and only 7% of them to the vault (Novak 1999). In both of these samples, the *face* seems to have been the primary target in domestic assaults.

There is a certain amount of evidence, then, that human gender violence tends to involve face-to-face conflict. This claim is supported by other clinical research in the United Kingdom (e.g., Fonseka 1974) and the United States (e.g., McDowell et al. 1992). At least two studies have concluded that trauma to a woman's face, if not the result of an automobile accident, should be considered evidence of domestic violence (Hartzell et al. 1996; Ochs et al. 1996). How well this conclusion holds cross-culturally has yet to be examined. What seems clear, in any case, is that women are hit in the face more often than are female chimpanzees.

We hypothesize that (1) a woman is more likely than a female chimpanzee to fight with an adult male face-to-face, and (2) this is because a woman has unique abilities to defend herself and even to go on the offensive against an adult male. Almost any weapon can reduce the physical advantage that a man would ordinarily have in a fight (e.g., Geldermalsen and Stuyft 1993; Burbank 1994). Modern weapons in particular tend to equalize the capacity of men and women to inflict physical harm. Indeed, with the abundance of firearms in North America, as many husbands are killed by their wives each year as wives are killed by their husbands (Wilson and Daly 1992). Even if women tend to murder men by ambush rather than direct confrontation, the ability of a lone female to challenge an adult male has no parallel in chimpanzees.

Linguistic ability, furthermore, allows women to argue, negotiate, and plead with their assailants, as well as to elicit support from relatives and other allies. Yet the use of words can be double-edged: verbal exchanges may exacerbate rather than ameliorate domestic conflict (e.g., Abraham 2000:136–137). In fact, one study in the United States suggested that a woman who pleads or argues with an intimate assailant actually increases her likelihood of being injured (Bachman and Carmody 1994). Whether this is the case in other social contexts is unknown, but the potentially crucial role of "verbal sparring" must be considered in the study of human gender violence.

In short, weapons and words make it possible for sexual partners to engage in *mutual combat*, a pattern that is unknown in chimpanzees. The prevalence of such combat, also known as "common couple violence" (Johnson 1995; Johnson and Ferraro 2000), has been demonstrated in a number of recent studies in the sociological literature (see review in McHugh 2005). For example, Bookwala et al. (2005) surveyed more than 6000 couples in the United States to examine gender differences in conflict strategies, physical aggression, and resulting injuries. Young couples were found to be more likely than middle-aged and older ones to "hit and throw things at each other and less likely to keep things to themselves or to have calm discussions during disagreements" (Bookwala et al. 2005:803). Interestingly, by comparison with men, women were "less likely to use calm discussions and more likely to engage in arguing/shouting heatedly during disagreements." At the same time, women were consistently more likely than men to be injured and to suffer more severe wounds (see also Brush 1990; Archer 2000, 2002).

Alternative Hypotheses

Our hypothesis focuses on the unique abilities of women to counter male aggression. An alternative approach, however, would shift the focus to men and to their distinctive tactics of aggression and manipulation. This kind of approach supplies us with at least three additional hypotheses:

1. Compared to male chimpanzees, men have a greater tendency to strike opponents of *either* sex in the face.
2. Men are especially prone to strike their sexual partners in the face to mar their appearance and thus stigmatize or shame them.
3. Men are able to confine females "structurally," thus preventing them from fleeing and making it more likely they will be struck in the face.

These hypotheses do not necessarily conflict with our own. All of them are plausible, and only empirical research can assess their validity. Here we briefly discuss each hypothesis just to bring out some of the issues to be addressed.

First, there is the possibility that men are more likely than male chimpanzees to attack any opponent's face. According to Walker (1997), human assailants in general tend to direct their blows toward the head and neck (see also Shepherd et al. 1990; Hussain et al. 1994; Brink et al. 1998). This selective targeting has two clear advantages (Walker 1997:160):

> From a strategic standpoint, the head and especially the face are attractive targets because injuries of this area can be very painful. Well placed blows to the head are also likely to produce bleeding and conspicuous bruises that serve as a highly visible symbol of the aggressor's social dominance.

Yet the point of Walker's argument is that boxing-style blows to the face are not typical of human combat throughout history, but seem to have become customary in certain times and places, perhaps under the influence of prize-fighting and other violent sports (1997:170–172; see also Brickley and Smith 2006). If Walker is correct, there may be a higher incidence of facial injuries in Western industrial societies than in other human groups. Because nearly all the available clinical data on domestic-violence injuries come from Western industrial settings, we must be cautious in assuming that these data are representative of humans in general. Any comparison with chimpanzees, furthermore, would have to be qualified accordingly.

A second hypothesis is that women are especially likely to be hit in the face because this physically marks and shames them. As Walker (1997:161) points out, batterers may be trying "to stigmatize their wives with highly visible, symbolically salient signs of their physical dominance." Moreover, facial wounds may serve to limit a woman's social and physical mobility by rendering her unattractive to other men and generally hampering her public performance. In this perspective, attacking the face is one manifestation of what Johnson and Ferraro (2000) call "intimate terrorism" (see also Johnson 1995; Dutton and Goodman 2005). Such terrorism, though not the only form of human gender violence, is a distinctly human pattern (Rodseth and Novak, Chapter 12 in this volume). Chimpanzees do not engage in it, if only because their sexual consortships are transitory and do not allow for a chronic pattern of private abuse.

This leads directly to the third hypothesis—that a woman is likely to be assaulted face-to-face because she is often *confined* and cannot simply flee her sexual partner as a female chimpanzee may do. In most societies, the confinement of women is "structural" in the broadest sense: it involves both physical structures (walls, houses) and social structures (laws, institutions). Of course, not all humans live within walls, and foraging women in particular are usually free of the physical constraints that are typically imposed on their sedentary counterparts (Rodseth and Novak, Chapter 12 in this volume). Even in the absence of permanent architecture, however, women are extremely unlikely to live independently, but must belong to a "household" (Rodseth and Novak 2000). This makes their situation radically different from that of female chimpanzees (and rather more like that of female gorillas). Unless women have independent access to strategic resources such as land, cattle, or money, they are likely to be beholden to their male aggressors. Thus prevented from fleeing, women tend to be "cornered" in a social as well as a physical sense.

Conclusion

Skeletal analysis is just one tool in the reconstruction of social behavior. Without a wide range of tools, we will be hard-pressed to understand what has happened and why. In a few cases, the remains of known chimpanzees from well-studied communities have been collected and analyzed, allowing behavioral observations to be integrated with skeletal findings (Jurmain 1989, 1997). Yet most museum collections are samples of anonymous individuals. The specimens come from scattered populations that have never been systematically observed or recorded. This is certainly the case for the Liberian series analyzed by Hatch (2006).

Despite these limitations, the LCOC sheds considerable light on at least one area of social life. Although teasing behavior out of bone is never easy, serious violence is likely to leave rather obvious physical traces. The incidence of healed trauma in the Liberian crania suggests an extremely high level of violence. Almost 60% of the 246 individuals in the sample had at least one healed lesion. This presents a stark contrast to the skeletal evidence from Central and East Africa, where violent conflict seems to have been much less frequent, less intense, or both. To account for these regional differences, we would have to examine the ecology, demography, and history of each of the skeletal populations. In northern Liberia, at least, the situation is clear enough. By the 1940s, trees were being extracted on a massive scale, and almost no forest remains in the Nimba region today (Sayer et al. 1992). Habitat destruction probably intensified competition among all the remaining chimpanzees, and human predation would have disrupted the hierarchies of specific chimpanzee communities.

Despite the regional differences in levels of violence, the distribution of wounds remains surprisingly similar across Africa. In the LCOC, as in comparable samples, males and females were about equally likely to have been injured by assault. At the same time, adult male chimpanzees consistently had more wounds to the facial region, whereas adult females were more likely to have injuries to the back of the head. This suggests that males are more likely to face their assailants, while females are more likely to turn away from them. Given her disadvantage in size and strength, a female chimpanzee has little to gain from a direct confrontation with an adult male and may have much to lose, especially if she is carrying an infant.

Women would seem to be in much the same situation. Sexual dimorphism in body mass, though greatly reduced in humans, still makes it difficult for most women to compete with men at the level of brute strength. In addition, the period of infant and juvenile dependency is greatly extended in humans, and this certainly constrains young mothers in their dealings with abusive partners. Yet women, unlike female chimpanzees, do not simply flee their assailants or huddle over their young. Indeed, in their injury patterns, victims of domestic violence (at least in a Western context) resemble *male* rather than female chimpanzees.

One interpretation of these patterns is that women, by comparison with female chimpanzees, perceive adult males as basically equal (though obviously dangerous) opponents. At the same time, physical and cultural barriers may prevent a woman from fleeing an abusive partner. Paradoxically, the walls that keep a woman secluded and vulnerable may also benefit her and her offspring by providing a refuge from external dangers. No such refuge is available to female chimpanzees, who must use their own bodies to shelter their young.

Acknowledgments

We would like to thank Martin Muller and Richard W. Wrangham for the opportunity to participate in the International Primatological Association conference, held June 25–30, 2006 in Entebbe, Uganda. We also thank Michele Morgan for coordinating access to the skeletal collection at the Peabody Museum of Natural History, Harvard University; the Bradford Royal Infirmary, Bradford, UK, and the Salt Lake City Police Department for facilitating the domestic violence research. A special thanks to Lars Rodseth for commenting on and contributing to this manuscript. Research was supported in part by a Marriner S. Eccles research grant (SN) and an Idaho State University GSRSC research grant F05–107 (MH).

Notes

1. Although museum records indicated sex for pre-reproductive juveniles, the records do not indicate how these assessments were made. Because secondary sex characteristics had not developed in the bony tissue, these crania were classified as indeterminate for sex.
2. Our adolescent age category corresponds to the chimpanzee skeletal standards used by Lovell (1987, 1990, 1991). In the behavioral literature, however, primatologists use slightly different criteria for age-group classification. Goodall's (1986:81) categories of "infancy" and "childhood," for example, would

be compressed into our "juvenile" category, whereas her stage of "early adolescence" closely corresponds to our category. Goodall also considers "old age" to be from about age 33 to death, while we rely on skeletal and dental markers that give us a cutoff between "young" and "old adults" at 20 years (Krogman 1930, 1969).

3. These zones were delineated by a coronal plane that ran from porion to porion—a craniometric landmark positioned at the superior margin of the external auditory meatus or auditory canal.

4. The frequencies of trauma reported here for Lovell (1987) have been recalculated to be comparable to frequencies presented in this study. Percentages of trauma as reported by Lovell (1990, 1991) may differ.

5. Jurmain increases his Gombe sample size from 11 to 14 in this study, although no additional trauma is reported. As a result, he calculates trauma to be found in 28.6% of the Gombe crania (Jurmain 1997:7).

6. It is unclear how Carter et al. (2008) calculated the trauma counts. Punctures and fractures are summarized separately (2008:396, Table 6). Based on the descriptions in the appendix (2008:398–402), cranial trauma was present in 9 individuals (47.4%). Moreover, if injuries to the mandible are considered, a total of 10 individuals (52.6%) displayed antemortem trauma in the skull. This frequency is significantly higher than what is reported by Lovell (1987), Jurmain (1989, 1997), and Jurmain and Kilgore (1998), and more like that seen in the Liberian series.

7. Based on the specimen descriptions in the appendix (Carter et al. 2008:398–402), we calculate that four of the adult male crania (25%) and all six of the adult female crania had antemortem trauma. However, this study was not included in our comparisons because some of the cranial lesions (especially those on the parietals) were not described in enough detail to determine whether the lesion occurred in the "face zone" or the "vault zone."

References

Abraham, M. *Speaking the Unspeakable: Marital Violence among South Asian Immigrants in the United States.* New Brunswick, N.J.: Rutgers University Press, 2000.

Allen, T., S. A. Novak, and L. L. Bench. "Pattern of Injuries: Accident or Abuse." *Violence against Women* 13 (2007): 802–816.

Archer, J. "Sex Differences in Aggression between Heterosexual Partners: A Meta-Analytic Review." *Psychological Bulletin* 126 (2000): 651–680.

———. "Sex Differences in Physically Aggressive Acts between Heterosexual Partners: A Meta-Analytic Review." *Aggression and Violent Behavior* 7 (2002): 313–351.

Bachman, R., and D. C. Carmody. "Fighting Fire with Fire: The Effects of Victim Resistance in Intimate Versus Stranger Perpetrated Assaults against Females." *Journal of Family Violence* 9 (1994): 317–331.

Boesch, C., and H. Boesch-Achermann. *The Chimpanzees of the Taï Forest: Behavioural Ecology and Evolution.* Oxford: Oxford University Press, 2000.

Bookwala, J., J. Sobin, and B. Zdaniuk. "Gender and Aggression in Martial Relationships: A Life-span Perspective." *Sex Roles* 52 (2005): 797–807.

Brickley, M., and M. Smith. "Culturally Determined Patterns of Violence: Biological Anthropological Investigation at a Historic Urban Cemetery." *American Anthropologist* 108 (2006): 163–177.

Brink, O., A. Vesterby, and J. Jensen. "Pattern of Injuries due to Interpersonal Violence." *Injury* 29 (1998): 705–709.

Brush, L. D. "Violent Acts and Injurious Outcomes in Married Couples: Methodological Issues in the National Survey of Families and Households." *Gender and Society* 4 (1990): 56–67.

Buikstra, J. "Healed Fractures in *Macaca mulatta:* Age, Sex, and Symmetry." *Folia Primatologia* 23 (1975): 140–148.

Burbank, V. K. *Fighting Women.* Berkeley: University of California Press, 1994.

D. A. Counts, J. K. Brown, and J. C. Campbell, eds. *To Have and to Hit: Cultural Perspectives on Wife Beating.* Urbana: University of Illinois Press, 1999.

Carter, M. L., H. Pontzer, R. W. Wrangham, and J. K. Peterhans. "Skeletal Pathology in *Pan troglodytes schweinfurthii* in Kibale National Park, Uganda." *American Journal of Physical Anthropology* 135 (2008): 389–403.

Crowell, N. A., and A. W. Burgess, eds. *Understanding Violence against Women.* Washington, D.C.: National Academy Press, 1996.

Dutton, M. A., and L. A. Goodman. "Coercion in Intimate Partner Violence: Toward a New Conceptualization." *Sex Roles* 52 (2005): 743–756.

Fonseka, S. "A Study of Wife-beating in the Camberwell Area." *British Journal of Clinical Practice* 28 (1974): 400–402.

Galloway, A. *Broken Bones: Anthropological Analysis of Blunt Force Trauma.* Springfield, Ill.: Charles C. Thomas, 1999.

Geldermalsen, A. A. V., and P. V. D. Stuyft. "Interpersonal Violence: Patterns in a Baeotho Community." *Journal of Tropical Medicine and Hygiene* 96 (1993): 93–99.

Goodall, J. *The Chimpanzees of Gombe: Patterns of Behavior.* Cambridge, Mass.: Belknap Press, 1986.

Graham-Kevan, N., and J. Archer. "Physical Aggression and Control in Heterosexual Relationships: The Effect of Sampling." *Violence and Victims* 18 (2003): 181–196.

Harley, G. W. "Roads and Trails in Liberia." *Geographical Review* 29 (1939): 447–460.

Harley, G. W. to E. A. Hooten, 1943 letter. Peabody Museum, Harvard University Archives.

Hartzell, K. N., A. A. Botek, and S. H. Goldberg. "Orbital Fractures in Women due to Sexual Assault and Domestic Violence." *Ophthalmology* 103 (1996): 953–957.

Hatch, M. A. "Skeletal Trauma Patterns in Liberian Chimpanzees *(Pan troglodytes verus)*." Unpublished MSc thesis, Idaho State University, 2006.

Humle T., and T. Matsuzawa. "Oil Palm Use by Adjacent Communities of Chimpanzees at Bossou and Nimba Mountains, West Africa." *International Journal of Primatology* 25 (2004): 551–581.

Hussain, K., D. B. Wijetunge, S. Grubnic, and I. T. Jackson. "A Comprehensive Analysis of Craniofacial Trauma." *Journal of Trauma* 36 (1994): 34–47.

Johnson, M. P. "Patriarchal Terrorism and Common Couple Violence: Two Forms of Violence against Women." *Journal of Marriage and the Family* 57 (1995): 283–294.

Johnson, M. P., and K. J. Ferraro. "Research on Domestic Violence in the 1990s: Making Distinctions." *Journal of Marriage and the Family* 57 (2000): 283–294.

Judd, M. A. "Continuity of Interpersonal Violence between Nubian Communities." *American Journal of Physical Anthropology* 131 (2006): 324–333.

Jurmain, R. "Trauma, Degenerative Disease, and Other Pathologies among the Gombe Chimpanzees." *American Journal of Physical Anthropology* 81 (1989): 333–342.

———. "Skeletal Evidence of Trauma in African Apes, with Special Reference to the Gombe Chimpanzees." *Primates* 38 (1997): 1–14.

Jurmain, R., and L. Kilgore. "Sex-related Patterns of Trauma in Humans and African Apes." In *Sex and Gender in Paleopathological Perspective,* eds. A. L. Grauer and P. Stuart-Macadam, pp. 11–26. Cambridge: Cambridge University Press, 1998.

Knight, B. *Forensic Pathology.* London: Edward Arnold, 1991.

Krogman, W. M. "Ectocranial and Endocranial Suture Closure in Anthropoids and Old World Apes." *American Journal of Anatomy* 46 (1930): 315–353.

———. "Growth Changes in the Skull, Face, Jaw, and Teeth of the Chimpanzee." In *The Chimpanzee, Vol. 1, Anatomy, Behavior, and Diseases of Chimpanzees,* ed. G. H. Bourne, pp. 104–164. Basel, Switzerland: Karger, 1969.

Lovell, N. C. "Skeletal Pathology of Wild-Shot Pongids: Implications for Human Evolution." Ph.D. dissertation, Cornell University, Ithaca, N.Y., 1987.

———. *Patterns of Injury and Illness in Great Apes.* Washington, D.C.: Smithsonian Institution Press, 1990.

———. "Patterns of Injury and Illness in Nonhuman Primates." *Yearbook of Physical Anthropology* 34 (1991): 117–155.

———. "Trauma Analysis in Paleopathology." *Yearbook of Physical Anthropology* 40 (1997): 139–170.

Martin, D. L. "Violence against Women in the La Plata River Valley (AD 100–1300)." In *Violent Times: Violence and Warfare in the Past,* eds. D. L. Martin and D. W. Frayer, pp. 45–75. Toronto: Gordon and Breach, 1997.

Matsuzawa, T., and G. Yamakoshi. "Comparison of Chimpanzee Material Culture between Bossou and Nimba, West Africa." In *Reaching into Thought: The Mind of the Great Apes,* eds. A. R. Russon, K. A. Bard, and S. T. Parker, pp. 211–232. Cambridge: Cambridge University Press, 1996.

McClusky, L. J. *"Here, Our Culture Is Hard": Stories of Domestic Violence from a Mayan Community in Belize.* Austin: University of Texas Press, 2001.

McDowell, J. D., D. K. Kassebaum, and S. E. Stromboe. "Recognizing and Reporting

Victims of Domestic Violence." *Journal of the American Dental Association* 123 (1992): 44–50.

McHugh, M. C. "Understanding Gender and Intimate Partner Abuse." *Sex Roles* 52 (2005): 717–724.

Muller, M. N. "Agonistic Relations among Kanyawara Chimpanzees." In *Behavioural Diversity in Chimpanzees and Bonobos,* eds. C. Boesch, G. Hohmann, and L. F. Marchant, pp. 112–124. Cambridge: Cambridge University Press, 2002.

Nishida, T. "Social Interactions between Resident and Immigrant Female Chimpanzees." In *Understanding Chimpanzees,* eds. P. G. Heltne and L. A. Marquardt, pp. 68–89. Cambridge: Cambridge University Press, 1989.

Nishida T., N. Corp, M. Hamai, T. Hasegawa, M. H. Hasegawa, K. Hosaka, K. D. Hunt, N. Itoh, K. Kawanaka, A. M. Oda, J. C. Mitani, M. Nakamura, K. Norikoshi, T. Sakamaki, L. Turner, S. Uehara, and K. Zamma. "Demography, Female Life History, and Reproductive Profiles among the Chimpanzees of Mahale." *American Journal of Primatology* 59 (2003): 99–121.

Novak, S. A. "Skeletal Manifestations of Domestic Assault: A Predictive Model for Investigating Gender Violence in Prehistory." Ph.D. dissertation, Department of Anthropology, University of Utah, Salt Lake City, 1999.

———. "Beneath the Façade: A Skeletal Model of Domestic Violence." In *Funerary Archaeology of Human Remains,* eds. R. Gowland and C. Knüsel, pp. 238–252. Oxford: Oxbow Press, 2006.

Ochs, H. A., M. C. Neuenschwander, and T. B. Dodson. "Are Head, Neck and Facial Injuries Markers of Domestic Violence?" *Journal of the American Dental Association* 127 (1996): 757–761.

Ohashi, G. "Bossou Chimpanzees Cross the National Border of Guinea into Liberia." *Pan Africa News* 13 (2006): 10–12.

Owsley, Douglas W., and Richard L. Jantz, eds. *Skeletal Biology in the Great Plains: Migration, Warfare, Health, and Subsistence.* Washington, D.C.: Smithsonian Institution Press, 1994.

Peters, W. "The Mosquitos of Liberia (Dipter: Culicidae), A General Survey." *Bulletin of Entymological Research* 47 (1956): 525–551.

Pusey, A. E., J. Williams, and J. Goodall. "The Influence of Dominance Rank on the Reproductive Success of Female Chimpanzees." *Science* 277 (1997): 828–830.

Randall, F. D. "The Skeletal and Dental Development and Variability of the Gorilla." *Human Biology* 16 (1944): 23–76.

Roberts C., and K. Manchester. *The Archaeology of Disease.* Ithaca, N.Y.: Cornell University Press, 2005

Rodseth, L., and S. A. Novak. "The Social Modes of Men: Toward an Ecological Model of Human Male Relationships." *Human Nature* 11 (2000): 335–366.

Rodseth, L., R. W. Wrangham, A. M. Harrigan, and B. B. Smuts. "The Human Community as a Primate Society." *Current Anthropology* 32 (1991): 221–254.

Sayer, J. A., C. S. Harcourt, and N. M. Collins, eds. *The Conservation Atlas of Tropical Forest—Africa.* London: Macmillan Publishing, 1992.

Schultz, A. H. "The Skeleton of the Chimpanzee." In *The Chimpanzee, Vol. 1, Anatomy, Behavior, and Diseases of Chimpanzees,* ed. G. H. Bourne, pp. 50–103. Basel, Switzerland: Karger, 1969.

Schwarz, G. T., and C. Dean. "Ontogeny of Canine Dimorphism in Extant Hominids." *American Journal of Physical Anthropology* 115 (2001): 269–283.

Shepherd, J. P., M. Shapland, N. X. Pearce, and C. Scully. "Pattern, Severity and Aetiology of Injuries in Victims of Assault." *Journal of the Royal Society of Medicine* 83 (1990): 75–78.

Shermis, S. "Domestic Violence in Two Skeletal Populations." *Ossa* 9–11 (1982): 143–151.

Smith, M. O. "Beyond the Palisades: The Nature and Frequency of Late Prehistoric Deliberate Violent Trauma in the Chickamauga Reservoir of East Tennessee." *American Journal of Physical Anthropology* 121 (2003): 303–318.

Smuts, B. B., and R. W. Smuts. "Male Aggression and Sexual Coercion of Females in Nonhuman Primates and Other Mammals: Evidence and Theoretical Implications." *Advances in the Study of Behavior* 22 (1993): 1–63.

Spedding, R. L., M. McWilliams, B. P. Nichol, and C. Deardon. "Markers for Domestic Violence in Women." *Journal of Accidental and Emergency Medicine* 16 (1999): 400–402.

Sugiyama, Y. "Demographic Parameters and Life History of Chimpanzees at Bossou, Guinea." *American Journal of Physical Anthropology* 124 (2004): 154–165.

Sugiyama, Y., and J. Koman. "Social Structure and Dynamics of Wild Chimpanzees at Bossou, Guinea." *Primates* 20 (1979a): 323–339.

———. "Tool-using and Mating Behavior in Wild Chimpanzees at Bossou, Guinea." *Primates* 20 (1979b): 513–524.

Walker, P. L. "Cranial Injuries as Evidence of Violence in Prehistoric Southern California." *American Journal of Physical Anthropology* 80 (1989): 313–323.

———. "Wife Beating, Boxing, and Broken Noses: Skeletal Evidence for the Cultural Patterning of Violence." In *Violent Times: Violence and Warfare in the Past,* eds. D. Martin and D. W. Frayer, pp. 145–179. Toronto: Gordon and Breach, 1997.

———. "A Bioarchaeological Perspective on the History of Violence." *Annual Review of Anthropology* 30 (2001): 573–596.

Whitehead, P., W. K. Sacco, and S. B. Hochgraf. *A Photographic Atlas for Physical Anthropology.* Englewood, Colo.: Morton Publishing, 2004.

Wilson, M., and M. Daly. "The Man Who Mistook His Wife for a Chattel." In *The Adapted Mind: Evolutionary Psychology and the Generation of Culture,* eds. J. H. Barkow, L. Cosmides, and J. Tooby, pp. 289–322. New York: Oxford University Press, 1992.

Wrangham, R. W., M. L. Wilson, and M. N. Muller. "Comparative Rates of Violence in Chimpanzees and Humans." *Primates* 47 (2006): 14–26.

Zihlman, A., D. Bolter, and C. Boesch. "Wild Chimpanzee Dentition and Its Implications for Assessing Life History in Immature Hominin Fossils." *Proceedings of the National Academy of Sciences* 101 (2004): 10541–10543.

14

Human Rape: Revising Evolutionary Perspectives

Melissa Emery Thompson

The role of biology in understanding human rape has been the subject of heated and not always scientific debate. Our study of pervasive sexual aggression in the human species can and should be informed by our less emotion-laden analyses of sexual dynamics in closely related species, and warrants objective analysis of scientific predictions within the context of our complex sociocultural environments. In this chapter, I explore hypotheses proposing that some types of human rape function as sexual coercion. Rape by any definition consists of verbal or physical coercion to engage in sexual activity. However, in the context of this book, and from an evolutionary perspective, sexual coercion is more restrictively viewed as a component of sexual selection. As such, rape is suggested to be a male reproductive strategy—part of a continuum of behaviors, comprising harassment and intimidation, in addition to forced copulation (Clutton-Brock and Parker 1995). Smuts and Smuts (1993) defined sexual coercion as "male use of force, or its threat, to increase the chances that a female will mate with the aggressor or to decrease the chances that she will mate with a rival, at some cost to the female."

There are significant obstacles to a scientific analysis of rape, including discrepancies in subjective assessment and reporting of rapes by victims. Empirically, however, a larger issue concerns categorization of a range of coercive sexual behaviors into the single behavioral category of rape. From a legal and social perspective, this practice is understandable. In terms of evaluating functional hypotheses, however, the available data suggest that at least two distinct types of rape should be recognized, depending on the preexisting social relationship between victim and attacker. As I will detail in this review, acquaintance rapes account for the majority of attacks but are less likely to involve physical force or injury to the victim and less likely to be reported to the police than are stranger rapes. And compared to men who rape acquaintances and intimate partners, men who rape strangers are more likely to exhibit symptoms of pathology and are overrepresented in prison populations. Therefore, although both types of rape merit earnest examination, they also are likely to fit separate explanatory models. Given its astonishing frequency and the lack of clear pathological correlates, acquaintance rape in particular merits evolutionary consideration within the spectrum of sexually coercive behaviors exhibited by other species.

Hypotheses about Rape

The simplest functional explanation for the occurrence of rape is that it is an adaptation to gain immediate opportunities to fertilize a fecund female (Thornhill and Palmer 2000). Randy Thornhill, the leading proponent of this hypothesis, uses scorpion flies as a model organism for understanding forced copulation. Male scorpion flies possess a "rape-specific adaptation," an appendage for which the only apparent purpose is to clamp onto a reluctant mate's forewings to prevent her from escaping during copulation. In an analogy to scorpion flies, Thornhill proposes that human males may possess a series of psychological rape-specific adaptations, among them the ability to assess the vulnerability and fecundity of potential victims and to modulate somatic and psychological arousal in a different manner than in consensual sex. Just as many scorpion flies can secure consensual matings without the use of their so-called rape appendage by presenting females with a "nuptial gift" of a dead insect, Thornhill argues that human males need only resort to coercive sex when they lack the opportunity, skill, or status to obtain copulation opportunities through normal means. The popular presentation of these hypothesis in *A Natural History of Rape* (Thornhill and Palmer 2000) received a considerable amount of attention—mostly negative. But as unsavory as Thornhill's idea may be to some feminist and cultural critics, it is legitimate as a *hypothesis;* we must at least consider whether rape may have evolved to directly enhance reproductive success through immediate access to fertilizations. Clearly, a rapist has a nonzero chance of reproducing as a result of his action, though this does not in itself establish rape as an adaptation. Apart from any other consideration, it must be asked whether he might have increased his fitness more by enacting a different strategy. As I will explore later in this chapter, there are a number of sound scientific reasons to reject Thornhill's hypothesis, though not to reject outright a biological perspective on the origin of rape.

A Natural History of Rape also considers an alternate evolutionary hypothesis for rape, that favored by Thornhill's co-author Craig Palmer. As originally proposed by evolutionary psychologist Donald Symons (1979), rape may result as a by-product of other psychological adaptations that increase male success at noncoercive reproduction. This hypothesis proposes that rape is an extreme part of a continuum of heterosexual behaviors and that differences in male and female reproductive capacity generate a conflict of interest with regard to sexual behavior. Whereas females' ability to reproduce is inherently limited and necessitates a large degree of parental investment, males' potential to inseminate females is theoretically limited only by access to new mates. Thus, it is argued that, compared to females, males have evolved increased susceptibility to visual arousal, higher sex drive, greater desire for sexual variety, and more willingness to engage in casual sex, together with a decreased ability to inhibit sexual impulses and lower discrimination in the choice of mates. Under the appropriate conditions, this combination of sexual proclivities may lead to coercion of a female whose adaptive psychological mechanisms favor more inhibited sexuality. Similarly, male psychological predispositions may lead to misinterpretations of sexual motivations between the sexes (Adams-Curtis and Forbes 2004). Evolutionary psychologists have garnered considerable empirical support for systematic differences in the sexual tendencies of men and women, though the magnitude of these differences is generally quite small (Buss 2004; Yost and Zurbriggen 2006). Moreover, this hypothesis of misapplied male sexual urges has weak explanatory potential in that evidence proposed in its favor will inevitably fit equally well with exclusively sociocultural explanations for rape.

A third evolutionary perspective has received relatively little attention. Smuts and Smuts (1993) reviewed a broad range of male aggression against females in primates and noted that, although some aggression clearly facilitates immediate mating opportunities, other forms of aggression serve instead to increase the male's probability of reproducing with that female in the future and/or reduce the probability that the female will seek matings from other males. As a pervasive form of male-female aggression, many instances of human rape may fit into this longer-term reproductive strategy. In contrast to the previously discussed hypotheses, the Smuts's proposal considers evolutionary conflicts of interest between males and females in light of the common pattern of long-term social or sexual relationships. As the present book reveals, male-female aggression is widespread among primates. However disagreeable this behavior may be, it appears to be a sexually selected male reproductive strategy in many nonhumans. Humans differ, however, from most primates in that male-female sexual relationships are typically long-term and biparental care is critical to infant survival. Coercive tactics, including but not limited to rape, may be a means for human males to secure, or protect, long-term sexual access to females (Wrangham and Peterson 1996; Hilton et al. 2000).

Smuts's relationship coercion hypothesis generates at least three predictions that distinguish it from Thornhill's hypothesis of a "rape-specific adaptation": (1) men who rape should not necessarily be low status or sexually unsuccessful; (2) a single act of forced sex does not necessarily lead to a high probability of conception, nor does the rapist necessarily expect it to, but may be more commonly associated with ongoing or anticipated male-female intimate relationships; and (3) forced sex should occur in the context of other coercive behaviors, such as verbal coercion, mate guarding or control of female sexual and social freedom, and nonsexual aggression.

A disturbing aspect of the relationship coercion hypothesis is that it implies a certain degree of female weakness and lack of control over their own sexuality, as well as a tendency to acquiesce to or be duped by male coercive tactics. On the one hand, this hypothesis does assume a generally weak bargaining position for women in one-on-one interactions. As in most other primate species, human males exhibit superior body size and strength to females. Furthermore, as even opponents of the biological view agree, this differential tends to be supported by male-male alliances and other sociocultural practices that reinforce male economic and political control over females. On the other hand, Smuts (1992; Smuts and Smuts 1993) argues that female vulnerability to male coercion varies with social standing, and such an evolutionary pressure has led to a variety of female counterstrategies that vary by context in their success. Female susceptibility to coercion is thus reduced when (a) female alliances are strong; (b) females have support from natal kin; (c) male alliances are relatively weak; (d) male status variance is minimal; and (e) female control of resources is enhanced (Smuts 1992). When contrasted to many of our primate relatives, human societies in general have characteristics that may support increased coercion by males (Wrangham 1980). Across and within cultures, rape incidence is increased by the presence of fraternal interest groups and reduced when women make relatively greater subsistence contributions (Otterbein 1979; Schlegel and Barry 1986; Lalumière et al. 2005). However, these conditions are hardly immutable or constant across cultures.

Many feminist and sociological scholars have argued broadly against an evolutionary approach to understanding rape, proposing instead that rape has its roots in psychological or cultural ills. A common thread is the assessment of rape behavior as pathological and/or a function of sexual frustration coupled with reduced psychosocial function (Figueredo 1992). Although various types of sexual deviance may be outcomes of such abnormal psychology, the available evidence quite clearly indicates that rape is more common than this hypothesis would suggest and, furthermore, that the psychological function of perpetrators of acquaintance rape differs in important ways from that of other criminals and sex offenders. Other arguments, too numerous and varied than I can give justice to in this forum, suggest that rape at its root is not a sexual behavior but part of a culturally mediated pattern of male patriarchy and subjugation of women. This hypothesis makes many of the same predictions as Smuts's coercion hypothesis about how rape behavior might vary cross-culturally, but proposes that rape is exclusively a product of culture, bearing at most superficial resemblance to the coercive reproductive strategies utilized by nonhuman animals. Other theorists argue by extension that rape may function as a homosocial act—that is, a means of male injury to other males' "property," such as in the context of war (Kimmel 2003).

The influence of feminist scholar Susan Brownmiller (1975) has been particularly strong in developing the argument that rape is ultimately a crime of violence and power rather than one of sexual desire. Although Brownmiller has been held up as a straw-woman by proponents of the evolutionary hypotheses, her writing reveals neither ignorance nor denial of biological influences on human sexuality (see also Lloyd 2003). In fact, the first few pages of her book *Against Our Will* discuss the sexual behavior of the primates and hypothesizes that concealed ovulation, male size dominance, and release of sexual behavior from hormonal control may have opened the door for males' potential to rape (though she assigns this to an "accident of biology" rather than to sexual selection). Brownmiller's and most feminist perspectives pose not that rape has no relation to sex, but that sex is secondary to violence and degradation as motivators for rape. Perhaps many readers would find compromise between the viewpoints of Smuts and Brownmiller, citing them as ultimate and proximate explanations, respectively, for the same behavior. From the female victims' perspective—as well as from the legal one—rape is undoubtedly a humiliating and emotionally, and often physically, violent experience. However, the distinction of evolutionary and sociocultural hypotheses is whether the motivations—whether conscious or unconscious—of male rapists are primarily reproductive (i.e., to increase sexual access to a female) or primarily dominating. In the case of the latter, it is then appropriate to ask what purposes (other than reproductive access) such demonstrations of power over a female might serve in a species that is endowed with the anatomical and social propensity for male dominance, and further why this dominance display tends to be expressed in sexual activity.

Testing hypotheses about the biological significance of rape is a complex and thorny prospect. As I will discuss, empirical study is hindered by highly variable legal and cultural definitions of rape, as well as the vast underreporting of sexual crimes. The majority of published reports on sexual assault concern proximate factors that may influence probability of raping or of being victimized. These types of studies are invaluable to public health and education concerns, but provide limited means of testing evolutionary hypotheses. Even theoretical debates are hindered by incomplete understanding of or misuse of evolutionary principles by both proponents and critics of the biological perspective. Acceptance of the strengths and limitations of evolutionary theory is essential to informed opinions about this debate and, more importantly, to structuring studies that may more adequately test evolutionary ideas.

Defending the Evolutionary Perspective

Evolutionary perspectives on human behaviors, particularly negative ones, often receive reactionary criticism of their perceived psychological and moral implications rather than their scientific merit. This frequently results from misunderstandings or misapplications of biological ideas, both by those who disdain an evolutionary approach, and, unfortunately, by some practitioners of the approach. Any discussion of the possible evolutionary underpinnings of rape should therefore begin with careful consideration of pitfalls common in evolutionary logic.

Naturalistic Fallacy

The naturalistic fallacy is the idea that biological explanations can help us to judge what is right or wrong—for example, that a Darwinian explanation for rape would soften moral judgments or undermine social or legal sanctions. Fear of this fallacy lies at the root of many responses to the evolutionary biology of human behavior. There is little rational basis for such fear, however. Evolution is recognized by biologists to occur without moral bias. Animals exhibit all manner of atrocities in the outcome of natural selection from the killing of infants and siblings to the consuming of a mother by her newborn young for nutritional advantage. Biologists who explain why the propensity for such behaviors has evolved see no moral lessons emerging from their analyses. Nor do the theorists behind the evolutionary hypotheses for rape propose that a biological explanation for the origin of rape reduces its weight as a crime. Admittedly, biological explanations might be invoked as part of the legal defense of an accused rapist, but evolutionary biology would argue strongly against such a stance. Fear of the naturalistic fallacy therefore does not justify a refusal to examine potential evolutionary influences on human behavior.

Genetic Determinism

Evolutionary explanations are often misinterpreted as ascribing biological influence on a behavior to genetically determined inevitability, or even an excuse, for that behavior. The nature/nurture dichotomy is no longer tenable at this stage of our understanding of genetics (Alexander 1979). Nonetheless, theorists on both sides of the debate often slip into arguments that dichotomize the influence of biology versus culture on our behavior. Evolutionarily derived traits are inherently shaped by the environment of the organism, and from the very anatomy of the organism to its behavior, the environment is critical in shaping each trait during an individual's life.

When we first learn about genetics, we often gloss over the gene–environment interaction and learn about simple systems such as how one gene controls whether we can roll our tongues or whether or not our ear lobes attach. We are quite satisfied with the genetic code as a blueprint for making us look like humans, with a pelvis shaped in a certain way and a brain that is very distinct from that of a chimpanzee. Similarly, the way we begin to understand the evolutionary process involves such simple models: organism A has a gene that makes him better adapted to the current environment than organism B, reproduces more successfully, and passes that gene to his offspring.

But realistic models are more complex. An overwhelming proportion of our genes act not to create a particular trait or behavior, but to manipulate the action of other genes. These genetic regulatory elements are direct conduits for the environment to influence the genome (Carroll 2005). In a similar vein, although a gene may always be responsible for producing a particular protein product, the amount of product produced, or the tissue in which it is produced, can lead to very different outcomes. In evolutionary terms, the ability to respond in different ways to different contexts is critical. Behavior is one large and very effective domain of variable response.

How genes influence behavior is not well understood. Nevertheless, psychological studies reveal suites of intricately correlated behavioral traits, suggesting common neurological origin. Imaging studies reveal how the brains of different people respond to different stimuli, responses that involve varying neurotransmitter levels, hormonal receptor densities, sensitivity to peripheral stimulation, and pathways of neural response, each of the elements that are directly influenced by gene expression.

Thus, it is very unlikely that there are many cases where a gene (or even a group of genes) controls a behavior as complex as rape. However, this does not negate the ability of selective forces to shape behavioral patterns just as they shape anatomical features. For in a socially complex species, variation in behavioral strategies may be a stronger determinant of reproductive success than variation in morphology.

Levels of Biological Explanation

Just as adaptations cannot be fully understood without the ecological context in which they evolved or the cultural context in which they are currently expressed, adaptive explanations are but one way to explain the existence of a behavior (or other trait). Tinbergen (1963) proposed that any behavior should be examined with four interrelated mechanisms in mind: (1) proximate causation: the stimuli that resulted in the behavioral response and the processes involved in its production; (2) developmental/ontogenetic influences: how production of the behavior has been influenced by environmental factors, including learning, during the individual's lifetime; (3) phylogenetic influences: how the behavior compares to that of closely related species; and (4) adaptive and nonadaptive explanations: whether and how the behavior may have been advantageous (i.e., increased survival and/or reproduction) during the animal's evolutionary history, with the caveat that some behaviors may be of recent origin or result from ultimate causes other than adaptation. This does not negate a biological understanding of these behaviors. Similarly, cultural processes are one alternative to an adaptive explanation, though this is not the only means of considering culture; social influences can be an important influence on both proximate and developmental aspects of behavior.

Behavioral Optimality

Many critics of evolutionary theories for human behavior argue that the behavior is not universally expressed or seems to be expressed in situations outside of those in which it is hypothesized to have evolved. This critique is founded on the mistaken impression that natural selection is claimed to produce traits that are expressed optimally or invariantly. This is hardly true for our anatomical characteristics and completely infeasible for traits as complex as social behavior. For instance, cross-cultural variation in the prevalence of a behavior is often touted as evidence against evolutionary hypotheses. Yet, the expression of consistent behavioral patterns across dramatically differing cultural contexts is a hallmark of a trait that warrants evolutionary examination. Similarly, as illustrated for many of the primates examined throughout this volume, behavioral strategies are often expressed differentially, depending on factors such as age, status, and previous experience. Furthermore, behavioral strategies are understood to be contingent on the constellation of strategies utilized by conspecifics. Very rarely would it be adaptive for all individuals to use the same strategy in pursuit of an end; thus, we see that sexual coercion is typically just one of many reproductive strategies used by male primates. Moreover, reproductive advantage is often predicted only for a small proportion of individuals utilizing the strategy. Some effective strategies can be risky and even result in harm to an individual's reproductive success if expressed under the wrong circumstances. While we look to our evolutionary past for the pressures under which a behavior was selected, we must similarly understand that selection continues to shape our behavioral patterns.

The Scope of the Problem

In the following sections, I will assess the hypotheses about rape's origin in light of the relevant data on the characteristics of rape incidence, attackers, and victims. A key problem is the lack of a uniform definition for rape (even among different states in the United States: Tobach and Reed 2003). Here,

I define rape as sexual intercourse that has been obtained by a man against a woman's consent through the use of threat or force. As many critics point out, this definition is somewhat limiting because it disregards a suite of seemingly related sexual crimes, including molestation of children, forced sodomy of male victims, and penetration of an unwilling victim with an object other than the penis. However, it is similarly problematic to assume that all such behaviors share the same cause or function (Drea and Wallen 2003). For instance, both evolutionary and feminist scholars might agree that forced copulation has more theoretical similarities to domestic violence than to the behavior most similar in form—consensual sex.

All data on rape are problematic for obvious reasons. With regard to sexual victimization, important data come from private and government surveys of women reporting their own sexual experiences. Such surveys reveal that a majority of women do not report rapes to the police or seek professional help from medical or crisis centers; many victims are also reticent to label their attacks as "rape". This means that statistics on the frequency of rape are universally underestimates.

Approximately 1 in 625 U.S. residents aged 12 and older experience a violent sexual offense each year (Greenfeld 1997; Fisher et al. 2000; Rennison 2002; Masho et al. 2005; Tjaden and Thoennes 2006). More than 85% of all rape and attempted rape victims are female. Over their lifetime, one in six women is raped. Nearly all perpetrators are male. There does not appear to be any consistent variation in victimization with regard to race, ethnicity, education level, or income. Two interesting trends emerge from this research: victims tend to be relatively young, and the majority of victims are assaulted by acquaintances, including intimate partners. These important issues will be addressed later.

Cross-cultural surveys of this detail are not available for rape specifically. The standard cross-cultural sample reports that among 95 societies, rape is commonplace or accepted practice in 18% and present in some frequency in additional 35% (Sanday 1981). Sanday categorizes the remaining 47% of societies as relatively rape-free, though this figure is questionable as the standard cross-cultural sample relies on reports of rape—among many other events—within available ethnographers' reports, which are not specifically focused on sexual issues and vary substantially in the duration of study. Because rape is a behavior that may frequently be hidden from the ethnographer (particularly male ethnographers or societies in which victims have no legal outlet), the absence of report by no means equates to absence of victimization.

Who Rapes?

Some researchers have argued that males who rape are likely to have psychopathologies, social inadequacies, experience of childhood sexual trauma, or limited sexual opportunities (Figueredo 1992; Thornhill and Palmer 2000). For example, Figueredo et al. (2000) reported that adolescent sex offenders tended to have various psychological, psychosocial, or learning disorders, in addition to histories of failed intimate relationships. With regard to sexual opportunities, rapists report frustration with their sexual opportunities, but characteristically have more sexual interactions and a greater number of partners than their peers (Lalumière et al. 2005). Perhaps more importantly, the samples pointing to psychopathology are skewed. For example, the Figueredo et al. (2000) sample only consisted of boys convicted of a sex crime, including child rape or incest, and subsequently referred to a clinic for sex offenders. Similarly, studies of convicted rapists tend to identify defects in measures of social competence and empathy (Stermac and Quinsey 1986; Lipton et al. 1987; Lalumière et al. 2005). Lalumière and colleagues (2005), who have studied convict populations extensively, report that rapists are similar to other violent offenders and tend to have extensive nonsexual criminal histories, and therefore propose that rape may be one common consequence of a generalized antisocial personality disorder. The prison samples used by all of these studies overrepresent men who attack strangers, due to a higher reporting and conviction rate; the more common pattern of acquaintance rape is typical of nonconvict samples (Koss et al. 1988; Malamuth 2003).

Thus, it is problematic to extrapolate the findings from convict populations to the larger incidence of acquaintance rape. Indeed, studies of men in nonprison populations fail to find a link between psychoticism or antisociality and sexual arousal in response to rape scenarios (Malamuth 1986; Lalumière and Quinsey 1996). In contrast to convict studies, a study of university students found no relationship between a coercive sexual history and social competence (Koralweski and Conger 1992). This contrast in studies of convicts and nonconvicts suggests that psychopathy is a leading risk factor for some rapists, but is not a strong predictor for the majority of sexual aggression—cases in which offenders do not end up in a prison population.

The hypothesis that rapists are sexually frustrated males who see force as the only way to obtain reproductive opportunities is clearly contradicted by the evidence. Contrary to the hypothesis that rapists lack alternative sexual outlets, acquaintance rapists (the largest subset of perpetrators) typically have active sex lives and are often married (Gebhard et al. 1965; Thornhill and Thornhill 1983; Kanin 1985; see also Lalumière and Quinsey 1996 for convicted rapists). In fact, having had relatively high success in sexual relationships can lead men to have higher expectations for sex and to respond more aggressively to refusal (Kanin 1985). Men who are relatively narcissistic have a higher tendency toward sexual aggression, express less empathy to rape victims, and are more likely to subscribe to "rape myths" that shift blame to the victim (Dean and Malamuth 1997; Bushman et al. 2003). Physiological and self-reported sexual arousal to rape depictions is also predicted by the degree to which men assign dominance as a key motivation for sex (Malamuth 1986).

Environmental factors likely play a significant role in the translation of sexual urges to sexual aggression. But within the United States, researchers find no difference in sexual coercion according to men's ethnicity, family income, or the size of the community in which they live (Koss and Gidycz 1985; Koss et al. 1987). One factor is clearly exposure to sexual aggression, particularly if this is portrayed as acceptable. Compared to men viewing a consensual sex scenario, men exposed to rape depictions are more likely to subsequently generate violent sexual fantasies (Malamuth 1981). Men exposed to rape depictions in which the victim shows signs of sexual arousal, as is common in aggressive pornography, are even less likely to be sympathetic to victims; they are more likely to view an unrelated depiction of rape as less traumatic for the victim and to show less motivation to punish the aggressor (Zillman and Bryant 1984; Malamuth and Check 1985). College men exhibit increased proclivity to rape when they perceive acceptance of rape myths among peers (Bohner et al. 2006). High among these rape myths is the assignment of alcohol consumption as a mitigating factor in sexual assault. In fact, arousal to depictions of coerced sex does not relate only to actual alcohol consumption but has been associated with perceived intoxication, such as after consumption of placebo beverages believed to be alcoholic (Briddell et al. 1978; Barbaree et al. 1983; Abbey et al. 1996). Higher prevalence of rape by young men may be a consequence of some of these societal factors, in addition to general proclivities toward risk-taking at this age (Daly and Wilson 1985; Lalumière et al. 2005). It should be noted, however, that rapists are not habitually younger than other sexual and nonsexual offenders (Lalumière et al. 2005).

Similar factors, such as the acceptance of interpersonal violence and degree of male social dominance over women, contribute to cross-cultural variation in rape prevalence (Sanday 1981). These data should be viewed with caution, however, as they may reflect increased perception of rape incidence rather than actual assault rates. Demographic variables may play a larger role in cultural variation. Based on Thornhill's hypothesis that men with limited sexual access commit rape (and the observation that men commit the majority of violent crimes, including rape), we might expect rape to be correlated with a high male-to-female sex ratio. Curiously, rape is reported to be more frequent in societies in which the sex ratio is low (i.e., relatively more women than men) (Barber 2000; Kimmel 2003). One cause of low sex ratio is increased incidence of tribal warfare, a condition that leads to higher relative social position of men over women, increased polygyny, and sometimes use of raiding to acquire (i.e., kidnap) wives (Sanday 1981; Kimmel 2003). Similarly, rape—as well as reduced control over other aspects of female sexuality—is reported most commonly in cultures where women make less of a contribution to subsistence, granting them relatively lower social status (Schlegel and Barry 1986). Such cross-cultural variation in rape incidence does not preclude evolutionary means as an ultimate explanation for rape and is in fact predicted by the relationship coercion model (Smuts 1992).

Who Are Rape Victims?

Rates of sexual victimization are highest among high school and college-aged women (ages 16–24; see Figures 14.1 and 14.2). In studies conducted since the 1950s, more than half of college-aged women reported that they had experienced attempts at sexual aggression or coerced sexual activity within a single year (Kirkpatrick and Kanin 1957; Kanin and Parcell 1977; Koss et al. 1987). Approximately 6 to 10% of college-aged women report that they experience legally defined rapes in a given year (Koss et al. 1987; Fisher et al. 2000).

The age distribution of rape victims is consistent across a great number of studies and has been advanced by Thornhill and Palmer (2000) as the single most compelling data point to support the hypothesis that rape is an alternate male avenue to fertilization opportunities. Their argument poses the idea that males target young adult females as objects of both coercive and noncoercive sexual interest because these females are most likely to conceive. Indeed, while adolescent and postmenopausal women are victimized, women of reproductive age are overrepresented among rape victims, suggesting that men may select targets that are capable of getting pregnant, or simply that they select women

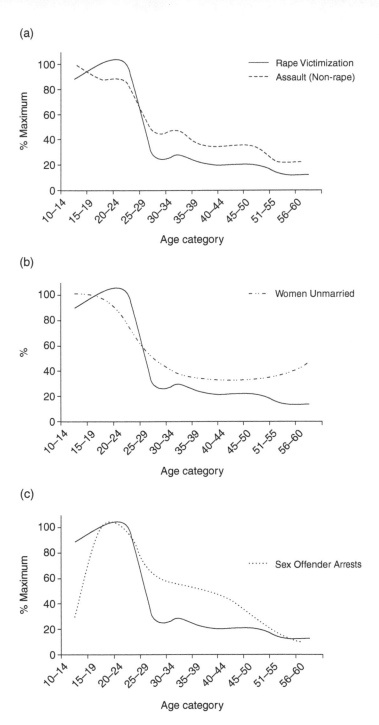

Figure 14.1 Rape victimization rates (bold lines) and indices of female vulnerability by age. Sources: *Rape and property crime victimization:* U.S., 2004 (Catalano 2005); *Women unmarried (or living apart from spouse:* U.S., 1998 (Lugaila 1998); *Sex offender arrests:* U.S., 2004 ("Crime in the United States" 2004).

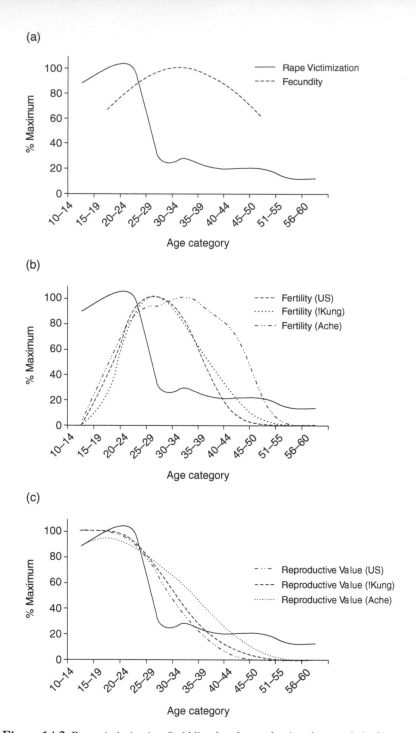

Figure 14.2 Rape victimization (bold lines) and reproductive characteristics by age. *Rape Victimization:* U.S., 2004 (Catalano 2005); *Fecundity:* Cross-cultural (Lipson and Ellison 1992, Ellison et al. 1993); *Age-specific fertility:* U.S., 2002 (Martin et al. 2003);

to whom they are otherwise attracted. It is also true that the age of highest victimization generally agrees with men's cross-culturally expressed preference for youthful, nulliparous mates (Buss 2004).

These age data are not as compelling, however, as Thornhill and Palmer suggest. Women who experience the highest victimization rates are also the most vulnerable. The age pattern of rape victims follows closely the pattern of nonsexual assaults in the United States (Figure 14.1a). The age groups most likely to be raped are also most likely to be unmarried (Figure 14.1b); thus these women are more likely to be living alone, and their routine activities place them in greater contact with potential attackers (Mustaine and Tewksbury 1998). College-aged women probably experience the highest victimization rates because they are the group most likely to be living away from natal kin but not yet with a domestic partner, increasing their vulnerability to a variety of crimes. Finally, the ages of the majority of rape victims coincide with the age at which men commit the most crimes, including rape and murder (Figure 14.1c; see Daly and Wilson 1988; "Crime in the United States," 2004). This is not just a factor of incarceration; sex offenders are more likely to reoffend if they are released at a relatively early age (Barbaree et al. 2003).

Although the data are problematic, the magnitude of the peak in victimization of young women does suggest that age of victims is a factor in male targeting of rape victims, something that warrants further exploration. However, as a reproductive ecologist, I find that the preference for youthful victims is inconsistent with the predictions put forth by Thornhill and Palmer (2000). This confusion results from a common misinterpretation of the terms *fertility* and *fecundity* in demographic context.

Female reproductive development is a remarkably slow process. The outward signs we use to judge reproductive maturity, such as the development of breasts and pubic hair, precede the development of reproductive capacity (Marshall and Tanner 1974). Although menarche is often used as the milestone of reproductive maturity, it is typically followed by years of nonconceptive cycling, referred to as *adolescent sterility* or more accurately *adolescent subfecundity* (Wood 1994). During this time, developing females experience irregular cycles and produce

---

!Kung: African hunter-gatherers (Howell 1979); Ache: South American hunter-gatherers (Hill and Hurtado, 1996); *Reproductive Value:* U.S, calculated from 2002 life tables (Kochanek et al. 2004) and birth data (Martin et al. 2003); !Kung/Ache calculated from respective demographic datasets (Hill and Hurtado 1996, Howell 1977)

relatively low levels of the hormones that facilitate ovulation and implantation. Although a girl may experience monthly menstrual periods, it can be expected that half of her cycles will be anovulatory for the first three to five years (Vihko and Apter 1984). In U.S. women, as well as in women in representative natural fertility cultures, the age curve of *fecundity*—probability of conception—does not peak until approximately ages 25–35 (Figure 14.2a; see Lipson and Ellison 1992)—after the peak in reproductive value and notably after the peak in rape victimization. *Fertility* curves indicating actual birth rates show a slightly earlier peak than fecundity curves (undoubtedly influenced by conscious control of fertility: Figure 14.2b). However, even these data suggest that men seeking women of prime conception age would target mates older than those reported to be the most frequent rape victims.

Evolutionary biologists have explained human male preference for young mates by noting that young females have high *reproductive value.* Men seeking long-term mates can theoretically monopolize a female's entire reproductive career; thus they should prefer mates that have the highest probability of future fertility. Similarly, men interested in their own genetic well-being should optimally prefer females who do not yet have offspring by previous mates. Curiously, this means that human males prefer young, subfecund mates over the mates that might give them the highest probability of immediate conception. By contrast, males of promiscuous primate species, who invest little in the care of offspring or in long-term mating partnerships, tend to avoid mating with young females (Anderson 1986). They appear to assess mating effort based on the immediate probability of fertilization, a clear contrast to the pattern seen in humans (Muller et al. 2006). Coercive males acting in accordance with Thornhill and Palmer's hypothesis should be expected to adopt the short-term mating strategy utilized by low-investment species. Yet this key piece of evidence advanced by Thornhill and Palmer (2000) does not support their argument. One explanation for the age pattern of rape victimization is simply that males are unable to subvert or contextualize their attraction to youth. However, it may be an indication that, if rape has an adaptive origin, it is not for immediate fertilization but for longer-term reproductive or social goals.

Can Rape Function as a Reproductive Strategy?

Two major questions are commonly asked in analyses of rape as a reproductive strategy. First, how could selection have acted on a behavior that has such a small probability of leading to successful conception? Second, why would a male seeking to increase his reproductive success through rape simultaneously inflict harm to the potential mother? It is important to evaluate the data behind each of these arguments, as well as to question whether the arguments really invalidate an evolutionary hypothesis. To be selected as a reproductive strategy, the behavior need not be the best approach, but a viable option for a given male in a given context. Long-term evolutionary success may depend less on an individual reproducing most prolifically than on its ability to consistently pass some copies of its genes to the next generation (Ellison 1994). In addition, reproductive success could result not as an immediate product of sexual coercion, but by increasing the probability of later fertile matings.

The common argument that rapes occur outside of the reproductive context (with males, children, and postmenopausal women) is not a particularly persuasive argument against the hypothesis that rape has an adaptive function for reproduction. It cannot be denied that the act of sex itself evolved for reproductive purposes, yet many millions of people engage in consensual sex in circumstances with little or no probability of conception (or even the hope that they will not conceive). Studies in the United States find that between 1 and 18% of women become pregnant as a result of their rape, equating to approximately 32,000 rape-pregnancies per year (Koss et al. 1987; Holmes et al. 1996; Gottschall and Gottschall 2003). This is likely to be an underestimate as many victims are offered emergency contraception, though it is comparable to rates of conception from a single act of consensual sex. It is apparent that forced intercourse can result in successful reproduction, though it is still questionable whether this small probability could select for the use of such a risky strategy.

Another objection to reproductive hypotheses is that violence seems a detriment to any chance of reproduction; after all, some rapists kill their victims. However, statistics suggest that severe injury to victims is actually quite rare. Although most victims suffer emotional trauma, only 5% of victims are severely injured and one-third report minor physical injuries (e.g., cuts, bruises, and chipped teeth) (Rennison 2002; Tjaden and Thoennes 2006). Again, there seems to be a conflict in evidence from stranger and acquaintance rape, suggesting that they may be functionally different phenomena with different underlying motivations. Victimization by stranger assailants is associated with increased use of power tactics, more frequent display of weapons, more severe injury, increased psychological trauma, and greater perceived life threat than acquaintance rape (Thornhill and Thornhill 1990; Cleveland et al. 1999; Ullman et al. 2006).

A Closer Examination of Acquaintance Rape

Surveys overwhelmingly find that the majority of rapes are committed by someone known to the victim (Jewkes and Abrahams 2002; Catalano 2005; Masho et al. 2005; Tjaden and Thoennes 2006). In fact, the majority of U.S. sexual assault victims report having a current or prior intimate relationship with their attacker (74%: Greenfeld 1997; >50%: O'Sullivan et al. 1998; 62%: Elliot et al. 2004). Approximately 1 in 13 American women report rape by their intimate partners (Tjaden and Thoennes 2006). Three-quarters of such victims do not report their rape to the police (Thornhill and Thornhill 1990; Rennison 2002). Rapes involving spouses or cohabiting partners are most likely to take place during the relationship rather than in response to a breakup, and most of these victims are raped more than once (Tjaden and Thoennes 2006).

Cross-population studies suggest that these U.S. statistics are not atypical. A meta-analysis of 48 population-based studies by the World Health Organization (Krug et al. 2002) found that between 10 and 69% of women reported being physically assaulted by their intimate partners. This physical abuse was coupled with sexual aggression in one-third to one-half of these cases. In a similar 18 country population survey, including Western and developing countries, between 6% (Japan) and 47% (Peru) of women with relationship histories had experienced rape or attempted rape by their intimate partner (Krug et al. 2002). In some developing countries, a large proportion of first sexual initiations are forced, frequently in the context of child marriage (Krug et al. 2002).

Thus, there appears to be a disturbing intrusion of forced sex into the context of "normal" male-female dating relationships. This suggests that rather than focusing on the pathological correlates of sexual aggression, a better understanding of the widespread rape phenomenon should consider why seemingly normal men in normal relationships should so often resort to force to obtain sexual access. Social scientists have argued that rape may serve different functions in relationships, from simply obtaining sex to reinforcing the nature of the relationship. This is likely to vary depending on the duration of the relationship, previous consensual sexual activity, and whether the man feels the relationship is threatened (Shotland 1992; Shotland and Goodstein 1992). This proximate model sounds remarkably like the Smuts's (Smuts and Smuts 1993) evolutionary argument for the origin of sexual coercion in primate species. In this model, force or the threat of force can be effective as a means of obtaining short- or long-term reproductive access to a female under circumstances in which male and female reproductive interests conflict.

This hypothesis predicts that sexual aggression is not the goal in and of itself but is used as one means to obtain a desired end. As previously mentioned, most victims of acquaintance rape do not experience physical injury, though presumably male attackers could inflict greater harm if so desired. Nearly all date rapists attempt to use a number of other strategies, including verbal aggression, before resorting to physical force, and the vast majority of rapes are preceded by mutually consensual intimate contact (Kanin 1985; Bownes et al. 1991; Ellis 1991). Compared with fewer than one-quarter of stranger rapes, nearly all acquaintance rapes involve some social interaction after the forced sex act (Bownes et al. 1991).

The relationship coercion hypothesis also predicts that sexual aggression would be more frequent if the male perceived a threat to his longer-term access to the female, rather than just resistance to the immediate copulation attempt. This prediction is supported by data that link decreases in sexual aggression to increases in relationship quality. Married women, for instance, experience lower levels of sexual aggression from their partners than unmarried women in long-term relationships (Cleveland et al. 1999; Tjaden and Thoennes 2006). In Flinn's (1988) landmark study of male aggression in a Caribbean village, men mate-guarded their partners most intensely, including closer proximity and increased aggression to the partner and other males, when the woman was more fecund and the relationship nonexclusive (see also Wilson and Daly 1992). Nonaggressive mate-guarding behaviors are positively associated with female-directed aggression, suggesting a common function (Shackelford et al. 2000).

A morally troubling prediction of Smuts and Smuts's hypothesis is that use of sexual aggression may be effective in continuing a male's sexual access to a female. This does seem to be the case in other ape species: chimpanzee females who receive physical aggression from a male are more likely to subsequently mate with that male and less likely to mate with other males (Muller et al. Chapter 8 in this volume); gorilla females whose infants are killed by a male frequently become part of the attacker's harem (Watts 1989). This does not require that females find these attackers sexually appealing (though selection of such males could be adaptive if the traits enabled their offspring to increase reproductive access through similar means: Weatherhead and Robertson 1979), but simply that the costs of acquiescence are lower than the costs of resistance. Among college-aged women, approximately 40% of rape victims report continuing to date their attackers (Wilson and Durrenberger 1982; Koss 1989). Women's positive expectations for a relationship correlated to self-blame and reduced anger in response to coercion (Macy et al. 2006). These victims did recognize that the assailant was to blame, but their assignment of blame was unrelated to the nature of the assailant's coercive actions and more strongly mediated by concerns over harming the relationship (Figueredo 1992; Macy et al. 2006).

Rape in Violent Conflict

Thus far in this discussion, I have given little mention to rape in the context of civil and gang warfare. In terms of absolute numbers, this may indeed be the most pervasive form of sexual victimization, affecting as many as half of women in some conflict areas, many the victims of repeated assault (World Health Organization 2004). This is a weighty topic that merits more than the cursory examination provided here. As I have emphasized, there are compelling reasons to distinguish different forms of rapes, such as stranger, acquaintance, and war rape in terms of our functional examination. In fact, the diversity of examples of sexual victimization in war suggests that "war rape" is also a category that includes distinct behavioral patterns. Many case studies provide little reason to suggest that rape is anything other than a weapon used to denigrate, wound, and terrorize the victimized group (Giller et al. 1991; Swiss and Giller 1993; World Health Organization 2004). However, the disorder of conflict also provides increased opportunities for individual males to victimize vulnerable acquaintances. In the Bosnian conflict, for example, many women interviewed by a United Nations human rights team knew the men who had raped them (United Nations 1993). In other cases, such as among the Yanomamö tribesmen of South America, the acquisition of wives is considered to be one desired side effect of a successful raid (Chagnon 1997). Such raiding, as well as forcible theft of brides from within the community, is reported from ethnographic populations on every major continent and from both polygynous and monogamous cultures (Ayres 1974). This behavior even extends to ritualized "mock" bride theft, in which the man is expected to abduct his new wife while she or her parents feign resistance until a suitable brideprice or other settlement can be made (Ayres 1974). Such displays suggest that the forcible abduction of wives may have been more widespread during human history.

Conclusions

As critics correctly note, we may never be able to firmly establish whether rape has an evolutionary history in the human species. The nature of the trait precludes controlled testing of its function and effectiveness, though available evidence gives us strong reasons to consider biological underpinnings. Statistics from a variety of sources reveal that, far from being random, rape is a crime overwhelmingly directed at young women by men casually or intimately known to them. Available evidence contradicts the controversial assertion of Thornhill and Palmer (2000) that men who are socially and sexually disadvantaged seek victims who are likely to conceive as a result of the attack. Both convict and community samples of coercive males have significant sexual experience. Furthermore, the age pattern of victimizations provides a closer fit for the relationship coercion hypothesis proposed by Smuts (1992), whereby males use sexual aggression, including but not limited to forced sex, as an attempt to secure and maintain long-term reproductive access to females. As predicted by Smuts, rape victimization varies in a predictable manner according to sociocultural context, including vulnerability of victims.

Acquaintance rapes significantly outnumber stranger rapes, and characteristics of assailants and the attacks themselves differ in predictable ways between these two contexts. This suggests that a single hypothesis is unlikely to comprehensively inform us about the ultimate cause of rape. Indeed, the phenomenon of stranger rape does not appear to warrant an adaptive explanation. Current data on the psychosocial characteristics of stranger rapists supports the hypothesis that stranger rape is frequently a pathological behavior (Figueredo 1992; Lalumière et al. 2005) or a by-product of an adaptation for heightened male sexuality (Palmer 1991).

With regards to acquaintance rape, the relationship coercion hypothesis is supported by the following evidence: (1) Acquaintance rapists are not characterized by major psychopathological or social competence deficits but tend to utilize other mate-guarding tactics and ascribe significance to being dominant over their partner; (2) acquaintance rapists rarely cause significant injury to their victims, and typically utilize less forceful strategies than stranger rapists prior to forcing intercourse; (3) sexual aggression frequently occurs in the context of ongoing relationships, including marriages, and both its incidence and its effectiveness are related to the quality of the relationship; and (4) a sizable proportion of women continue their relationships with their attackers following an acquaintance rape. The eventual reproductive consequences of sexual aggression are not known, but the available data suggest that such a strategy, though undesirable, may frequently be successful in securing long-term access to a desired female. This is particularly true when social contexts limit female economic opportunities or support networks.

The coercion hypothesis is consistent with feminist and sociological perspectives in arguing that rape is intrinsically related to male social and physical dominance over females. Critics from these fields frequently argue that examinations of rape cannot be informed by studies of other species from which humans diverge in so many aspects of their behavior. In fact, human sexual behavior does differ in significant ways from that of our closest relatives. However, the most parsimonious perspective is to assume that our sexual strategies have been influenced in comparable ways by the forces of natural and sexual selection. Sexual conflicts of interest and the use of aggression toward reproductive ends are pervasive among animal species, and there is little reason to believe our ancestors would have been immune from these pressures. Does an evolutionary perspective help us better understand how to prevent rape? At the very least it may give us a clearer picture of the enormity of the problem we are dealing with. At best it may help reveal the most salient socioecological factors that exacerbate sexual aggression against women (Rosenfeld, Chapter 17 in this volume). Increased awareness of these factors should facilitate our ability to morally sanction and punish sexually coercive behavior, a unique and elegant counteradaptation.

Acknowledgments

Thank you to Martin Muller, Richard Wrangham, Diane Rosenfeld, and Ian Gilby for sharing ideas and for comments on the chapter.

References

Abbey, A., L. T. Ross, D. McDuffie, and P. McAuslan. "Alcohol, Misperception, and Sexual Assault: How and Why Are They Linked?" In *Sex, Power, Conflict: Evolutionary and Feminist Perspectives*, eds. M. Buss and N. M. Malamuth, pp. 138–161. New York: Oxford University Press, 1996.

Adams-Curtis, L., and G. B. Forbes. "College Women's Experiences of Sexual Coercion: A Review of Cultural, Perpetrator, Victim, and Situational Variables." *Trauma, Violence and Abuse* 5 (2004): 91–122.

Alexander, R. *Darwinism and Human Affairs.* Seattle: University of Washington Press, 1979.

Anderson, C. M. "Female Age: Male Preference and Reproductive Success in Primates." *International Journal of Primatology* 7 (1986): 305–325.

Ayres, B. "Bride Theft and Raiding for Wives in Cross-Cultural Perspective." *Anthropological Quarterly* 47 (1974): 238–252.

Barbaree, H. E., W. Marshall, E. Yates, and L. Lightfoot. "Alcohol Intoxication and Deviant Sexual Arousal in Male Social Drinkers." *Behaviour Research and Therapy* 21 (1983): 365–373.

Barbaree, H. E., R. Blanchard, and C. M. Langton. "The Development of Sexual Aggression through the Life Span: The Effect of Age on Sexual Arousal and Recidivism among Sex Offenders." *Annals of the New York Academy of Science* 989 (2003): 59–71.

Barber, N. "The Sex Ratio as a Predictor of Cross-National Variation in Violent Crime." *Cross-Cultural Research* 34 (2000): 264–282.

Bohner, G., F. Siebler, and J. Schmelcher. "Social Norms and the Likelihood of Raping: Perceived Rape Myth Acceptance of Others Affects Men's Rape Proclivity." *Personality and Social Psychology Bulletin* 32 (2006): 286–297.

Bownes, I. T., E. G. O'Gorman, and A. Sayers. "Rape—A Comparison of Stranger and Acquaintance Assaults." *Medical Science and Law* 2 (1991): 102–109.

Briddell, D. W., D. C. Rimm, G. R. Caddy, G. Krawitz, D. Sholis, and R. J. Wunderline. "Effects of Alcohol and Cognitive Set on Sexual Arousal to Deviant Stimuli." *Journal of Abnormal Psychology* 87 (1978): 418–430.

Brownmiller, S. *Against Our Will: Men, Women and Rape.* New York: Simon and Schuster, 1975.

Bushman, B. J., A. M. Bonacci, M. van Dijk, and R. F. Baumeister. "Narcissism, Sexual Refusal, and Aggression: Testing a Narcissistic Reactance Model of Sexual Coercion." *Personality Processes and Individual Differences* 84 (2003): 1027–1040.

Buss, D. M. *Evolutionary Psychology: The New Science of the Mind.* Boston: Allyn and Bacon, 2004.

Carroll, S. B. "Evolution at Two Levels: On Genes and Form." *Public Library of Science Biology* 3 (2005): 1159–1166.

Catalano, S. M. *National Crime Victimization Survey (NCJ 210674).* Washington, D.C.: U.S. Department of Justice, Bureau of Justice Statistics, 2005.

Chagnon, N. A. *Yanomamö.* Fort Worth, Tex.: Harcourt Brace College Publishers, 1997.

Cleveland, H. H., M. P. Koss, and J. Lyons. "Rape Tactics from the Survivors' Perspective: Contextual Dependence and Within-Event Independence." *Journal of Interpersonal Violence* 14 (1999): 532–547.

Clutton-Brock, T., and G. Parker. "Sexual Coercion in Animal Societies." *Animal Behaviour* 49 (1995): 1345–1365.

Crime in the United States. Washington, D.C.: U.S. Department of Justice, Federal Bureau of Investigation, 2004.

Daly, M., and M. Wilson. "Competitiveness, Risk-taking, and Violence: The Young Male Syndrome." *Ethology and Sociobiology* 6 (1985): 59–73.

———. "Evolutionary Social Psychology and Family Homicide." *Science* 242 (1988): 519–524.

Dean, K. E., and N. M. Malamuth. "Characteristics of Men Who Aggress Sexually and of Men Who Imagine Aggressing: Risk and Moderating Variables." *Journal of Personality and Social Psychology* 72 (1997): 449–455.

Drea, C. M., and K. Wallen. "Female Sexuality and the Myth of Male Control." In *Evolution, Gender, and Rape,* ed. C. B. Travis, pp. 29–60. Cambridge, Mass.: MIT Press, 2003.

Elliot, D. M., D. S. Mok, and J. Briere. "Adult Sexual Assault: Prevalence, Symptomatology, and Sex Differences in the General Population." *Journal of Traumatic Stress* 17 (2004): 203–211.

Ellis, L. "The Drive to Possess and Control as a Motivation for Sexual Behavior: Applications to the Study of Rape." *Social Science Information* 30 (1991): 663–675.

Ellison, P. T., S. F. Lipson, M. T. O'Rourke, G. R. Bentley, A. M. Harrigan, C. Panter-Brick, and V. Vitzthum, "Population Variation in Ovarian Function." *The Lancet* 342 (1993): 433–434.

Ellison, P. T. "Extinction and Descent." *Human Nature* 5 (1994): 155–165.

Figueredo, A. "Does Rape Equal Sex Plus Violence?" *Behavioral and Brain Sciences* 15 (1992): 384–385.

Figueredo, A., B. D. Sales, K. P. Russell, J. V. Becker, and M. Kaplan. "A Brunswikian Evolutionary-Developmental Theory of Adolescent Sex Offending." *Behavioral Sciences and the Law* 18 (2000): 309–329.

Fisher, B. S., F. T. Cullen, and M. G. Turner. *The Sexual Victimization of College Women (NCJ 182369).* Washington, D.C.: U.S. Department of Justice, Office of Justice Programs, 2000.

Flinn, M. V. "Mate Guarding in a Caribbean Village." *Ethology and Sociobiology* 9 (1988): 1–28.

Gebhard, P., J. Gagnon, W. Pomeroy, and C. Christenson. *Sex Offenders.* New York: Harper and Row, 1965

Giller, J. E., P. J. Bracken, and S. Kabaganda. "Uganda: War, Women, and Rape." *The Lancet* 337 (1991): 604.

Gottschall, J. A., and T. A. Gottschall. "Are Per-Incident Rape-Pregnancy Rates Higher Than Per-Incident Consensual Pregnancy Rates?" *Human Nature* 14 (2003): 1–20.

Greenfeld, L. A. *Sex Offenses and Offenders (NCJ-163392).* Washington, D.C.: U.S. Department of Justice, Bureau of Justice Statistics, 1997.

Hill, K., and A. M. Hurtado. *Ache Life History: The Ecology and Demography of a Foraging People.* New York: Aldine de Gruyter, 1996.

Hilton, N. Z., Harris, G. T., and M. E. Rice. "The Functions of Aggression by Male Teenagers." *Journal of Personality and Social Psychology* 79 (2000): 988–994.

Holmes, M. M., H. S. Resnick, D. G. Kilpatrick, and C. L. Best. "Rape-Related Pregnancy: Estimates and Descriptive Characteristics from a National Sample of Women." *American Journal of Obstetrics and Gynecology* 175 (1996): 320–325.

Howell, N. *Demography of the Dobe! Kung.* New York: Aldine de Gruyter, 1979.

Jewkes, R., and N. Abrahams. "The Epidemiology of Rape and Sexual Coercion in South Africa: An Overview." *Social Science and Medicine* 55 (2002): 1231–1234.

Kanin, E. J. "Date Rapists: Differential Sexual Socialization and Relative Deprivation." *Archives of Sexual Behavior* 14 (1985): 219–231.

Kanin, E. J., and Parcell, S. R. "Sexual Aggression: A Second Look at the Offended Female." *Archives of Sexual Behavior* 6 (1977): 67–76.

Kimmel, M. "An Unnatural History of Rape." In *Evolution, Gender, and Rape,* ed. C. B. Travis, pp. 221–234. Cambridge, Mass,: MIT Press, 2003.

Kirkpatrick, C., and E. Kanin. "Male Sex Aggression on a University Campus." *American Sociological Review* 22 (1957): 52–58.

Kochanek, K. D., S. L. Murphy, R. N. Anderson, and C. Scott. "Deaths: Final Data for 2002." *National Vital Statistics Survey* 53 (2004), National Center for Health Statistics.

Koralweski, M. A., and J. C. Conger. "The Assessment of Social Skills among Sexually Coercive College Males." *The Journal of Sex Research* 29 (1992): 169–188.

Koss, M. P. "Hidden Rape: Sexual Aggression and Victimization in a National Sample of Students in Higher Education." In *Violence in Dating Relationships,* eds., M. A. Pirog-Good and J. E. Stets, pp. 145–168. New York: Praeger, 1989.

Koss, M. P., T. E. Dinero, C. A. Seibel, and S. L. Cox. "Stranger and Acquaintance Rape: Are There Differences in the Victim's Experience?" *Psychology of Women Quarterly* 12 (1988): 1–24.

Koss, M. P., and C. A. Gidycz. "Sexual Experiences Survey: Reliability and Validity." *Journal of Consulting and Clinical Psychology* 53 (1985): 422–423.

Koss, M. P., C. A. Gidycz, and N. Wisniewski. "The Scope of Rape: Incidence and Prevalence of Sexual Aggression and Victimization in a National Sample of Higher Education Students." *Journal of Consulting and Clinical Psychology* 55 (1987): 162–170.

Krug, E. G., L. L. Dahlberg, J. A. Mercy, A. B. Zwi, and R. Lozano. *World Report on Violence and Health.* Geneva, Switzerland: World Health Organization, 2002.

Lalumière, M. L., G. T. Harris, V. L. Quinsey, and M. E. Rice. *The Causes of Rape: Understanding Individual Differences in Male Propensity for Sexual Aggression.* Washington, D.C.: American Psychological Association, 2005.

Lalumière, M. L., and V. L. Quinsey. "Sexual Deviance, Antisociality, Mating Effort, and the Use of Sexually Coercive Behaviors." *Personality and Individual Differences* 21 (1996): 33–48.

Lipson, S. F., and P. T. Ellison. "Normative Study of Age Variation in Salivary Progesterone Profiles." *Journal of Biosocial Science* 24 (1992): 233–244.

Lipton, D. N., E. C. McDonel, and R. M. McFall. "Heterosocial Perception in Rapists." *Journal of Consulting and Clinical Psychology* 55 (1987): 17–21.

Lloyd, E. A. "Violence against Science: Rape and Evolution." In *Evolution, Gender, and Rape,* ed. C. B .Travis, pp. 235–262. Cambridge, Mass.: MIT Press, 2003.

Lugaila, T. A. *Current Population Reports: Marital Status and Living Arrangements (P20–514).* Washington, D.C.: U.S. Department of Commerce, Census Bureau, 1998.

Macy, R. J., P. S. Nurius, and J. Norris. "Responding in Their Best Interests: Contextualizing Women's Coping with Acquaintance Sexual Aggression." *Violence against Women* 12 (2006): 478–500.

Malamuth, N. M. "Rape Fantasies as a Function of Exposure to Violent Sexual Stimuli." *Archives of Sexual Behavior* 10 (1981): 33–47.

———. "Predictors of Naturalistic Sexual Aggression." *Journal of Personality and Social Psychology* 50 (1986): 953–962.

———. "Criminal and Noncriminal Sexual Aggressors: Integrating Psychopathy in a Hierarchical-Mediational Confluence Model." *Annals of the New York Academy of Science* 989 (2003): 33–58.

Malamuth, N. M., and J. V. Check. "The Effects of Aggressive Pornography on Beliefs in Rape Myths: Individual Differences." *Journal of Research in Personality* 19 (1985): 299–320.

Marshall, W., and J. Tanner. "Puberty." In *Scientific Foundations of Paediatrics,* eds. J. A. Davis and J. Dobbing, pp. 124–151. Philadelphia: W. B. Saunders Co., 1974.

Martin, J. A., B. E. Hamilton, P. D. Sutton, S. J. Ventura, D. Menacker, and M. L. Munson. "Births: Final Data for 2002." *National Vital Statistics Survey 52* (2003), National Center for Health Statistics.

Masho, S. W., R. K. Odor, and T. Adera. "Sexual Assault in Virginia: A Population-Based Study." *Women's Health Issues* 15 (2005): 157–166.

Muller, M. N., M. Emery Thompson, and R. W. Wrangham. "Male Chimpanzees Prefer Mating with Old Females." *Current Biology* 16 (2006): 2234–2238.

Mustaine, E. E., and R. Tewksbury. "Predicting Risks of Larceny Theft Victimization: A Routine Activity Analysis Using Refined Lifestyle Measures." *Criminology* 36 (1998): 829–858.

O'Sullivan, L. F., E. S. Byers, and L. Finkelman. "A Comparison of Male and Female College Students' Experiences of Sexual Coercion." *Psychology of Women Quarterly* 22 (1998): 177–195.

Otterbein, K. F. "A Cross-cultural Study of Rape." *Aggressive Behavior* 5 (1979): 425–435.

Palmer, C. T. "Human Rape: Adaptation or By-Product?" *The Journal of Sex Research* 28 (1991): 365–386.

Rennison, C. M. *Rape and Sexual Assault: Reporting to Police and Medical Attention, 1992–2000 (NCJ 194530).* Washington, D.C.: U.S. Department of Justice, Office of Justice Programs, 2002.

Sanday, P. R. "The Socio-Cultural Context of Rape: A Cross-Cultural Study." *Journal of Social Issues* 37 (1981): 5–27.

Schlegel, A., and H. Barry. "The Cultural Consequences of Female Contributions to Subsistence." *American Anthropologist* 88 (1986): 142–150.

Shackelford, T., V. Weekes-Shackleford, G. LeBlanc, A. Bleske, H. Euler, and S. Hoier. "Female Coital Orgasm and Male Attractiveness." *Human Nature* 11 (2000): 299–306.

Shotland, R. "A Theory of the Causes of Courtship Rape: Part 2." *Journal of Social Issues* 48 (1992): 127–143.

Shotland, R., and L. Goodstein. "Sexual Precedence Reduces the Perceived Legitimacy of Sexual Refusal: An Examination of Attributions Concerning Date Rape and Consensual Sex." *Personality and Social Psychology Bulletin* 18 (1992): 756–764.

Smuts, B. "Male Aggression against Women: An Evolutionary Perspective." *Human Nature* 3 (1992): 1–44.

Smuts, B. B., and R. W. Smuts. "Male Aggression and Sexual Coercion of Females in Nonhuman Primates and Other Mammals: Evidence and Theoretical Implications." *Advances in the Study of Behavior* 22 (1993): 1–63.

Stermac, L. E., and V. L. Quinsey. "Social Competence among Rapists." *Behavioral Assessment* 8 (1986): 171–185.

Swiss, S., and J. E. Giller. (1993). "Rape as a Crime of War: A Medical Perspective." *Journal of the American Medical Association* 270 (1993): 612–615.

Symons, D. *The Evolution of Human Sexuality.* New York: Oxford University Press, 1979.

Thornhill, N. W., and R. Thornhill. "An Evolutionary Analysis of Psychological Pain Following Rape: II. The Effects of Strange, Friend, and Family-Member Offenders." *Ethology and Sociobiology* 11 (1990): 177–193.

Thornhill, R., and N. W. Thornhill. "Human Rape: An Evolutionary Analysis." *Ethology and Sociobiology* 4 (1983): 137–173.

Thornhill, R., and C. T. Palmer. *A Natural History of Rape: Biological Bases of Sexual Coercion.* Cambridge, Mass.: MIT Press, 2000.

Tinbergen, N. "On Aims and Methods of Ethology." *Zeitschrift fur Tierpsychologie* 20 (1963): 410–429.

Tjaden, P., and N. Thoennes. *Extent, Nature, and Consequences of Intimate Partner Violence: Findings from the National Violence against Women Survey (NCJ 210346).* Washington, D.C.: U.S. Department of Justice and Centers for Disease Control, Office of Justice Programs, 2006.

Tobach, E., and Reed, R. "Understanding Rape." In *Evolution, Gender, and Rape,* ed. C. B. Travis, pp. 105–138. Cambridge, Mass.: MIT Press, 2003.

Ullman, S. E., H. H. Filipas, S. M. Townsend, and L. L. Starzynski. "The Role of Victim-Offender Relationship in Women's Sexual Assault Experiences." *Journal of Interpersonal Violence* 21 (2006): 798–819.

United Nations. *Report on the Situation of Human Rights in the Territory of the Former Yugoslavia (E/CN.4/1993/50).* Geneva, Switzerland, 1993.

Vihko, R., and D. Apter. "Endocrine Characteristics of Adolescent Menstrual Cycles: Impact of Early Menarche." *Journal of Steroid Biochemistry* 20 (1984): 231–236.

Watts, D. P. "Infanticide in Mountain Gorillas: New Cases and a Reconsideration of the Evidence." *Ethology* 81 (1989): 1–18.

Weatherhead, P. J., and R. J. Robertson. "Offspring Quality and the Polygyny Threshold: 'The Sexy Son Hypothesis'." *The American Naturalist* 113 (1979): 201–208.

Wilson, M. I., and M. Daly. "The Man Who Mistook His Wife for a Chattel." In *The Adapted Mind: Evolutionary Psychology and the Generation of Culture,* eds. J. Barkow, L. Cosmides, and J. Tooby, pp. 289–322. New York: Oxford University Press, 1992.

Wilson, W., and R. Durrenberger. "Comparison of Rape and Attempted Rape Victims." *Psychological Reports* 50 (1982): 198–207.

Wood, J. W. *Dynamics of Human Reproduction: Biology, Biometry, Demography.* New York: Aldine de Gruyter, 1994.

World Health Organization. "Sexual Violence in Conflict Settings and the Risk of HIV." *The Global Coalition on Women and AIDS Information Bulletin Series* 2 (2004).

Wrangham, R. W. "An Ecological Model of Female-Bonded Primate Groups." *Behaviour* 75 (1980): 262–299.

Wrangham, R. W., and D. Peterson. *Demonic Males: Apes and the Origins of Human Violence.* Boston: Houghton-Mifflin Co., 1996.

Yost, M. R., and E. L. Zurbriggen. "Gender Differences in the Enactment of Sociosexuality: An Examination of Implicit Social Motives, Sexual Fantasies, Coercive Sexual Attitudes, and Aggressive Sexual Behavior." *The Journal of Sex Research* 43 (2006): 163–173.

Zillman, D., and J. Bryant. "Effects of Massive Exposure to Pornography." In *Pornography and Sexual Aggression,* eds. N. M. Malamuth and E. Donnerstein, pp. 115–138. New York: Academic Press, 1984.

IV

FEMALE COUNTERSTRATEGIES

15

"Friendship" with Males: A Female Counterstrategy to Infanticide in Chacma Baboons of the Okavango Delta

Ryne Palombit

The fundamental conflict of reproductive interests between males and females, originally articulated by Trivers (1972) and Parker (1979), may have diverse evolutionary consequences at multiple levels: genetic, morphological, physiological, social, ecological, life historical (Arnqvist and Rowe 2005; Chapman 2006). This volume addresses one dramatic manifestation of this conflict: coercive behavior directed at females by adult males. This phenomenon draws our attention immediately to selection on females for adaptive strategies countering the costs of male coercion.

In this chapter, I examine evidence for one such proposed counterstrategy in one particular system: "friendship" in a population of chacma baboons. "Friendship" refers to a cohesive social relationship between an anestrous female and an adult male (Figure 15.1). Smuts (1985) originally developed the concept in her classic study of olive baboons *(Papio hamadryas anubis)*. Though sometimes labeled differently by various researchers, friendships have been observed in other olive baboon populations (Ransom and Ransom 1971), as well as in yellow baboons *(P. h. cynocephalus)* (Altmann 1980) and chacma baboons *(P. h. ursinus, P. h. griseipes)* (Seyfarth 1978; Anderson 1983; Palombit et al. 1997; Weingrill 2000). Friendship is a general feature of multimale, multifemale baboon societies, though it is not limited to baboons (Manson 1994; Palombit 1999) and does not necessarily have a unitary functional explanation.

This chapter focuses on research conducted for over a decade on a population of chacma baboons inhabiting the Okavango Delta in northwestern Botswana. These baboons live in relatively large groups (mean = 75 individuals)

Figure 15.1 A "friendship" in the Okavango chacma baboons, comprising an adult male, and an anestrous female (with her dependent infant). Photograph by Ryne Palombit.

comprising multiple adult males, adult females, and their offspring. As with baboons in other parts of Africa, groups are organized around linear dominance hierarchies and matrilineal relationships among female kin (for additional details concerning the study group and site, see Kitchen et al., Chapter 6 in this volume). This population is of special interest because a diverse array of data collectively suggests that male-female friendships are a female counterstrategy to a particular manifestation of sexual conflict: infanticide.

The "Coercive" Problem: Infanticide by Males

Male infanticide is the killing of unweaned infants by conspecific adult males. Clarke et al. (Chapter 3 in this volume) summarize the increasing evidence supporting the sexual selection hypothesis that infanticide is a reproductive strategy of males: killing an unrelated infant prematurely terminates its mother's lactational amenorrhea, returning her to ovulatory status much sooner, thereby making her available for fertilization to an infanticidal male whose own mating opportunities are constrained by intrasexual competition.

At first glance, male infanticide might seem to depart in two ways from Smuts and Smuts's (1993) original definition of sexual coercion as "use by a male of force, or threat of force, that functions to increase the chances that a female will mate with him at a time when she is likely to be fertile, and to decrease the chances that she will mate with other males, at some cost to the female." First, the potential mating opportunities afforded by male aggression are less proximate when directed at anestrous females than at currently cycling females. If, however, loss of an infant accelerates a mother's availability as a potential mate to the male, infanticide can in principle be examined within a sexual coercion framework. Nevertheless, it is worthwhile to appreciate that the more extended temporal course for coercion of anestrous females may involve different conditions, decision making, and proximate mechanisms than coercion of cycling females.

A second issue is that infanticide may not be usefully viewed as aggression to females or perhaps even as "aggression" per se. The target of infanticide is the infant. In primates, with the apparent exception of chimpanzees and mountain gorillas, aggression to mothers during infanticidal attacks is relatively infrequent or erratic, being usually a response to maternal defense rather than an inherent tactical feature of the attack itself. Also, it is rarely injurious (van Schaik 2000; Palombit 2003). Indeed, some primarily anecdotal evidence suggests that males may preferentially improve the probability of success by attacking when an infant's mother is unlikely to be involved (i.e., she is some distance away). Several workers have also argued that infanticide "is best understood as a motivational system separate from general aggressiveness" (Hrdy et al. 1995; van Schaik 2000:46), partly in light of evidence that, at least in rodents, the neuroendocrine bases of infanticide are distinct from those mediating conspecific aggression (vom Saal 1984; Schneider et al. 2003).

Whether or not inclusion of infanticide improves our analysis of "sexual coercion," there is no doubt that infanticide is an arresting example of "sexual conflict" in the broader sense (Arnqvist and Rowe 2005), in which a male reproductive strategy imposes costs on female fitness. The cost becomes particularly acute in organisms with slow rates of reproduction such as in primates. The potential fitness consequences of infanticide are so significant that it is rightly regarded as being as important as "traditional" socioecological factors, such as predation, in models of primate social evolution in general and female biology in particular (Kappeler and van Schaik 2002; Hrdy 1979; Agrell et al. 1998; Ebensperger 1998). Confusion of paternity through female promiscuity may perhaps be the best known of these female counterstrategies (Clarke et al., Chapter 3, in this volume), but protective association with a male defender has received increasingly focused scrutiny (van Schaik and Kappeler 1997; Palombit 2000).

Infanticide in the Okavango Chacma Baboons

Infanticide is seldom observed partly because attacks are typically unpredictable, opportunistic, and rapid, but also because opportunities for successful infanticide are rare. Sometimes, however, infanticide occurs at relatively high rates. The chacma baboons of the Okavango Delta are one such population. Enlarging the original data sets of Busse and Hamilton III (1981) and Collins et al. (1984), Palombit et al. (2000) reported that infanticide accounts for at least 38% of infant mortality, though relative rates as high as 75% may occur in certain years. A subsequent demographic analysis over a longer period not only confirmed this figure, but demonstrated that infanticide was comparatively more important than any other cause of death, including disease and predation (Cheney et al. 2004).

The patterning of infanticide in this population is more consistent with Hrdy's sexual selection hypothesis than alternative models (such as the generalized aggression hypothesis or the social pathology hypothesis). Infanticide is usually perpetrated by an adult male who has recently immigrated into the group and attained the alpha position in the dominance hierarchy (Figure 15.2). Targeted infants are not only those in the group at the time of alpha male turnover, but also those born within one gestation period after the new

Figure 15.2 An infanticidal male with his victim. Photograph by Ryne Palombit.

alpha's immigration. His status as a new immigrant largely eliminates the possibility of killing his own offspring. And his dominance rank positions him to benefit reproductively since (1) the mother's resumption of menstrual cycling is greatly accelerated by loss of her infant; and (2) alpha male chacma baboons largely monopolize periovulatory matings (Bulger 1993). The full significance of this benefit is made clear by the brief tenure of alpha males, averaging only seven to eight months. In other words, an alpha male chacma baboon has near exclusive sexual access to females, but only for a brief period. This scenario intensifies the potential benefit of infanticide. And the benefit is realized: following the majority of infanticides, mothers were observed to copulate with infanticidal males during subsequent conceptive cycles (Palombit et al. 2000).

Occasionally, a long-term resident male in the group may rise to the alpha position. We might expect that these males would be less inclined toward infanticide than a new immigrant male who has never copulated with the females of the group. But infanticide following such "resident male" turnover has been observed, and in all cases the male had not been observed to previously consort with the female during the cycle in which she conceived the infant he killed. The noteworthy ability of the alpha male to monopolize copulations may be responsible: mating may be so skewed in his favor that other resident males run little risk of killing their own offspring. Because the sample size of "resident male alpha turnovers" is relatively small, however, this supposition awaits further tests.

Demographic data from the Okavango not only suggest the potential reproductive value of infanticide to males, but also illuminate the dire problem it poses to females. Consider that a new adult male immigrates into the group every few months on average (Palombit 2003), that alpha male tenure averages only seven months, and that gestation lasts about six months (Gilbert and Gillman 1951). In principle, then, a female could be faced with a steady succession of infanticidal alpha males: following the death of her infant, she would resume cycling within a month or two, mate with the male and conceive a month or so after that, only to give birth six months later when, once again, a new immigrant male is now alpha, who then kills her infant, and so on. Females do succeed in rearing infants, of course, for at least four reasons. First, not all new immigrant males attain alpha status rapidly or at all (Cheney et al. 2004). Second, not all alpha males become "infanticidal." Rather, evidence of infanticide and high infant mortality is associated with one-third to one-half of new alpha males, indicating that infanticide may be pursued facultatively by males, depending on currently unclear criteria (Palombit et al. 2000). Third, older infants are more likely to survive wounds inflicted by infanticidal males (Palombit et al. 2000). Finally, the threat of infanticide may be mitigated to variable degrees by female counterstrategies, notably an association with a male defender.

The Counterstrategy: Recognizing "Friendships"

A necessary first step is reliably differentiating a friendship from the social relationships an anestrous female has with the other adult males in her group. Heterosexual friendships are, by and large, highly conspicuous associations, even to casual observers. Since social relationships are a function of the nature, patterning, and diversity of social interactions between individuals (Hinde 1983), the distinctiveness of friendships is best appreciated through consideration of multiple behavioral dimensions.

A useful initial indicator of an association is pronounced spatial proximity. A commonly used technique utilizes the so-called Composite Proximity Score (or C Score), which sums the amount of time two individuals are found within variable distances of one another. For example:

$$C = 1(T_{0\text{-}2\,m}) + 0.25(T_{2\text{-}6\,m})$$

where $T_{0\text{-}2m}$ and $T_{2\text{-}6m}$ are the percentages of time that a male and female spend within 0–2 m and 2–6 m, respectively (Palombit et al. (1997) (corresponding constants of 1 and 0.25 are weighting factors for each spatial category—see Smuts [1985]). A C-score is calculated for each male-female dyad in the group, and then the distribution of scores for a particular female is examined: a discontinuous distribution indicates a distinctive spatial association of this female with a particular male (or two) (Figure 15.3). Thus, friendships are male-(anestrous) female dyads characterized by a high degree of spatial proximity.

A striking feature of the Okavango chacma baboons is that this propinquity typically becomes conspicuous only after a female gives birth to an infant, usually quickly thereafter (Palombit et al. 1997). In the period immediately preceding the birth of an infant *and* also, in the relevant cases, immediately following the death of an infant, a female's spatial relations do not clearly reveal "special relationships" with any adult males in particular. Friendships are tied closely to the presence of a dependent infant.

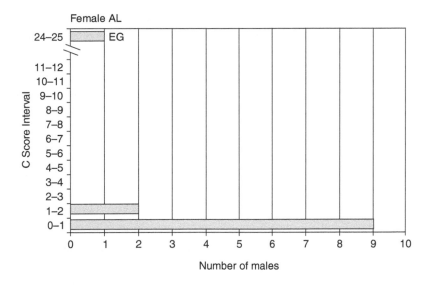

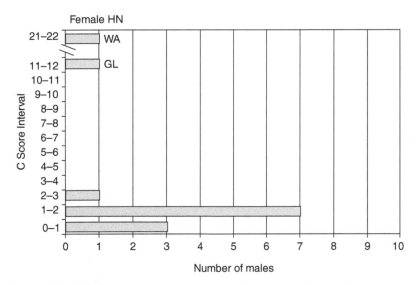

Figure 15.3 The distribution of composite proximity scores (C scores) for two representative females, AL and HN. The number of males in the group (on the *x*-axis) with a C score falling within the range given (on the *y*-axis) is shown for each female. "EG," "WA," and "GL" designate the names of male friends identified based on discontinuities in the distributions.

There is much more to friendship than simple spatial association, however, as reflected by the additional social interactions of friends. One of these is "infant handling," the prolonged, relaxed touching and manipulation of an infant attended or carried by its mother (Figure 15.4). A female's baby is frequently handled by multiple adult females (Silk et al. 2003), but when it comes to males, her friend handles an infant at nearly three times the rate of all other (non-friend) males, even after controlling for the greater propinquity of friends. Indeed, for nearly two-thirds of lactating females, friends are the *only* males who handle infants. The functional significance of infant handling remains unclear, though this behavior is not "agonistic buffering," in which a male uses an infant to shield himself from aggression from another male (e.g., Busse 1984a). The relevance of infant handling for distinguishing friendships is that it reflects a pronounced level of maternal tolerance ("trust") of a particular male and his attitude toward her infant.

Allogrooming is also significantly more common between friends. Approximately 7% of the time that friends are in close proximity is devoted to grooming, compared to no more than 2% (and usually far less) for nonfriend males. As with infant handling, the male friend is the *only* male grooming partner for many lactating females.

Figure 15.4 "Infant handling" by an adult male friend. Photograph by Ryne Palombit.

Finally, there is a distinct vocal dimension of friendships. The importance of vocalizations in regulating social behavior was first revealed in this population by study of female-female interactions. Adult females sometimes emit low-amplitude, tonal grunts when approaching one another (Cheney et al. 1995). Compared to "quiet approaches," this grunting increases the probability of both sustained proximity and ensuing affinitive interaction (such as grooming) between the females, especially if they have disparate dominance ranks. Grunts by males have a similar function, but with one important qualification (Palombit et al. 2000). In male approaches to females, grunts by males are also associated with elevated rates of affinitive interaction, if the female is cycling or pregnant. When the female is lactating, however, grunts have this mollifying effect *only* when they are performed by an approaching male friend (Figure 15.5). In other words, a female with infant shuns affinitive interaction with a nonfriend male, whether he grunts or not. Thus, females appear to classify males on the basis of their friendship status and accordingly moderate their response to a signal that normally paves the way for friendly interaction.

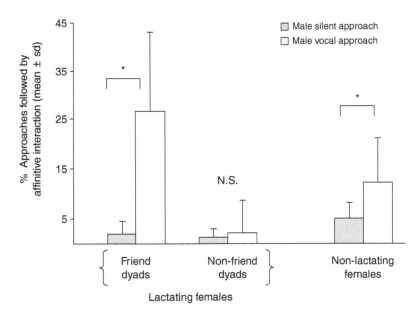

Figure 15.5 The effect of male grunting on subsequent affinitive interaction with females.

* Indicates significantly higher rates of affinitive interaction following vocal approaches by males ($p < 0.05$).

Who Are the Friends?

The variables we have described set friendships clearly apart from lactating females' relationships with other males in the group, but which individuals engage in friendship behavior? The vast majority of female chacma baboons have friendships with an adult male during at least some part of their lactation period. For example, of 19 births monitored among 14 females by Palombit et al. (1997), in only two cases did a parturient female fail to establish a friendship (in one case, the infant survived to weaning; in the other case, it was killed a few hours after birth). Females of all dominance ranks have male friends, although competition among females may sometimes influence the distribution of friendships in the group and over time. Similarly, both multiparous and primiparous mothers engage in friendship behavior. The modal pattern is for a lactating female to have one male friend at a time. In Palombit et al.'s (1997) study, for example, only 3 of 12 lactating females had multiple male friends simultaneously. None had more than two. In summary, any female who gives birth to an infant is likely to develop a friendship, regardless of her dominance and parity status.

Male friends are almost always fully adult, nonnatal individuals (in one case, a lactating female's friend was her young adult son). Males of all dominance rank may be potentially involved in friendships; thus most males participate in friendships at some point in their lives. The current alpha male is rarely a friend to females, however, unless his tenure exceeds one gestation period (i.e., unless he copulated with the lactating female during her previous conceptive cycle). Although lactating females avoid the current alpha male as friend, about 60% of the friend dyads involve *former* alpha males (Palombit et al. 1997), and females may preferentially seek out such males as associates.

The Adaptive Value for Females: Do Male Friends Make a Difference?

A useful starting point for understanding the purpose of friendships concerns the *costs* of friendships. Relationships are usefully viewed as elements of an individual's social world that it must develop, maintain, and "service" (Kummer 1978). If females benefit from associating with a male friend, we predict that they will invest in these relationships. Two behavioral measures reflect this investment.

First, the "Hinde Index" (Hinde 1977) usefully assesses which partner in a dyad is responsible for the close proximity (≤ 2 m) they share. In the current context, the index is obtained by subtracting the percentage of withdrawals due to the female from the percentage of approaches due to her action. The resulting index varies from −100, signifying complete male responsibility, to + 100, reflecting female responsibility. Hinde indices for chacma baboon friends are almost all positive, and the average value is strongly positive. Thus, it is the female who is responsible for the fact that friends are often near one another. A second assay of investment concerns grooming. Because allogrooming is a primary mechanism for establishing and maintaining social relationships in primates (Dunbar 1988; Harcourt 1988), its distribution may indicate the "value" individuals ascribe to their different social relationships (*sensu* Simpson 1991). Notably, nearly 90% of the grooming exchanged between friends is done by the female (Figure 15.6). Male friends rarely reciprocate grooming. In this sense, too, the female commitment to maintaining the relationship is conspicuously greater than the male's involvement. The existence of these costs of friendships to females implies compensatory benefits. Three hypotheses address the adaptive benefits of friendships to females:

Figure 15.6 Adult female grooming her male friend. Photograph by Ryne Palombit.

The *infanticide protection hypothesis* argues that males protect their female friends' infants from infanticidal attack.

The *female harassment hypothesis* posits that male friends protect lactating females from aggressive interactions with higher-ranking females. Dominant females may harass mothers (Altmann 1980) or handle roughly or temporarily "kidnap" their infants (e.g., Wasser 1983).

The *future male caretaker hypothesis* suggests that there are no immediate, protection-related benefits of friendship to females. Rather, a female establishes a friendship with a particular male in order to promote the development of a relationship (even perhaps attachment) between her infant and the male (Ransom and Ransom 1971; Seyfarth 1978). The proposed benefit of this bond comes later, when the youngster is an older infant or a juvenile, and the male may reduce its travel costs through occasional carrying (Anderson 1992), defend it from predators or from nonlethal aggression from other immatures or adult females (Altmann and Altmann 1970; Buchan et al. 2003), increase its access to food (Altmann 1980; Packer 1980), or accelerate its socialization and attainment of independence (Altmann 1980).

Four independent lines of evidence—demography, behavioral observations, playback experiments, and hormonal profiles—strongly suggest that the benefit of male friends to females relates more to infanticide protection than alternative contingencies. This evidence can be described with reference to these hypotheses' respective predictions, as follows.

Prediction 1: Friendship reduces infanticide risk to mothers.

Directly demonstrating this fitness benefit of male associates is difficult because there are not enough data to compare rates of infanticide suffered by females who have male friends and those *very* few females who do not. Another way to test this prediction, however, is to rephrase it:

Prediction 2: Friendship status enhances a male's willingness to invest in the anti-infanticide defense of particular females and their infants.

This prediction can be tested in two ways. First, we can compare the behavior of males during potential (or actual) infanticidal attacks by a new alpha male. Such comparisons are necessarily qualitative because these attacks do not lend themselves to systematic collection of behavioral data. Typically, many individuals are rushing about, covering relatively long distances, calling loudly, chasing, seeking cover, fleeing. It is important to understand that the commotion surrounding these attacks does not mean that infanticide is a side effect of ongoing aggression escalating "out of control" to fatally injure bystanders (*sensu* Bartlett et al. 1993). On the contrary, infanticidal attacks are typically sudden and precipitous, and are not preceded by other aggression. The pronounced arousal of other baboons is a *response* to the attack (e.g., chasing a male who has clutched an infant in his mouth). Even with the limitations on observations, two general features of infanticidal defense emerge.

One feature is that male friends are not the *only* respondents to these attacks. In a majority of cases (at least 60%), other nonfriend males and females are unambiguously aroused and move rapidly to the site of an attack, vocalize (e.g., scream or grunt), or even chase the attacker (Palombit et al. 2000). In one such attack, for example, an infant was saved when its 5-year-old sister snatched it up and carried it quickly away from the pursuing alpha male.

Another feature is the qualitative difference in the responses of adult males: males observed to directly engage the attacker (e.g., approach, chase, or fight him) or the infant (e.g., carry it) are always the friend of the infant's mother (Table 15.1; Figure 15.7). Although these males may be actively confronting a rival male who is arguably superior in fighting ability, their defense may still be effective in part because the distribution of injuries in escalated male-male fights is largely independent of the interactants' rank (Drews 1996). In other words, a middle-ranking male may still inflict a serious injury during a prolonged fight with a physically superior alpha. As predicted by the infanticide protection hypothesis, friends are more willing than other males to participate in the energetically costly and risky forms of anti-infanticide defense.

Table 15.1 Qualitative differences in male response to potentially infanticidal attacks by alpha male (summarized from Palombit et al. 2000).

Response Dimension	Friend Males	Non-friend Males
Move rapidly to attack site?	+	+
Remain at attack site for prolonged period in state of "aroused vigilance"	+	+
Vocalize in response to attack	+	+
Chase attacker	+	+
Initiate chase of attacker	+	−
Direct threats, appeasement signals, vocalizations at attacker at close range	+	−
Initiate and maintain close proximity to attacker	+	−
Physically fight attacker	+	−
Carry infant for extended travel	+	−

Figure 15.7 A vigilant adult male watches an infant he is protecting. Photograph by Ryne Palombit.

A second way to test Prediction 2 is through field playback experimentation, which provides supplemental data on male defensive predispositions under more controlled conditions. The experimental rationale derives from the fact that female cercopithecines typically emit screams when attacked by others, and these calls may function in part to recruit defensive aid from conspecifics (Gouzoules et al. 1984; Cheney and Seyfarth 1990; Bergman et al. 2003). The response of male chacma baboons to naturally occurring screams from out-of-sight females does not always involve movement toward the caller. Sometimes males respond simply by looking and visually scanning—albeit intently—for variable amounts of time.

Experimental playbacks exploit this natural variation in male responsiveness by selectively presenting female screams to male friends or nonfriend males under uniform conditions that control for numerous potentially relevant variables (e.g., immediately preceding social interaction) (Palombit et al. 1997). With this technique, we can now obtain a better sense of how males vary in their solicitude toward females in distress. Such experiments reveal that (1) a particular female scream elicits a stronger response from her male friend than a control male of similar rank and friendship status (i.e., the control male also has a female friend in the group, but not the one whose scream he hears); and (2) any given male responds more strongly to his female friend's scream than to the scream of another lactating female of similar dominance rank (Figure 15.8a). Visual scanning is likely to reflect the listener's investment in obtaining further information about the specific circumstances surrounding the vocalization (Marler et al. 1992) and is generally predictive of imminent changes in behavior (Rowell and Olson 1983).

Consequently, the results support the conclusion suggested by the *ad libitum* observations of infanticidal attacks that friendship with a female heightens a male's predisposition to protect her. Although this finding was partly unanticipated by the experimental design, in two trials a male subject did move immediately and quickly in the direction of the playback speaker—these males were both friends of the (playback) caller. Thus, as predicted by the infanticide protection hypothesis, a female's male friend seems to "care" more than other males about her safety and security. This hypothesis also specifies the temporal patterning of this male solicitude:

Prediction 3: A male's willingness to invest in defense is strictly tied to the presence of the female friend's infant.

The playbacks revealed striking support for this prediction: the solicitude of male friends vanishes with the infant. At various times during our study, infants died (usually, but not always, due to infanticide). When we repeated the same playbacks shortly after the deaths of these infants, male friends responded far less strongly; in fact, the magnitude of their responses was significantly *weaker* than that of control males (Figure 15.8b). Thus, as predicted by the infanticide protection hypothesis, a female's male friend seems to "care" more than other males about her safety and security, but only when she has an infant. Similarly, as noted earlier, a female ceases investment in her friendship behavior shortly following the death of her infant. For both sexes, then, the function of friendships is closely tied to neonates.

The empirical support for Predictions 2 and 3 is inconsistent with the future male caretaker,hypothesis, which holds that the advantage does not lie in immediate protection by males at the time of the friendship, but rather in the bonds that form later between males and youngsters.

The female harassment hypothesis, however, cannot be rejected on the basis of Predictions 2 and 3 since it also predicts that male friends will come to

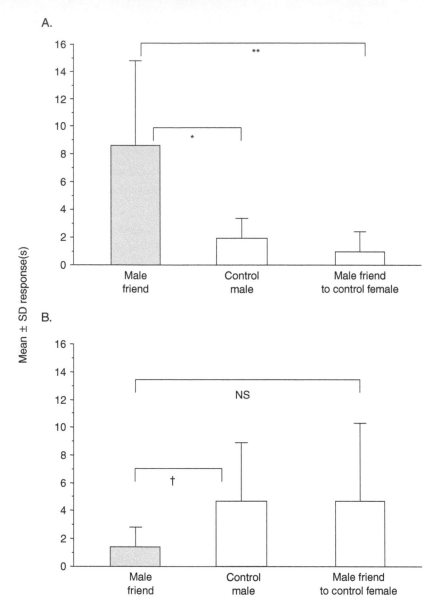

Figure 15.8 Experimental audio playback of female screams to adult male chacma baboons. The first two bars show the responses to the same stimulus scream of her male friend (black bar) or a control male of similar dominance rank and friendship status; that is, he has a female friend (white bar). The hatched bar is the response of the original male friend to the scream of a control (lactating) female. Part a presents results when the male friend's female partner had a living infant: male friends showed significantly (p<0.01) stronger responses than control males to the same scream (*), and male friends responded significantly stronger to the screams of their female friends than control females (**).

the aid of females in distress. To further test between the Infanticide Protection and the Female Harassment hypotheses, two additional predictions focus on their respective sources of danger.

Prediction 4: Male friends will respond more strongly than control males to attacks on females by the potentially infanticidal alpha male (infanticide protection hypothesis) or by higher-ranking females (female harassment hypothesis).

A final series of playback experiments reveal that it is the infanticidal contingency that most clearly differentiates the responses of friend versus nonfriend males. These experiments utilized a stimulus comprising a female victim's screams (as above), but now combined with the threat calls of the (simulated) attacker (Palombit et al. 1997). In these trials, male friends respond more strongly than control males when the aggressor was an infanticidal alpha male rather than a high-ranking female (or even a high-ranking, but noninfanticidal [beta-ranking] adult male). In other words, males seem to perceive that a potentially infanticidal attack threatens greater danger to their friends (and their infants) than does female harassment (or attack by nonalpha males), and they respond accordingly.

Prediction 5: Lactating females will exhibit avoidance of the potentially infanticidal alpha male (infanticide protection hypothesis) or higher-ranking females (female harassment hypothesis).

Lactating females unambiguously avoid new immigrant alpha males, as expressed by rapid retrieval of infants and flight from proximity of the male, often accompanied by highly intense displaying, such as screaming and the "tail-up" display (Busse 1984b; Palombit et al. 2000). This conspicuous pattern of avoidance is much less common among other females or among lactating females whose infants happen to be elsewhere at the time. Another striking feature of

---

Part b presents results of playbacks conducted within 1 to 4 weeks of the death of an infant: at this time, male friends showed significantly ($p < 0.05$) weaker responses to control males (†), and the male friends' response to friends' screams did not differ significantly from their response to the control female's scream. The response measure is the duration of the subject's looking in the direction of the playback speaker in the 20s after the vocal stimulus is played back minus the duration of looking in the same direction in the 20s preceding playback (Mean ± SD for pooled subjects.) All tests were the relevant paired comparisons as indicated, performed with two-tailed, Wilcoxon signed ranks tests for 30 playbacks (Part A) or 18 playbacks (Part B) distributed equally across conditions See Palombit et al., 1997 for details).

this avoidance is that the new alpha is often simultaneously directing affinitive or conciliatory signals such as lip-smacking or grunting at these fleeing, scream-ing females; these signals fail to mollify mothers carrying infants.

Behavioral data do not support the analogous prediction of the female harass-ment hypothesis, however: lactating females approach and avoid higher-ranking, unrelated females at the same rate they do when they are cycling. Finally, the fact that high-ranking females all the way up to the group's alpha female pursue friendships as reliably and vigorously as low-ranking individuals do (see later in this chapter) suggests that protection from harassment from other females is not the primary benefit offered by male friends.

Thus, behavioral data from the Okavango chacma baboons support the in-fanticide protection hypothesis more than the two alternative hypotheses. Re-cent hormonal data further support the infanticide protection hypothesis. Glucocorticoid hormones, such as cortisol, have long been considered physio-logical assays of stress (Sapolsky 1993; Palme et al. 2005). It is not surprising that immigration of a new male chacma baboon elevates glucocorticoid levels among group members. Not only does he destabilize the male dominance hi-erarchy by his arrival, but he pursues alpha status through increased aggression and threat displays, such as loud "wahoo" calling contests and vigorous chases (Kitchen et al. 2003; Fischer et al. 2004). Thus, circulating cortisol levels rise in males (including the immigrant) at this time (Bergman et al. 2005). It is the corresponding hormonal responses among females that are of interest in test-ing the infanticide protection hypothesis, which posits that:

Prediction 6: Compared to other females, females at risk of infanticide will experi-ence greater stress following immigration of a new male, and therefore show higher levels of circulating cortisol.

Alternatively, however, we might argue that the social disturbance caused by intensified male-male competition would affect females generally, or, if any-thing, cycling females in particular, for they among females are disproportion-ately targeted in the protracted, aggressive chases that seem to advance a new male's rise to alpha status (Kitchen et al., Chapter 6 in this volume):

Prediction 7: Immigration of a new male will be followed by cortisol increases in all females, or more in cycling females.

The hormonal data support Prediction 6 but reject Prediction 7. It is not cycling, but lactating (and pregnant) females who exhibit a significant rise in glucocorticoids following immigration of a new male (Beehner et al. 2005). These data demonstrate that this hormonal spike is not due to changes in rates of (noninfanticidal) aggression from males, female-female grooming, female-female aggression, and male-male aggression. The implication is clear: females at risk of infanticide from the new male undergo peculiarly high stress upon his arrival. This is borne out by observed *additional* increases in glucocorticoid concentration among lactating females when a new alpha commits infanticide (Engh et al. 2006). Finally, and most importantly, the infanticide protection hypothesis predicts that:

Prediction 8: Compared to mothers with male friends, lactating females without male friends are at higher risk of infanticide, and therefore will exhibit higher levels of stress-related hormones.

The data demonstrate that the few lactating females who do not have male friends not only sustain higher circulating levels of cortisol generally, but they also experience this predicted higher cortisol spike at alpha male turnover (Beehner et al. 2005). In light of the fact that males rarely—if ever—groom their female friends, this hormonal difference cannot be attributed simply to less grooming from a male friend.

Is Infanticide Protection from Friends Worth Competing For?

Competition among females for males is relatively rare in mammals, and consequently little studied (Berglund et al. 1993; Andersson 1994). Sexual selection theory (Darwin 1871) predicts that such female-female competition is most likely in two contexts: (1) a "sex role reversed" mating system (Petrie 1983), which certainly does not apply to the chacma baboon; and (2) when males provide a resource or service crucial to female reproduction. It is difficult to imagine a male service more critical to a female chacma baboon than preventing the death of her infant, particularly in light of the clear demographic impact of infanticide. Thus, more so than its alternatives, the infanticide protection hypothesis generates several additional predictions:

Prediction 9: Females will compete for male friends.

Female competition implies that males vary in their effectiveness as protectors of infants and hence in their value as friends. A male's dominance rank and paternity status are likely focal points of female-female competition because they imply greater competitive ability and defensive motivation, respectively:

Prediction 10: Females will compete for access to male friends who are high ranking and/or fathers of their infants.

Finally, we ultimately need to demonstrate the benefit of female-female competition:

Prediction 11: Competitively superior females will gain access to better male protectors and therefore suffer less risk of infanticide than less successful females.

Beginning with Prediction 9, several lines of evidence suggest that such competition occurs among lactating female chacma baboons.

Differential Access to the Male Friend. First, sometimes two mothers maintain friendships with the same male simultaneously (Figure 15.9). In such cases, the higher-ranking female maintains greater levels of close proximity and interaction (e.g., allogrooming) with him than her subordinate counterpart. The magnitude of this disparity is especially pronounced when the dominant is a young female and the subordinate is an old female. This may be partly because primiparous females appear to be more vulnerable than their multiparous counterparts to infanticide (Palombit et al. 2000; Cheney et al. 2004). That is, younger females not only have more to gain from friendships, but they may also have more to lose from unsuccessful competition with other females.

Competitive Exclusion from a Male Friend. Second, the pattern of differential access to the male friend results from competitive "displacement" of the lower-ranking female by the dominant rival. We can observe this progression

Figure 15.9 Two lactating females who are simultaneously engaged in friendships with the same adult male. Photograph by Ryne Palombit.

in situations when one lactating female has an exclusive friendship with a particular male, and then another female gives birth and establishes a friendship
with him as well. The consequences for the first female's friendship depend
on the dominance ranks of the females. If the first female is low-ranking, and
she is then "joined" by a higher-ranking mother, the subordinate female experiences a quick and dramatic decline in her propinquity and social interaction
with the male friend. The converse does not occur, however: a dominant's
friendship remains largely unchanged if a second, newly parturient low-ranking
female begins associating with the male (Palombit et al. 2001).

These changes in friendships are the result of female behavior rather than
of male friend behavior. These females do not usually physically fight or aggressively interfere with each other, but dominants supplant subordinates from
male friends in much the same way that they supplant them from foods. In
other words, dominant females compel low-ranking rivals to withdraw from the
contested resource, the male friend. Males apparently do little to influence this
outcome. In many cases, a low-ranking female is entirely supplanted from her
friendship with a male; that is, she abandons the relationship and pursues an
alternative strategy.

Attributes of Male Friends. Is there evidence that certain male characteristics
are particularly "valued" in female competition for male friends (Prediction
10)? One initial piece of evidence is the fact that two females' simultaneous
sharing of a single male friend does not simply derive from a lack of males. In
all cases where we observed this phenomenon, there were three to four long-
term resident, nonnatal adult males who were unengaged in friendships at the
time (though each of them participated in friendships at other times). Thus,
when establishing friendships, lactating females overlook "unbonded" adult
males in favor of males who already possess a male friend. Notably, subordinate
females are just as likely as dominants to choose a male that already has a friend,
even though sharing a friend imposes higher competitive costs on them. This
pattern already implies variation in male quality.

Two features of males stand out in this regard: dominance rank and probability of paternity of the female's infant. The importance of male rank in female competition is suggested by the positive correlation between friends' ranks:
high-ranking females acquire high-ranking males as friends (Figure 15.10).
There are several potential advantages of high rank in a male protector against
infanticide. First, the physical superiority of dominant males may make them
more effective defenders. Second, as described earlier, the vigor and speed of

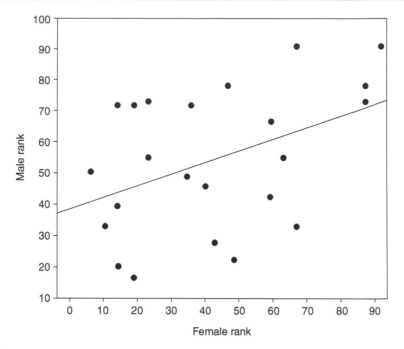

Figure 15.10 Association between the dominance ranks of male and female friends (rs = +0.42, p < 0.05). Rank is expressed as the percentage of individuals dominated.

the male friend's response to infanticidal attack may facilitate indirect defensive participation from other baboons. High rank in the male friend may facilitate this involvement by nonfriends, since rank and the relative fighting abilities of opponents influence monkeys' decisions about intercession in ongoing aggression (Harcourt 1992; Noë and Sluijter 1995; Silk 1999). Finally, of course, dominant males may have a higher probability of paternity of a female's infant and, therefore, a greater predisposition to defend it.

The potential importance of paternity is indicated by the fact that in at least 68% of friendships, the male had previously copulated with the female in the cycle in which she conceived her current infant (Palombit et al. 1997). Since females typically mate with only one or two adult male partners per cycle (Bulger 1993), this result suggests that females preferentially target possible fathers as friends. Without direct genetic analyses, however, it is difficult to specify how paternity influences male defense or friendship behavior, but one unexpected result of the experimental playbacks is intriguingly suggestive. There is a positive correlation between the magnitude of a male friend's response to the female's scream and the dominance rank he held when his female friend conceived her infant (6–10 months earlier) ($r_s = +0.70$, N = 10, p < 0.05). There was no such correlation between the male's response and his rank at the time of the actual playback, however. Hauser (1986) similarly found that playback of infant vervet monkey *(Cercopithecus aethiops)* screams elicited stronger responses from males with a presumed higher probability of paternity. Thus, to the extent that male rank is correlated with mating success, this result may reflect the importance of paternity in male protectiveness. If paternity influences chacma male protectiveness, it becomes a resource that females should seek to acquire and, if necessary, compete for.

Benefits of Competition: In light of these patterns, we can test Prediction 11 by answering this question: If high-ranking males are better protectors, and high-ranking females usually acquire them as friends, then are high-ranking females less vulnerable to infanticide? Demographic data collected to date do not suggest improved survivorship among the infants of high-ranking females (Cheney et al. 2004). If anything, *middle-ranking* mothers seem to suffer less infanticide than either their high- or low-ranking rivals. If we divide the female hierarchy into equal thirds, we find that approximately 12% of infanticide victims were born to mid-ranking females, compared to 53% and 36% being the offspring of high- and low-ranking females, respectively.

The implications of this demographic result for understanding female competition for friends is unclear. The fact that a decade of demographic data have not yet revealed any clear-cut reproductive advantage of high rank to females generally (Cheney et al. 2004) recommends further study of this long-lived, slowly reproducing primate. Conversely, high- and low-ranking females may be more vulnerable to infanticide. High-ranking females may be more vulnerable because they enjoy priority of access to feeding sites, as does the (infanticidal) alpha male, thereby bringing them together frequently. Low-ranking females may be more susceptible because their (lower-ranking) male friends are less effective deterrents. In any case, longitudinal data are needed to test this prediction.

Is Infanticide Protection Worth Competing For? Infanticide protection in chacma baboons raises a fundamental question: why is competition for male friends necessary at all? A "nondepreciable" resource should not stimulate competition (Kleiman and Malcolm 1981), and it is not immediately obvious why a male friend cannot guard several mothers (and their infants) at the same time, especially in light of the female responsibility for proximity maintenance.

After all, analogous protection against predation has been considered a service males can provide to multiple females simultaneously (van Schaik and van Noordwijk 1989; Rose and Fedigan 1995).

So, why do females compete for males? There are at least two possible reasons. First, females may act "spitefully" to lower the fitness of rivals by depriving them of a critical resource that is, in fact, shareable (Knowlton and Parker 1979; Berglund et al. 1993). Current data do not allow a test of this hypothesis in the chacma baboon. Alternatively, and more plausibly, the substantial investment female chacma baboons make in friendships suggests that social, spatial, and temporal access to a particular male, in order to "service" a relationship with him, is not necessarily equally shareable among several females at once. In other words, the opportunity to develop the friendship itself—which seems to crucially influence male defense—becomes the effectively depreciable resource for which females compete.

The Costs and Benefits of Friendships to Males

The adaptive significance of friendships as a female counterstrategy cannot be fully understood until we consider males. From a male perspective, it does not matter that females stand to benefit reproductively from their association. If friendship-based protection is costly to males, there must be compensatory fitness advantages for them in order for females to garner infanticide protection.

One might reasonably argue that at one level the costs of friendships to male chacma baboons appear trivial, or at least considerably less substantive than those impacting females. Even the casual observer of chacma baboon friendships is struck by the conspicuous female contribution to the relationship—through maintaining proximity and allogrooming—and the contrasting modicum that males seem to invest in this regard. In this situation, however, experimental playbacks usefully distill out the significant commitment of male friends. That is, although indifference may seem to characterize a male's daily interactions with an attentive female friend, playbacks under controlled conditions reveal that he does "care" disproportionately for her (and her infant) in circumstances that are potentially dangerous. The consequential costs of this solicitude are suggested by the nature of these males' direct defense during infanticidal attacks. Vigilance, active protection, and risk of injury are costs to males that direct our attention to potential benefits, which can be grouped into three hypothetical alternatives.

The parental effort hypothesis suggests that males are the fathers of the infants they protect, and thus they directly enhance their individual fitness through friendships. Observations that male friends are usually one of several former sexual consorts of their female partners in their previous conceptive cycle support this hypothesis. Thus, there is clear potential for paternity to motivate baboon social behavior.

The mating effort hypothesis suggests that males are unrelated to the infants of their current female friends, but friendship increases the probability of siring *future* offspring with these females through preferential mating (Smuts 1985). This scenario seems unlikely to operate in the Okavango chacma baboons in light of the apparent capacity of alpha males to competitively exclude other males from periovulatory matings, but only genetic data can conclusively test the hypothesis.

A third set of hypotheses can be subsumed under the diverse "Social Advantages" males may derive directly or indirectly from association with females. For example, friendship with an anestrous female may facilitate an immigrating male's integration into a group (Smuts 1985). Although there is one recorded case of a pregnant female chacma baboon befriending an immigrant male in this manner (Palombit unpublished data), this hypothesis cannot generally explain chacma baboon friendships because new immigrants are virtually never chosen as friends by lactating females (almost certainly because they constitute the primary risk of infanticide). Alternatively, male and female friends may mutually support each other in aggressive interactions. For example, the aid male rhesus macaques *(Macaca mulatta)* receive from their female friends crucially helps them achieve and maintain high rank (Smuts and Smuts 1993). There is currently no evidence, however, that male status in the Okavango chacma baboons depends in any meaningful way on assistance from females or, for that matter, from other males. In this population, coalitionary support is rare among females (even among kin) (Silk et al. 1999) and is virtually nonexistent among adult males. Finally, friendship with a female may facilitate a male's access to her infant to use as an "agonistic buffer" when another (higher-ranking) male threatens or attacks him (Deag and Crook 1971; Ransom and Ransom 1971). This hypothesis has not been studied intensively, but there is evidence that males may be able to use friends' infants as buffers more frequently than other infants (Busse 1984a; Palombit unpublished data). What is lacking, however, is a clear understanding of the quantitative fitness benefits of agonistic buffering to males and of whether these compensate for the costs of protecting these infants.

In summary, the fitness consequences for males—both costs and benefits— remain one of the least understood aspects of friendship behavior, not only in chacma baboons, but in baboons and primates generally (van Schaik and Paul 1996). At present, the parental effort hypothesis seems most plausible, but additional data are needed for direct tests. To this end, the relevant paternity analyses of DNA from the Okavango chacma baboon population are currently underway.

Conclusions and Future Directions

Taken together, these diverse data suggest that heterosexual friendships in this population are best understood as a female counterstrategy to infanticide. The Okavango chacma baboons are not unique in this regard. Weingrill (2000) came to the same conclusion in his study of chacma baboons in the Drakensberg Mountains of South Africa. More generally, there is accumulating evidence that female association with a male protector has evolved as an anti-infanticide strategy in a variety of mammals, birds, and insects (Palombit 2000). Of course, infanticide avoidance is not the only adaptive reason for male-female bonds generally (e.g., Gubernick and Teferi 2000), and even perhaps for baboon friendships in particular. As noted, infanticide does not appear to be nearly as important in male reproductive strategies among East African baboons (Palombit 2003), though friendships are widespread and prominent. In olive baboons, the benefits of friendships to females appear to relate to protection from nonlethal harassment (Smuts 1985; Palombit unpublished data).

Many of the questions remaining about chacma baboon friendships, infanticide, and their functional bases require much more study. Even the nature of the friendship itself warrants greater scrutiny. One intriguing question, for example, concerns whether baboon friendships involve distinctive psychoneuroendocrine foundations of the kind characterizing pair bonds in neotropical monkeys (e.g., Mendoza and Mason 1997), which appear to constitute adult attachment relationships (Mason and Mendoza 2002) *sensu* Bowlby (1977). Anecdotal behavioral evidence (see especially Smuts 1985, 1999), preliminary hormonal data, and their highly consequential fitness implications raise compelling possibilities that baboon friendships may indeed constitute strong, emotionally salient bonds for their participants, but this question awaits further study. Are perhaps these relationships as important to individuals as the matrilineal dominance relationships usually emphasized in descriptions of baboon societies? Indeed, it is the transient, but reproductively significant, nature of friendships makes them valuable foci of new explorations of social cognition (e.g., Cheney and Seyfarth 2004).

One particularly important question concerns how variation in the nature of friendships affects male defense, infanticide risk, and, especially, female lifetime reproductive success. Reproductive data accumulate slowly in baboons, but ancillary questions can be examined. For example, in light of the apparent differences in infant mortality due to infanticide, do friendships involving high- and low-ranking mothers differ systematically from those of middle-ranking females? The same question can be usefully extended across baboon populations; that is, do friendships vary in meaningful ways with infanticide risk intraspecifically?

Our understanding of the costs of friendships for females and especially males is relatively poor. It is difficult to measure the costs of anti-infanticide defense for males, but other costs, such as female-female competition, also merit attention. A female displacing a lower ranking mother from her friendship potentially imposes a significant cost on the relevant male if he is the father of the supplanted female's infant and if weakened friendship status heightens infanticide risk. If paternal or mating benefits underpin male involvement in friendships, it is surprising that males apparently do little to forestall this competitive outcome.

Intimately related to this understanding of costs is a related question: how does friendship compare with other anti-infanticide strategies available to females? Even the precise nature and limits of the alternatives are unclear. In this sense, a prominent association with a male has been argued to be a relatively *undesirable* anti-infanticide strategy for females because it effectively "advertises" the distribution of paternity, thereby *increasing* a female's vulnerability to infanticide (Harcourt and Greenberg 2001). This potential cost is probably insignificant in this chacma baboon population, however. Infanticide is usually perpetrated by a new immigrant male whose lack of any copulatory history in the group whatsoever makes the distribution of friendships in his new group largely superfluous to paternity-based contingencies. The cost is potentially more relevant when a long-term resident male rises to alpha rank, but even in this case, the argued ability of alpha male chacma baboons to skew mating in their favor once again implicates sexual history as more important. The degree of reproductive skew is, in fact, another deficiency in our knowledge, and it bears directly on the question of how much opportunity female chacma baboons have to avoid infanticide by confusing paternity (O'Connell and Cowlishaw 1994;

Henzi 1996;Cowlishaw and O'Connell 1996). Although confusion of paternity may not offer anti-infanticide benefits to female chacma baboons, Palombit et al. (2001) have suggested two other possible counterstrategies. First, older, dominant mothers may spend more time with female kin, at the expense of time spent with the male friend. Second, low-ranking females "displaced" from access to a male friend by a higher ranking female may become peripheral, avoiding all males generally. We know little about the adaptive significance of these possible counterstrategies, however. Presently, friendships appear to be the best option available to females confronted by this significant risk of infanticide.

In summary, then, study of friendships in the Okavango chacma baboons suggests that although males are a preeminent problem for female reproductive success, they are also, in a different context, the (best?) solution.

Acknowledgments

I thank Martin Muller and Richard Wrangham for inviting this contribution and for helpful suggestions on the chapter. I am grateful to the Office of the President and the Department of Wildlife and National Parks of the Republic of Botswana for sponsorship of the research, to the late Mokopi Mokopi for his contribution to data collection, and to Dorothy Cheney for fruitful discussion of the data presented here. Okavango Tours and Safaris and Gametrackers generously provided logistical support in the field. The research was supported by grants from the National Science Foundation, the National Institutes of Health, the Wenner-Gren Foundation, the L.S.B. Leakey Foundation, the Center for Human Evolutionary Studies (Rutgers University), the Research Foundation of the University of Pennsylvania, and the Institute for Research in the Cognitive Sciences (University of Pennsylvania).

References

Agrell, J., J. O. Wolff, and H. Ylönen. "Counter-Strategies to Infanticide in Mammals: Costs and Consequences." *Oikos* 83 (1998): 507–517.

Altmann, J. *Baboon Mothers and Infants*. Cambridge, Mass.: Harvard University Press, 1980.

Altmann, S. A., and J. Altmann. *Baboon Ecology*. Chicago: University of Chicago Press, 1970.

Anderson, C. M. "Levels of Social Organization and Male-Female Bonding in the Genus *Papio*." *American Journal of Physical Anthropology* 60 (1983): 15–22.

———. "Male Investment under Changing Conditions among Chacma Baboons at Suikerbosrand." *American Journal of Physical Anthropology* 87 (1992): 479–496.

Andersson, M. *Sexual Selection.* Princeton, N.J.: Princeton University Press, 1994.

Arnqvist, G., and L. Rowe. *Sexual Conflict.* Princeton, N.J.: Princeton University Press, 2005.

Bartlett, T. Q., R. W. Sussman, and J. M. Cheverud. "Infant Killing in Primates: A Review of Observed Cases with Specific Reference to the Sexual Selection Hypothesis." *American Anthropologist* 95 (1993): 958–990.

Beehner, J. S., J. Bergman, D. L. Cheney, R. M. Seyfarth, and P. L. Whitten. "The Effect of New Alpha Males on Female Stress in Free-Ranging Baboons." *Animal Behaviour* 69 (2005): 1211–1221.

Berglund, A., C. Magnhagen, A. Bisazza, B. König, and F. Huntingford. "Female-Female Competition over Reproduction." *Behavioral Ecology* 4 (1993): 184–187.

Bergman, T. J., J. C. Beehner, D. L. Cheney, and R. M. Seyfarth. "Hierarchical Classification by Rank and Kinship in Baboons." *Science* 302 (2003): 1234–1236.

Bergman, T. J., J. C. Beehner, D. L. Cheney, R. M. Seyfarth, and P. L. Whitten. "Correlates of Stress in Free-Ranging Male Chacma Baboons, *Papio hamadryas ursinus.*" *Animal Behaviour* 70 (2005): 703–713.

Bowlby, J. "The Making and Breaking of Affectional Bonds. I. Aetiology and Psychopathology in the Light of Attachment Theory." *British Journal of Psychiatry* 130 (1977): 201–210.

Buchan, J. S., S. C. Alberts, J. B. Silk and J. Altmann. "True Paternal Care in a Multi-Male Primate Society." *Nature* 425 (2003): 179–181.

Bulger, J. B. "Dominance Rank and Access to Estrous Females in Male Savanna Baboons." *Behaviour* 127 (1993): 67–103.

Busse, C. D. "Tail Raising by Baboon Mothers toward Immigrant Males." *American Journal of Physical Anthropology* 64 (1984a): 255–262.

———. "Triadic Interactions among Male and Infant Chacma Baboons." In *Primate Paternalism,* ed. D. M. Taub, pp. 186–212. New York: Van Nostrand Reinhold, 1984b.

Busse, C. D., and W. J. Hamilton III. "Infant Carrying by Male Chacma Baboons." *Science* 212 (1981): 1281–1283.

Chapman, T. "Evolutionary Conflicts of Interest between Males and Females." *Current Biology* 16 (2006): R744–R754.

Cheney, D. L., and R. M. Seyfarth. *How Monkeys See the World.* Chicago: University of Chicago Press, 1990.

———. "The Recognition of Other Individuals' Kinship Relationships." In *Kinship and Behavior in Primates,* eds. B. Chapais and C. M. Berman, pp. 347–364. Oxford: Oxford University Press, 2004.

Cheney, D. L., R. M. Seyfarth, J. Fischer, J. Beehner, T. Bergman, S. E. Johnson, D. M. Kitchen, R. A. Palombit, D. Rendall, and J. B. Silk. "Factors Affecting Reproduction and Mortality among Baboons in the Okavango Delta, Botswana." *International Journal of Primatology* 25 (2004): 401–428.

Cheney, D. L., R. M. Seyfarth, and J. B. Silk. "The Role of Grunts in Reconciling Opponents and Facilitating Interactions among Adult Female Baboons." *Animal Behaviour* 50 (1995): 249–257.

Collins, D. A., C. D. Busse, and J. Goodall "Infanticide in Two Populations of Savanna Baboons." In *Infanticide: Comparative and Evolutionary Perspectives,* eds. G. Hausfater and S. B. Hrdy, pp. 193–215. New York: Aldine, 1984.

Cowlishaw, G., and S. M. O'Connell. "Male-Male Competition, Paternity Certainty and Copulation Calls in Female Baboons." *Animal Behaviour* 51 (1996): 235–238.

Darwin, C. *The Descent of Man, and Selection in Relation to Sex,* 1st ed. London: J. Murray, 1871.

Deag, J. M., and J. H. Crook. "Social Behaviour and 'Agonistic Buffering' in the Wild Barbary Macaque, *Macaca sylvana* L." *Folia Primatologica* 15 (1971): 183–200.

Drews, C. "Contexts and Patterns of Injuries in Free-Ranging Male Baboons *(Papio cynocephalus)." Behaviour* 133 (1996): 443–474.

Dunbar, R. I. M. *Primate Social Systems.* New York: Cornell University Press, 1988.

Ebensperger, L. A. "Strategies and Counterstrategies to Infanticide in Mammals." *Biological Reviews of the Cambridge Philosophical Society* 73 (1998): 321–346.

Engh, A. L., J. C. Beehner, T. J. Bergman, P. L. Whitten, R. R. Hoffmeier, R. M. Seyfarth, and D. L. Cheney. "Female Hierarchy Instability, Male Immigration and Infanticide Increase Glucocorticoid Levels in Female Chacma Baboons." *Animal Behaviour* 71 (2006): 1227–1237.

Fischer, J., D. M. Kitchen, R. M. Seyfarth, and D. L. Cheney. "Baboon Loud Calls Advertise Male Quality: Acoustic Features and Their Relation to Rank, Age, and Exhaustion." *Behavioral Ecology and Sociobiology* 56 (2004): 140–148.

Gilbert, C., and J. Gillman. "Pregnancy in the Baboon *(Papio ursinus)." South African Journal of Medical Science* 16 (1951): 115–124.

Gouzoules, S., H. Gouzoules, and P. Marler. "Rhesus Monkey *(Macaca mulatta)* Screams: Representational Signalling in the Recruitment of Agonistic Aid." *Animal Behaviour* 32 (1984): 182–193.

Gubernick, D. J., and T. Teferi. "Adaptive Significance of Male Parental Care in a Monogamous Mammal." *Proceedings of the Royal Society of London Series B Biological Sciences* 26 (2000): 147–150.

Harcourt, A. H. "Alliances in Contests and Social Intelligence." In *Machiavellian Intelligence: Social Expertise and the Evolution of Intellect in Monkeys, Apes, and Humans,* eds. R. W. Byrne and A. Whiten, pp. 132–152. Oxford: Clarendon Press, 1988.

———. "Coalitions and Alliances: Are Primates More Complex Than Non-Primates?" In *Coalitions and Alliances in Humans and Other Animals,* eds. A. H. Harcourt and F. B. M. de Waal, pp. 445–471. Oxford: Oxford University Press, 1992.

Harcourt, A. H., and J. Greenberg. "Do Gorilla Females Join Males to Avoid Infanticide? A Quantitative Model." *Animal Behaviour* 62 (2001): 905–915.

Henzi, S. P. "Copulation Calls and Paternity in Chacma Baboons." *Animal Behaviour* 51 (1996): 233–234.

Hinde, R. A. "On Assessing the Bases of Partner Preferences." *Behaviour* 62 (1977): 1–9.

Hinde, R. A., ed. *Primate Social Relationships: An Integrated Approach.* Sunderland, Mass.: Sinauer, 1983.

Hrdy, S. B. "Infanticide among Animals: A Review, Classification, and Examination of the Implications for Reproductive Strategies of Females." *Ethology and Sociobiology* 1 (1979): 13–40.

Hrdy, S. B., C. H. Janson, and C. P. van Schaik. "Infanticide: Let's Not Throw out the Baby with the Bath Water." *Evolutionary Anthropology* 3 (1995): 151–154.

Kappeler, P. M., and C. P. van Schaik. "Evolution of Primate Social Systems." *International Journal of Primatology* 23 (2002): 707–740.

Kitchen, D. M., R. M. Seyfarth, J. Fischer, and D. L. Cheney. "Loud Calls as Indicators of Dominance in Male Baboons *(Papio cynocephalus ursinus).*" *Behavioral Ecology and Sociobiology* 53 (2003): 374–384.

Kleiman, D. G., and J. R. Malcolm "The Evolution of Male Parental Investment in Mammals." In *Parental Care in Mammals,* eds. D. J. Gubernick and P. H. Klopfer, eds., pp. 347–387. New York: Plenum, 1981.

Knowlton, N., and G. A. Parker. "Evolutionary Stable Strategy Approach to Indiscriminate Spite." *Nature* 279 (1979): 419–421.

Kummer, H. "On the Value of Social Relationships to Nonhuman Primates: A Heuristic Scheme." *Social Science Information* 17 (1978): 687–705.

Manson, J. H. "Mating Patterns, Mate Choice, and Birth Season Heterosexual Relationships in Free-Ranging Rhesus Macaques." *Primates* 35 (1994): 415–433.

Marler, P., C. S. Evans, and M. D. Hauser "Animal Signals: Motivational, Referential, or Both?" In *Nonverbal Vocal Communication: Comparative and Developmental Approaches,* eds. H. Papousek, U. Jürgens, and M. Papousek, pp. 66–86. New York: Cambridge University Press, 1992.

Mason, W. A., and S. P. Mendoza. "Generic Aspects of Primate Attachments: Parents, Offspring and Mates." *Psychoneuroendocrinology* 23 (1998): 765–778.

Mendoza, S. P., and W. A. Mason. "Attachment Relationships in New World Primates." *Annals of the New York Academy of Sciences* 807 (1997): 203–209.

Noë, R., and A. A. Sluijter. "Which Adult Male Savanna Baboons Form Coalitions?" *International Journal of Primatology* 16 (1995): 77–105.

O'Connell, S. M., and G. Cowlishaw. "Infanticide Avoidance, Sperm Competition and Mate Choice: The Function of Copulation Calls in Female Baboons." *Animal Behaviour* 48 (1994): 687–694.

Packer, C. "Male Care and Exploitation of Infants in *Papio anubis.*" *Animal Behaviour* 28 (1980): 512–520.

Palme, R., S. Rettenbacher, C. Touma, S. M. El-Bahr, and E. Möstl. "Stress Hormones in Mammals and Birds: Comparative Aspects Regarding Metabolism, Excretion, and Noninvasive Measurement in Fecal Samples." *Annals of the New York Academy of Sciences* 1040 (2005): 162–171.

Palombit, R. A. "Infanticide and the Evolution of Pair Bonds in Nonhuman Primates." *Evolutionary Anthropology* 7 (1999): 117–129.

———. "Male-Female Social Relationships and Infanticide in Animals." In *Male Infanticide and Its Implications,* eds. C. P. van Schaik and C. H. Janson, pp. 239–268. Cambridge: Cambridge University Press, 2000.

———. "Male Infanticide in Wild Savanna Baboons: Adaptive Significance and Intraspecific Variation." In *Sexual Selection and Reproductive Competition in Primates: New Perspectives and Directions,* ed. C. Jones, pp. 364–411. New York: American Society of Primatologists, 2003.

Palombit, R. A., D. L. Cheney, J. Fischer, S. Johnson, D. Rendall, R. M. Seyfarth, and J. B. Silk "Male Infanticide and Defense of Infants in Wild Chacma Baboons." In *Infanticide by Males and Its Implications,* eds. C. P. van Schaik and C. H. Janson, pp. 123–152. Cambridge: Cambridge University Press, 2000.

Palombit, R. A., D. L. Cheney, and R. M. Seyfarth. "The Role of Male Grunts in Facilitating Interactions with Females in Wild Chacma Baboons *(Papio cynocephalus ursinus)." Behaviour* 136 (1999): 221–242.

———. "Female-Female Competition for Male "Friends" in Wild Chacma Baboons *(Papio cynocephalus ursinus)." Animal Behaviour* 61 (2001): 1159–1171.

Palombit, R. A., R. M. Seyfarth, and D. L. Cheney. "The Adaptive Value of Friendships to Female Baboons: Experimental and Observational Evidence." *Animal Behaviour* 54 (1997): 599–614.

Parker, G. A. "Sexual Selection and Sexual Conflict." In *Sexual Selection and Reproductive Competition in Insects,* eds. M. S. Blum and N. A. Blum, pp. 123–166. New York: Academic, 1979.

Petrie, M. "Mate Choice in Role-Reversed Species." In *Mate Choice,* ed. P. Bateson, pp. 167–179. Cambridge: Cambridge University Press, 1983.

Ransom, T. W., and B. S. Ransom. "Adult Male-Infant Relations among Baboons *(Papio anubis)." Folia Primatologica* 16 (1971): 179–195.

Rose, L. M., and L. M. Fedigan. "Vigilance in White-Faced Capuchins, *Cebus capucinus,* in Costa Rica." *Animal Behaviour* 49 (1995): 63–70.

Rowell, T. E., and D. K. Olson. "Alternative Mechanisms of Social Organization in Monkeys." *Behaviour* 86 (1983): 31–54.

Sapolsky, R. M. "Neuroendocrinology of the Stress-Response." In *Behavioral Endocrinology,* eds. J. Becker, S. M. Breedlove, and D. Crews, pp. 287–324. Cambridge, Mass.: MIT Press, 1993.

Schneider, J. S., M. K. Stone, K. E. Wynne-Edwards, T. H. Horton, J. Lydon, B. O'Malley, and J. E. Levine. "Progesterone Receptors Mediate Male Aggression toward Infants." *Proceedings of the National Academy of Sciences of the United States of America* 100 (2003): 2951–2956.

Seyfarth, R. M. "Social Relationships among Adult Male and Female Baboons. II. Behaviour throughout the Female Reproductive Cycle." *Behaviour* 64 (1978): 227–247.

Silk, J. B. "Male Bonnet Macaques Use Information about Third-Party Rank Relationships to Recruit Allies." *Animal Behaviour* 58 (1999): 45–51.

Silk, J. B., D. Rendall, D. L. Cheney, and R. M. Seyfarth. "Natal Attraction in Adult Female Baboons *(Papio cynocephalus ursinus)* in the Moremi Reserve, Botswana." *Ethology* 109 (2003): 627–645.

Silk, J. B., R. M. Seyfarth, and D. L. Cheney. "The Structure of Social Relationships Among Female Savanna Baboons." *Behaviour* 136 (1999): 670–703.

Simpson, M. J. A. "On Declaring Commitment to a Partner." In *The Development and Integration of Behaviour: Essays in Honour of Robert Hinde*, ed. P. Bateson, pp. 271–293. Cambridge: Cambridge University Press, 1991.

Smuts, B. B. *Sex and Friendship in Baboons.* New York: Aldine, 1985.

———. *Preface to reprinted edition of Sex and Friendship in Baboons.* Cambridge, Mass.: Harvard University Press, 1999.

Smuts, B. B., and R. W. Smuts. "Male Aggression and Sexual Coercion of Females in Nonhuman Primates and Other Mammals: Evidence and Theoretical Implications." *Advances in the Study of Behavior* 22 (1993): 1–63.

Strum, S. C. "Life with the Pumphouse Gang: New Insights into Baboon Behavior." *National Geographic* 147 (1974): 672–691.

Trivers, R. L. "Parental Investment and Sexual Selection." In *Sexual Selection and the Descent of Man, 1871–1971*, ed. B. Campbell, pp. 136–179. Chicago: Aldine, 1972.

van Schaik, C. P. "Infanticide by Male Primates: The Sexual Selection Hypothesis Revisited." In *Infanticide by Males and Its Implications*, eds. C. P. van Schaik and C. H. Janson, pp. 27–60. Cambridge: Cambridge University Press, 2000.

van Schaik, C. P., and P. M. Kappeler. "Infanticide Risk and the Evolution of Male-Female Association in Primates." *Proceedings of the Royal Society of London Series B Biological Sciences* 264 (1997): 1687–1694.

van Schaik, C. P., and A. Paul. "Male Care in Primates: Does It Ever Reflect Paternity?" *Evolutionary Anthropology* 5 (1996): 152–156.

van Schaik, C. P., and M. A. van Noordwijk. "The Special Role of Male *Cebus* Monkeys in Predation Avoidance and Its Effect on Group Composition." *Behavioral Ecology and Sociobiology* 24 (1989): 265–276.

vom Saal, F. S. "Proximate and Ultimate Causes of Infanticide and Parental Behavior in Male House Mice." In *Infanticide: Comparative and Evolutionary Perspectives*, eds. G. Hausfater and S. B. Hrdy, pp. 401–424. New York: Aldine, 1984.

Wasser, S. K. "Reproductive Competition and Cooperation Among Yellow Baboons." In *Social Behavior of Female Vertebrates*, ed. S. K. Wasser, pp. 349–390. New York: Academic, 1983.

Weingrill, T. "Infanticide and the Value of Male-Female Relationships in Mountain Chacma Baboons *(Papio cynocephalus ursinus)*." *Behaviour* 137 (2000): 337–359.

16

The Absence of Sexual Coercion in Bonobos

Tommaso Paoli

Bonobos *(Pan paniscus)* and chimpanzees *(Pan troglodytes)* share a similar fission-fusion society, characterized by male philopatry and female dispersal, in which communities of up to 140 individuals form temporary subgroups (or "parties") that fluctuate in size, composition, and duration. Despite these common features, three decades of research have revealed clear differences between the two species in their social behavior, particularly in the nature of social bonds within and between the sexes. Bonobos exhibit more frequent male-female and female-female associations (Kuroda 1980; Badrian and Badrian 1984; White 1996; de Waal, 2001), a reduced level of aggression within and between groups, and extensive nonconceptive sexuality (de Waal 1987; Wrangham 1993; de Waal 1995; Manson et al. 1997; Paoli et al. 2006a). Bonobos also exhibit co-dominance between the sexes (defined as female dominance over some males: Hemelrijk et al. 2003; Wrangham 1993; Paoli et al. 2006b). This phenomenon has led many authors to describe bonobos as "egalitarian" (Kuroda 1980; Kano, 1992; Wrangham and Peterson 1996; Furuichi 1997; Fruth et al. 1999;Hohmann and Fruth, 2003), but recent data suggest that "tolerant" is a more accurate term (Paoli et al. 2006b). In either case, bonobos can confidently be described as less despotic than chimpanzees (Boehm 1999).

In a controversial review, Stanford (1998:406) argued that the purported diversity "between the social systems of bonobos and chimpanzees may not accord well with field data." Some authors supported this position, attributing published differences between the two species to unequal sampling (see comments by Fruth and McGrew in Stanford 1998). Others were more skeptical, noting that many behavioral differences between the two species are readily observable in both the wild and captivity (see comments by de Waal, Kano, and Parish in Stanford 1998). Doran et al. (2002) reviewed a range of data on wild populations of both species and concluded that bonobos and chimpanzees can be distinguished on the basis of at least five features. Bonobos are characterized by (1) more varied intergroup encounters (including sexual interactions between communities) and different mechanisms of social integration for female immigrants, (2) absence of male dominance and a strong tendency for females to maintain feeding priority, (3) greater female sociality (reduced tendency for females to travel alone and less disparity in male and female ranging behavior), (4) weaker male-male bonds (less male-male association and reduced importance of male coalitions for establishing and maintaining rank and defending territories) and (5) a novel use of socio-sexuality. In light of the data reported so far, it can be fairly assumed that bonobos present a unique behavioral repertoire among primates: female bonding without female kinship, male kinship without male bonding, and relatively little ability of males to dominate females (White 1996).

Bonobo Sexuality

Bonobo sexual behavior differs from that of chimpanzees in several ways. First, almost all studies of bonobo sexuality report both ventroventral and ventrodorsal mountings and copulations. In contrast, chimpanzee mating is typically ventrodorsal (Tutin 1979; Tutin and McGinnis 1981,; Blount 1990). Second, bonobos frequently use sexual behavior to ease tension and defuse potential conflict (Kano and Mulavwa 1984; Kuroda 1984; Thompson-Handler et al. 1984; de Waal 1987, 1992; Blount 1990; Paoli et al. 2007). Third, sexual interactions can involve individuals of all sex and age classes: socio-sexual contact occurs between males (e.g., rump-rump rubbing, in which two males dorsodorsally rub their anal are: Kano 1980), between females (see the next paragraph), and between immature individuals and adults.

Bonobo females are well known for their genitogenital-rubbing (or GG-rubbing: Kano 1980; Kuroda 1980), a homosexual mounting behavior in which two participants grasp each other and swing their hips laterally while keeping the front tips of the vulvae, where the clitorises protrude, in close contact. Initially observed in captivity by Savage-Rumbaugh and Wilkerson (and christened "homosexual copulation"), this behavior occurs regularly in the wild (Kano 1980; Thompson-Handler et al. 1984; Hohmann and Fruth 2000).

Anestis (2004) recently claimed to have documented GG-rubbing in a small group of captive adolescent female chimpanzees, but the term seems inappropriate in this case. Genital contacts in female chimpanzees differ in several respects from GG-rubbing in both captive and wild bonobos, particularly in the variety of positions assumed and the context in which the sessions occur. Most remarkably, chimpanzee females were found to exchange genital contacts ventroventrally in a small fraction of sessions; they never lifted their partners off the ground, and they showed no evidence of muscle contractions or clenching resulting from rubbing. Thus, GG-rubbing appears to be characteristic of female *Pan paniscus*, and many hypotheses have been proposed to elucidate its biological meaning, including signaling dominance, expressing relationship quality, regulating social tension, and indicating female proceptivity (for an extensive review, see Hohmann and Fruth 2000). Paoli et al. (2006a), suggest an additional hypothesis: given that GG-rubbing is an immediate and intense way of testing the willingness to interact "fairly" by the exposure of a vulnerable part of the body, it may be a means of social assessment (Wrangham 1993), possibly complementary to affinitive interactions (cf. baboon "diddling": Whitham and Maestripieri 2003).

In addition to these differences in sexual behavior, many studies have reported physiological differences in the ovarian cycles of chimpanzees and bonobos, with an increased period of maximal swelling in bonobos (Thompson-Handler et al. 1984; Dahl 1986; Furuichi 1987; Blount 1990; Kano 1992). Recent data (Paoli et al. 2006a) showed a maximum swelling duration of 13.4 ± 0.7 S.E. days (37.6% of the perineal cycle), which is only slightly longer than that reported for chimpanzees (ranging from 9.6 to 12.5 days in the literature: reviewed in Paoli et al. 2006a). Minimally, a discrepancy of 1–3 days seems to exist, and the presence of this "prolonged" perineal tumescence has been related to the extended female attractivity (Kano 1989; Stanford 1998; Furuichi and Hashimoto 2004) and hypersexuality (de Waal 1987; Wrangham 1993) of female bonobos.

Does Sexual Coercion Exist in Bonobos?

To date there have been multiple contradictory results from both captive and wild populations on the ability of male bonobos to monopolize females. Some wild studies have indicated a positive relationship between male dominance rank and copulation rates (see Kano 1996), whereas others reported the absence of such an effect (Gerloff et al. 1999; Furuichi and Hashimoto 2004). In captivity males have been described as not monopolizing copulations (Stevens et al. 2001), and even where an unequal distribution of copulations has been observed, the alpha male was not always responsible for the majority (Marvan et al. 2006; Paoli and Palagi 2008). It has been suggested that the high status of females in bonobo groups may enhance the potential for female choice (Furuichi 1992; Kano 1996; Fruth et al. 1999), which is facilitated by prolonged and frequent swellings (Furuichi and Hashimoto 2004; Paoli et al. 2006a). Thus, it is possible that high-ranking females choose their mating partners with little or no objection from other males.

Although our knowledge of bonobo behavior has steadily increased over the past two decades, males have never been reported to employ coercive aggression against females in the immediate context of courtship (Wrangham 1993; Kano 1996; Parish 1994, 1996; Hohmann and Fruth 2003). In order to determine whether unpublished observations of such behavior exist, I developed a simple questionnaire which was distributed to both captive and wild bonobo researchers. Specifically, I asked participants to provide the number of coercion cases per male, the total observation time (hours), the number of adult males and females in the group, and the actor's rank.

Some researchers were able to provide detailed data, whereas others had anecdotal, albeit interesting, observations. In addition, I included personal observations of the bonobo group hosted at the Apenheul Primate Park (The Netherlands; details on the group and study methods can be found in Paoli et al., 2006a, 2006b), thus obtaining the data in Table 16.1.

Table 16.1　Lack of male sexual coercion in captive study groups.

Bonobo Group	# Adult Males and Females	Observation Time	# Male Sexual Coercion Cases
Apenheul	3 ♂, 5 ♀ (individuals at the beginning of the study)	Group: about 1700 hours	0
San Diego Zoo[a]	2 ♂, 4 ♀	Group: about 114 hours	0
San Diego Wild Animal Park[a]	3 ♂, 4 ♀	Group: about 114 hours	0[b]

[a] Data from personal communication by Amy Pollick.
[b] 5 cases described, but without use of force, thus more correctly defined as strong advances. 2 cases for the highest ranking male and 3 cases for the lowest ranking.

Jo Thompson provided a fascinating anecdotal account of the behavior of bonobos in the Lukuru project area, Democratic Republic of Congo: "Although I have witnessed countless incidents of sexual interaction, I have observed no cases of coercion. I gave thought to the context of a coercive act . . . incidents where force was applied for sexual activity. During heightened social excitement or conflict situations associated with a grimacing expression, sexual activity occurs to appease individuals or ease tension in the group. In all cases that I have observed, individuals either approach each other, turn towards each other, or engage in the activity without force."

Another remarkable comment comes from Monique Fortunato's study of the Columbus Zoo bonobo colony: "I have witnessed males of all rank sexually present to females to such an extent that I would almost call it "harassment", but I have never observed male sexual coercion in bonobos. In contrast, I have seen some 'sexual coercion' by females to males—not in the strictest sense, in which there are beatings and/or forced copulation, but there is one female in particular that makes very strong advances to males for oral or manual stimulation".

A similar description of "coercion" by females was provided by Amy Pollick in her study of the San Diego Zoo group: "I saw plenty of sexual coercion in the San Diego Zoo bonobos—mostly just one female (Lolita) towards one male (Junior). Lolita (second-ranking female) must have done it 40 times in 114 hours of observation, whereas the third-ranking female just 3 times."

In the captive Apenheul group, I have never observed any case of sexual coercion immediately preceding copulation by either males or females. Males generally performed strong advances toward females during periods of high excitement (mostly feeding times; see also Paoli et al. 2007) but they never used their physical strength to force females into a sexual contact.

As previously mentioned, it has also been suggested that male aggression against females may not be followed immediately by copulation, but can still increase the mating success of the aggressor, either by making females more likely to mate with him in the future (intimidation) or by discouraging females from mating with other males (coercive mate guarding). To test for such an effect in the Apenheul group, I looked at the correlation between rates of male-female aggression and overall copulatory frequency across individual male-female dyads (for aggression data see Table 2 in Paoli et al., 2006b; copulatory rates are based on 450 hours of focal sampling (Altmann 1974) in one observation session). No correlation was found (Matman's rowwise correlation test: $K_r = -1$, $tau_{rw} = -0.038$, n.s.; Figure 16.1), and thus even these less immediate forms of coercion seem to be absent in bonobos.

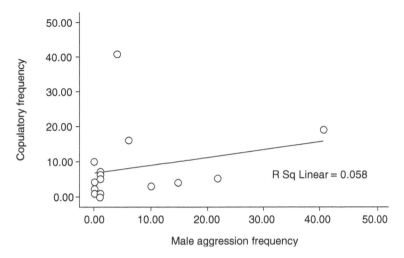

Figure 16.1 Correlation between male aggression against females and copulatory frequency across male-female dyads.

If male aggression against females functions as sexual coercion, then it is expected to intensify in reproductive contexts: specifically, males should direct aggression predominantly toward females that are sexually attractive (e.g., showing maximum swelling) (Smuts and Smuts 1993; Muller et al., Chapter 1 in this volume). To test this hypothesis, I performed a comparison of rates of aggression by males toward females across all male-female dyads, as a function of female swelling status (for this analysis aggression rates are based on focal observations in one study period, including three adult males). Interestingly, I found that males directed significantly lower rates of aggression toward females during the maximum swelling phase (randomization test for paired samples: $t=2.074$, $n=15$, $p<0.05$; Figure 16.2).

This finding contrasts with those of Hohmann and Fruth (2003), who reported a higher frequency of male aggression toward estrous females in a wild community of bonobos at Lomako. However, Hohmann and Fruth used pooled data and deviation from expectation, which is a less reliable approach from an ethological perspective. Furthermore, there was no indication of an immediate causal relationship between male-female aggression and copulation in the same dyads in the following 15 minutes. Nor were there any observations of forced copulation.

As a further test of coercion in the Apenheul group, I examined male aggression rates toward maximally swollen females as a function of female reproductive state (parous *vs.* nulliparous; see Table 3 in Paoli et al. 2006a), as done by Muller

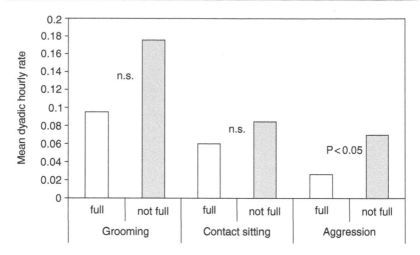

Figure 16.2 Grooming and contact sitting rates in male-female dyads and aggression by males toward females during maximum swelling phase (white bars) and nonmaximum swelling/detumescence (black bars).

and colleagues (2007, Chapter 8) for Kanyawara chimpanzees. In this case, a randomization test for two independent samples showed that rates of male aggression did not differ significantly between male-female$_{parous}$ dyads and male-female$_{nulliparous}$ dyads ($t = -1562$, $n_1 = 9$, $n_2 = 6$, n.s.), though the mean dyadic hourly rate was slightly higher for nulliparous females (male-female$_{parous}$ = 0.007 ± 0.021 SD; male-female$_{nulliparous}$ = 0.054 ± 0.089 SD). This differs from the pattern in Kanyawara chimpanzees, where parous swollen females received significantly higher rates of male aggression compared to nulliparous females. Thus, it seems that in bonobos, male aggression does not predictably intensify in reproductive contexts.

In light of this overall evidence, if we use the definition of sexual coercion provided by Smuts and Smuts (1993): "use by a male of force, or threat of force, that functions to increase the chances that a female will mate with him at a time when she is likely to be fertile, and to decrease the chances that she will mate with other males, at some cost to the female"), these new data support the idea that this behavior is virtually nonexistent in bonobos.

Some Hypotheses on the Absence of Sexual Coercion in Bonobos

Why do male bonobos apparently not engage in sexual coercion of females? Smuts and Smuts (1993) suggested that female-female alliances can be an important source of protection against sexual coercion, and other authors have reported that female bonobos are more attracted to other females than to males as social partners (White 1988; Idani 1991; Kano 1992; Palagi et al. 2004). Thus, it is possible that in bonobos strong female alliances discourage male aggression (the *female-defense alliance hypothesis:* Parish 1996). However, Hohmann and Fruth (2003) reported that females did not support each other during charges against males. They proposed that male sexual coercion is scarce in bonobos not because of the risk of retaliation from females, but because of the support that a female may receive from other females *and* males. The authors noted that when a female headed an attack against resident or strange males, she was always supported by both males *and* females.

On the other hand, it is unquestionable that bonobo females in both the wild and captivity are characterized by strong social bonds. An interesting hypothesis suggested by Takahata et al. (1996) is that "female bonobos may show maximal swelling in order to exchange GG-rubbing with other females, rather than to copulate with males." In support of this idea, Paoli et al. (2006a) found that GG-rubbing among females was significantly more frequent when focal females were in the maximum swelling phase. Moreover, in this study females were found to participate in GG-rubbing at higher rates than they participated in heterosexual copulations during maximum swelling. In light of this evidence, an interesting hypothesis can be suggested: if the perineal swelling is a means of attractivity also among females, then the occurrence of a prolonged maximum swelling phase might function as a source of protection against male coercion. Given that females tend to GG-rub frequently in the swelling phase and that they likely enhance their social bonds by doing so (Kano 1992; Wrangham 1993; Furuichi 1997; Paoli et al. 2006a), they might hamper male coercive attempts and facilitate relatively quiet copulatory interactions. This idea is supported by the significantly lower frequency of male aggression against females during the maximum swelling phase that I observed in the Apenheul group. However, in light of the small sample size of females in this study and the contrasting evidence provided by Hohmann and Fruth (2003), additional investigation is needed to obtain a larger and more detailed bonobo data set.

Given the prolonged swelling period of female bonobos, males may benefit more by associating with females throughout their entire perineal cycle than restricting their sexual initiative to the maximum swelling phase (Hohmann and Fruth 2003). In the Apenheul group, rates of female-male contact sitting and grooming exchange were independent of the female swelling status (contact sitting: randomization test for paired samples $t=0.592$, n$=15$, n.s.; grooming: randomization test for paired samples $t=0.162$, n$=15$, n.s.; see Figure 16.2). Males may benefit by permanently associating with females because in bonobos sexual swellings are not reliable indicators of ovulation. Fully one-third of ovulatory events in bonobos fall outside the maximum swelling phase period (Reichert et al. 2002). Male reproductive success may thus be increased by mating with females independently of their swelling phase, given that female bonobos are receptive throughout their reproductive cycle. Paoli et al. (2006a), for example, found no significant increase in copulation rates in the maximum swelling phase (mean hourly copulatory rate: maximum swelling phase$=0.22\pm0.083$ SD; nonmaximum swelling and detumescence$=0.154\pm0.083$ SD).

Moreover, it is noteworthy that across all male-female dyads, no statistical difference was found between aggressive interactions initiated by males and by females (randomization test for paired samples: $t=-0.184$, n$=15$, n.s.), with a mean dyadic hourly frequency slightly higher for aggression directed by females to males (males to females: 0.009 ± 0.016 SD; females to males: 0.015 ± 0.025 SD). These and previously published data (Paoli et al. 2006b; Paoli and Palagi submitted) support the idea of co-dominance among males and females, a condition that likely diminishes male ability to coerce females.

Conclusions

It is difficult to isolate a single reason for the lack of sexual coercion by male bonobos. Wrangham (1999) proposed that the generally less aggressive temperament of male bonobos compared to male chimpanzees could derive from the dominance of males by powerful female-female coalitions, or the greater importance of mothers than other males as allies for individual males.

Female bonobos are co-dominant to males, and they do not "fear" one-on-one aggressive interactions with males. This is likely critical to understanding why, even in the absence of female coalitionary support against males *(female-defense alliance hypothesis)*, the occurrence of male sexual coercion is so rare in this species. Male deference toward females is probably related to multiple aspects of bonobo sociality. First, sexual interactions in bonobos occur independently of the female reproductive state, with females always easily accessible to mating. Second, the use of sexuality goes beyond conception and assumes in this species a role in communication and tension reduction (e.g., GG-rubbings and mountings: de Waal 1987; Paoli et al. 2007). Third, close association between males and females is observed throughout the female reproductive cycle.

What can account for these novel aspects of the bonobo social system? From an evolutionary perspective, the availability of large patches of high-quality fallback foods and feed-as-you-go foraging that characterize bonobos (Wrangham 2000) may have provided an ecological basis for the occurrence of mixed-sex parties regardless of female swelling phase (Furuichi 1997; Gerloff et al. 1999). Thus, selection may have favored a generally less aggressive male temperament in bonobos to facilitate continuous male-female association. In the bonobo social system, where females generally enjoy high status and feeding priority (Wrangham 1993; Parish 1994; Furuichi 1997; Paoli and Palagi 2008), the costs of group living are reduced for females, who also benefit from male protection in the mixed-sex party (Hohmann and Fruth 2003). As a result, female reproductive success may be increased. However, from the male perspective, the combination of continual mating access and proceptive females (whose attractiveness is granted by a long-lasting perineal swelling: about 38% of the perineal cycle; see Paoli et al. 2006a) may have undermined the use of coercion and, given the nondeferent temperament of females, rendered attempts at coercion costly. In addition, the misleading perineal signal characterizing bonobo females (Reichert el al. 2002) appears to support the male strategy of deference and close association with females (which I call the *deferent/close-association hypothesis*), which is also effective from the point of view of their reproductive success.

In this overall scenario characterized by flexibility, co-dominance, hypersexuality, prolonged perineal swelling, and permanent male-female association, selection may have achieved an equilibrium point where the reproductive success of both sexes is maximized by male deference and female continuous proceptivity.

References

Altmann, J. "Observational Study of Behaviour Sampling Methods." *Behaviour* 49 (1974): 227–265.

Anestis, S. F. "Female Genito-Genital Rubbing in a Group of Captive Chimpanzees." *International Journal of Primatology* 25 (2004): 477–488.

Badrian, A., and N. Badrian. "Social Organization of *Pan paniscus* in the Lomako Forest, Zaire." In *The Pygmy Chimpanzee: Evolutionary Biology and Behavior,* ed. R. L. Susman, pp. 325–346. New York: Plenum Press, 1984.

Blount, B. G. "Issues in Bonobo *(Pan paniscus)* Sexual Behavior." *American Anthropologist* 92 (1990): 702–714.

Boehm, C. *Hierarchy in the Forest: The Evolution of Egalitarian Behavior.* Cambridge, Mass.: Harvard University Press, 1999.

Dahl, J. F. "Cyclic Perineal Swelling during the Intermenstrual Intervals of Captive Female Pygmy Chimpanzees *(Pan paniscus)*." *Journal of Human Evolution* 15 (1986): 369–385.

de Waal, F. B. M. "Tension Regulation and Nonreproductive Functions of Sex on Captive Bonobos *(Pan paniscus)*." *National Geographic Research* 3 (1987): 318–335.

———. "Appeasement, Celebration, and Food Sharing in the Two *Pan* Species." In *Topics in Primatology Vol. I—Human Origins,* eds. W. C. McGrew, P. Marler, and M. Pickford, pp. 37–50. Tokyo: University of Tokyo Press, 1992.

———. "Bonobo Sex and Society: The Behavior of a Close Relative Challenges Assumptions about Male Supremacy in Human Evolution." *Scientific American* (1995): 82–88.

———. "Apes from Venus: Bonobos and Human Social Evolution." In *Tree of Origin: What Primate Behavior Can Tell Us about Human Social Evolution,* ed. F. B. M. de Waal, pp. 41–68. Cambridge, Mass.: Harvard University Press, 2001.

Doran D. M., W. L. Jungers, Y. Sugiyama, J. G. Fleagle, and C. P. Heesy "Multivariate and phylogenetic approaches to understanding chimpanzee and bonobo behavioural diversity." In *Behavioural Diversity in Chimpanzees and Bonobos,* eds. C. Boesch, G. Hohmann, and L. F. Marchant, pp. 14–34. Cambridge: Cambridge University Press, 2002.

Fruth B., G. Hohmann, and W. C. McGrew. "The *Pan* Species." In *The Non Human Primates,* eds. P. Dolhinow and A. Fuentes, pp. 64–72. Mountain View, Calif.: Mayfield Publishing Company, 1999.

Furuichi, T. "Agonistic Interactions and Matrifocal Dominance Rank of Wild Bonobos *(Pan paniscus)* at Wamba." *International Journal of Primatology* 18 (1987a): 855–875.

———. "Sexual Swelling, Receptivity, and Grouping of Wild Pygmy Chimpanzee Females at Wamba Zaire." *Primates* 28 (1987b): 309–318.

———"Dominance Relations among Wild Bonobos *(Pan Paniscus)* at Wamba, Zaire." Paper presented at the 14th Congress of the International Primatological Society, Strasbourg, 1992.

Furuichi, T., and C. Hashimoto. "Sex Differences in Copulation Attempts in Wild Bonobos at Wamba." *Primates* 45 (2004): 59–62.

Gerloff, U., B. Hartung, B. Fruth, G. Hohmann, and D. Tautz. "Intracommunity Relationships, Dispersal Pattern and Paternity Success in a Wild Living Community of Bonobos *(Pan paniscus)* Determined from DNA Analysis of Faecal Samples." *Proceedings of the Royal Society B* 266 (1999): 1189–1195.

Hemelrijk, C. K., J. Wantia, and M. Daetwyler. "Female Co-Dominance in a Virtual World: Ecological, Cognitive, Social and Sexual Causes." *Behaviour* 140 (2003): 1247–1273.

Hohmann, G., and B. Fruth "Use and Function of Genital Contacts among Female Bonobos." *Animal Behaviour* 60 (2000): 107–120.

———. "Intra- and Inter-Sexual Aggression by Bonobos in the Context of Mating." *Behaviour* 140 (2003): 1389–1413.

Idani, G. "Social Relationships between Immigrant and Resident Bonobo *(Pan paniscus)* Females at Wamba." *Folia Primatologica* 57 (1991): 83–95.

Kano, T., and M. Mulavwa. "Feeding Ecology of *Pan paniscus* at Wamba." In *The Pygmy Chimpanzee: Evolutionary Biology and Behavior,* ed. R. L. Susman, pp. 233–274. New York: Plenum Press, 1984.

Kano, T. "Social Behavior of Wild Pygmy Chimpanzees *(Pan paniscus)* of Wamba: A Preliminary Report." *Journal of Human Evolution* 9 (1980): 243–260.

———. "The Sexual Behavior of Pygmy Chimpanzees." In *Understanding Chimpanzees,* eds. P. G. Heltne and L. A. Marquard, pp. 176–183. Cambridge, Mass.: Harvard University Press, 1989.

———. *The Last Ape.* Palo Alto, Calif.: Stanford University Press, 1992.

———. "Male Rank Order and Copulation Rate in a Unit-Group of Bonobos at Wamba, Zaire." In *Great Ape Societies,* eds. W. C. McGrew, L. F. Marchant, and T. Nishida, pp. 135–155. Cambridge: Cambridge University Press, 1996.

Kuroda, S. "Social Behavior of the Pygmy Chimpanzee." *Primates* 21 (1980): 181–197.

Kuroda, S. "Interaction over Food among Pygmy Chimpanzees." In *The Pygmy Chimpanzee: Evolutionary Biology and Behavior,* ed. R. L. Susman, pp. 301–324. New York: Plenum Press, 1984.

Manson, J. H., S. Perry, and A. R. Parish. "Nonconceptive Sexual Behavior in Bonobos and Capuchins." *International Journal of Primatology* 18 (1997): 767–786.

Marvan, R., J. M. G. Stevens, A. D. Roeder, I. Mazura, M. W. Bruford, and J. R. de Ruiter. "Male Dominance Rank, Mating and Reproductive Success in Captive Bonobos *(Pan paniscus).*" *Folia Primatologica* 77 (2006): 364–376.

Matsumoto-Oda, A., and R. Oda. "Changes in the Activity Budget of Cycling Female Chimpanzees." *American Journal of Primatology* 46 (1998): 157–166.

Palagi, E., T. Paoli, and S. M. Borgognini Tarli. "Reconciliation and Consolation in Captive Bonobos *(Pan paniscus).*" *American Journal of Primatology* 62 (2004): 15–30.

Paoli, T., and E. Palagi. "What Does Agonistic Dominance Imply in Bonobos?" In *The Bonobos: Behavior, Ecology and Conservation,* eds. T. Furuichi and J. Thompson, pp 39–54. New York: Springer, 2008.

Paoli, T., E. Palagi, and S. M. Borgognini Tarli. "Perineal Swelling, Intermenstrual Cycle, and Female Sexual Behavior in Bonobos" *(Pan paniscus). American Journal of Primatology* 68 (2006a): 333–347.

———. "Reevaluation of Dominance Hierarchy in Bonobos *(Pan paniscus)."* *American Journal of Physical Anthropology* 130 (2006b): 116–122.

Paoli, T., G. Tacconi, S. M. Borgognini Tarli, and E. Palagi. "Influence of Feeding and

Short-Term Crowding on the Sexual Repertoire of Captive Bonobos *(Pan panis-cus)." Annales Zoologici Fennici* 44 (2007): 81–88.

Parish, A. R. "Sex and Food Control in the "Uncommon Chimpanzee": How Bonobo Females Overcome a Phylogenetic Legacy of Male Dominance." *Ethology and Sociobiology* 15 (1994): 157–179.

———. "Female Relationships in Bonobos *(Pan paniscus):* Evidence for Bonding, Cooperation, and Female Dominance in a Male-Philopatric Species." *Human Nature* 7 (1996): 61–69.

Reichert, K. E., M. Heistermann, J. K. Hodges, C. Boesch, and G. Hohmann. "What Females Tell Males about Their Reproductive Status: Are Morphological and Behavioural Cues Reliable Signals of Ovulation in Bonobos, *Pan paniscus?" Ethology* 108 (2002): 583–600.

Savage-Rumbaugh, E. S., and B. J. Wilkerson. "Socio-Sexual Behavior In *Pan paniscus* and *Pan troglodytes:* A Comparative Study." *Journal of Human Evolution* (1978): 327–344.

Smuts, B., and R. W. Smuts RW. "Male Aggression and Sexual Coercion of Females in Nonhuman Primates and Other Mammals: Evidence and Theoretical Implications." *Advances in the Study of Behavior* 22 (1993): 1–63.

Stanford, C. "The Social Behavior of Chimpanzees and Bonobos." *Current Anthropology* 39 (1998): 399–420.

Stevens, J. M. G., H. Vervaecke, and L. van Elsacker. "Sexual Strategies in *Pan paniscus:* Implications of Female Dominance." *Primate Report Special Issue* 60 (2001): 42–43.

Takahata, Y., H. Ihobe, and G. Idani. "Comparing Copulations of Chimpanzees and Bonobos: Do Females Exhibit Proceptivity or Receptivity?" In *Great Ape Societies,* eds. W. C. McGrew, L. F. Marchant, and T. Nishida, pp. 146–155. Cambridge: Cambridge University Press, 1996.

Thompson-Handler, N., R. K. Malenky, and N. Badrian "Sexual Behavior of *Pan Paniscus* under Natural Conditions in the Lomako Forest, Equateur, Zaire." In *The Pygmy Chimpanzee: Evolutionary Biology and Behavior,* ed. R. L. Susman, pp. 347–368. New York: Plenum Press, 1984.

Tutin, C. E. G., and P. R. McGinnis "Chimpanzee Reproduction in the Wild." In *Reproductive Biology of the Great Apes: Comparative and Biomedical Perspectives,* ed. C. E. Graham, pp. 239–264. New York: Academic Press, 1981.

Tutin, C. E. G. "Mating Patterns and Reproductive Strategies in a Community of Wild Chimpanzees *(Pan troglodytes schweinfurthii)." Behavioral Ecology and Sociobiology* 6 (1979): 29–38.

White, F. J. "Party Composition and Dynamics in *Pan paniscus." International Journal of Primatology* 9 (1988): 179–193.

———. "Comparative Socio-ecology of *Pan paniscus.*" In Great Ape Societies, eds. W. C. McGrew, L. F. Marchant, and T. Nishida, pp. 29–41. Cambridge: Cambridge University Press, 1996.

Whitham, J. C., and D. Maestripieri. "Primate Rituals: The Function of Greetings between Male Guinea Baboons." *Ethology* 109 (2003): 847–859.

Wrangham, R. W., and D. Peterson. *Demonic Males.* New York: Houghton Mifflin, 1996.

Wrangham R. W. "The Evolution of Sexuality in Chimpanzees and Bonobos." *Human Nature* 4 (1993): 47–79.

———. "The Evolution of Coalitionary Killing." *Yearbook of Physical Anthropology* 42 (1999): 1–30.

———. "Why Are Male Chimpanzees More Gregarious Than Mothers? A Scramble Competition Hypothesis." In *Primate Males,* ed. P. Kappeler, pp. 248–258. Cambridge: Cambridge University Press, 2000.

17

Sexual Coercion, Patriarchal Violence, and Law

Diane L. Rosenfeld

This chapter considers how evolutionary perspectives on male sexual coercion can usefully inform legal policy. The underlying conceit is that by analyzing the law's treatment of male sexual coercion through an evolutionary lens, we can help explain the law's failure to effectively prevent or redress much violence against women. I briefly explain the current state of law in the United States and suggest ways in which a legal approach that focuses on the sexual coercion underlying much violence against women might advance us toward the goal of gender equality.

I refer to sexual coercion using Smuts's definition: "male use of force, or its threat, to increase the chances that a female will mate with the aggressor or to decrease the chances that she will mate with a rival, at some cost to the female" (Smuts and Smuts 1993: 2–3). I apply Smuts's theories regarding intragender alliances throughout this chapter, arguing that male-male alliances to refrain from interfering with one another's property (including, notably females) are reflected within the law. The chapter has four proposals.

The first is that an evolutionary perspective reveals the vast prevalence of male sexual coercion in almost all human societies—informing us that this behavior is commonplace rather than aberrational. Yet, the laws of these societies fail to provide recourse for most forms of sexual coercion, recognizing only the most extreme (and uncommon) behaviors.

My second proposal is that law reflects—rather than interrupts—the role of male sexual coercion in establishing and maintaining a patriarchal agreement not to intervene in another male's coercive tactics. The three stages of law relevant to this analysis are, first, the passage of legislation through a law-making body; second, the enforcement of laws; and third, the interpretation of laws through the courts. Women have had the most success in the passage of legislation to protect them against male aggression; less success at compelling law enforcement—meaning police and prosecutors—to implement the laws; and possibly the least success in judicial interpretation of protective laws.

The third proposal offers a theory of "patriarchal violence" that explains the variability in the forms and severity of male sexual coercion as that which is necessary to preserve a patriarchal social order. Although others have theorized about patriarchy and violence against women (MacKinnon 1989; Smuts 1995), my contribution is to offer the term *patriarchal violence* to connect the occurrence of male sexual coercion in primates and in humans in a way that describes the functionality of such behavior and explains its possible source of variation. Support for this theory comes from an analysis of the two most important Supreme Court cases involving gender violence, which found that democratically achieved legislative gains, enabling women to challenge male sexual violence, exceeded constitutional limitations.

My fourth proposal is that the potential of female-female alliances to resist male sexual coercion would be strengthened by confronting directly the ways in which male sexual coercion fractures female bonding, as well as incorporating collective physical resistance to male aggression in a bonobo-like fashion.

First Proposal: The prevalence and variation of male sexual coercion indicates that it is commonplace rather than aberrational, and for the law to effectively prevent and punish sexual coercion, it must recognize its prevalence and functionality.

Violence against women is a worldwide problem, with women in every country living under the threat of physical, mental, and sexual abuse at the hands of men. The World Health Organization's (WHO) Multi-Country Study on Women's Health and Domestic Violence against Women in 2005 showed that women are more at risk of experiencing violence in intimate relationships than anywhere else, "challenging the notion that home is a safe haven" (WHO 2005:9). The vast majority of abuse occurs at the hands of a partner rather than a stranger; over 75% of women physically or sexually abused since the age of 15 reported abuse by a partner, and between 20 and 75% of women had experienced emotional abuse. The study found that men who physically abuse their partners also exhibit higher rates of controlling behavior than men who do not, including isolation (keeping her from seeing or communicating with friends and family), sexual jealousy, and wanting to know where she is at all times. In addition, the study found "a great variation in the degree to which such behaviour is acceptable (normative) in different cultures" (WHO 2005:19). Because these milder forms of coercive behavior often escalate, we need to learn to recognize the significance of even subtle and seemingly benign instances of controlling and sexually coercive behaviors.

The WHO report showed that not only are women experiencing violence from the people who should pose the least threat to them, but they are also subjected to violence at a time when one would least expect it to occur: during pregnancy. Between one-quarter and one-half of the women who were physically abused during pregnancy were kicked or punched in the abdomen. The study reported that over 90% of these women were abused by the biological father of the child the woman was carrying. Many said they were beaten for the first time during pregnancy (WHO 2005:26–27).

Although the United States was not included in the WHO study, other research shows that the prevalence of rape, domestic violence, and domestic violence homicide in the United States remains distressingly high. In 2001, more than half a million American women were victims of nonfatal violence committed against them by an intimate partner, and as many as 324,000 women experience intimate partner violence against them during pregnancy. Studies also reveal that pregnant and recently pregnant women are more likely to die as a result of domestic homicide than any other cause. On average in the United States, more than three women per day are murdered by their husbands, boyfriends, ex-husbands, estranged husbands, or ex-boyfriends (Bureau of Justice Statistics 2001; Family Violence Prevention Fund 2007).

Despite the prevalence of male sexual coercion, the law does not reflect its commonality. Thus the most frequent occurrences are either not prosecuted (acquaintance rapes) or inadequately enforced (typical domestic violence cases). Although society devotes resources and time to stranger rapes, it pays very little attention to acquaintance rapes, even though these account for the vast majority of rapes (Emery Thompson, Chapter 14 in this volume). Most colleges offer self-defense programs to women based on the paradigm of stranger rape, whereas estimates are that one in four college women will experience a rape or attempted rape by an acquaintance or intimate partner during her time at school (Karjane et al. 2005).

As for domestic violence, its everyday occurrence receives little public attention except when associated with a homicide case. As a result, wife-beating is presented as though it were an unusual event. Tracy and Crawford (1999) argue that this treatment of violence against women as pathological rather than commonplace contributes to the law's failure to counteract it. "As long as wife-beating is viewed as a bizarre event, with blame placed on both batterer and victim—rather than as a remnant of our ancestral past, when it was 'beneficial'—it will continue to occur. Effective remedies cannot be undertaken at the individual level unless societies choose to censure wife beating" (Tracy and Crawford 1999:39).

Legal scholar Duncan Kennedy has likewise argued that the popular understanding of sexual abuse as pathological rather than normal tends to sustain "the view that [sexually coercive acts] play no significant structural role in the relations between 'normal' men and women" (1992:1321). Kennedy states:

> The combination of the limits of the formal law and the actual workings of the legal system has the result that men can and do commit large numbers of sexual abuses of women without any official sanction. Although, by hypothesis, most people would regard the conduct as clearly wrong, and injurious, there is no punishment and no redress. The crucial point . . . is that *some* abuse, what I will call the "tolerated residuum," is plausibly attributed to contestable social decisions about what abuse is and how important it is to prevent it. The law defines murder quite clearly, and the "system" devotes substantial resources to catching and punishing perpetrators. It defines rape much less clearly, and devotes less resources to it, some of the time, than to less important crimes. (Kennedy 1992:1321).

Kennedy goes on to consider the "consequences of setting up the legal system to condemn sexual abuse of women by men in the abstract, but at the same time operating the system so that many, many instances of clearly wrongful abuse are tolerated . . . men and women gain and lose from the practices of abuse, *whether or not* they themselves are actually abusers or victims." One very clear consequence of this is that women operate and live within this structure, bargaining in the shadow of a legal system that does not provide real protection against male sexual abuse (Kennedy 1992; Rosenfeld 2004).

Yet, low-level coercion forms the basis for the more extreme commission of sexual violence against women. Male coercion is understood to occur on a continuum ranging from simple gender-role enforcement to domestic violence homicide. I refer to this as a "continuum of male entitlement" over an intimate partner. This coercive and violent behavior is widely understood to be motivated by a desire to exercise control over the victim (Duluth Project 1989).

At the least serious end of the continuum is behavior based on gendered expectations of male supremacy and female subordination exemplified in stereotypes of American housewives from the 1950s. This behavior enforces the idea that a man should be served his meals and is generally exempt from household chores and other "women's work." The continuum moves on to emotional abuse and economic abuse, which can include isolation, an important tactic of abusers (Duluth Project 1989). A batterer will typically criticize his partner's relationships with her family and friends gradually, until she is cut off from resources and alliances that might help her escape her partner's abuse. Similarly, an abusive partner might gradually gain control of his partner's finances either by insisting she is incapable of managing them herself or by forcing her to quit her job and become financially dependent on him. In either case, the abusive partner deliberately reduces his partner's agency by forcing this financial and economic dependence.

Physical and sexual assault occur next on the continuum. This violence is deliberate, habitual, and often premeditated. As with the case of emotional and financial abuse, a batterer uses chronic physical abuse to reduce his partner's agency and ensure that she becomes dependent on him and submissive to him. Thus, the assumption that a man will only physically abuse his partner when he is moved by extreme emotion or exceptional circumstances is mistaken. Indeed, the common claim of batterers that they "just lost control" (Adams 2007:171–172) should be rejected, for most batterers do not similarly lose control with other people in their lives, such as bosses or strangers (Rosenfeld 1994). Far from an aberration in the batterer's behavior or an exceptional case of loss of control, chronic physical abuse should be viewed as a tactic that batterers use to exert control over their female partner's behavior.

Legal recognition of the prevalence and functionality of male sexual coercion is necessary to the design of an effective deterrent to such behavior. Evolutionary perspectives about the cross-cultural prevalence of sexual coercion could help legal policymakers see the larger context of this everyday behavior. Thus, we can begin to address it as a social norm rather than an aberration, and refocus our legal efforts to prevent and punish it accordingly.

Second Proposal: Law reflects—rather than interrupts—a patriarchal agreement among males not to intervene in each other's sexually coercive tactics.

Human patriarchy has its beginning in the forest ape social world, a system based on males' social dominance and coercion of females. . . . Language would eventually generate both marriage and the patriarchal rules that favor married men. Men, following an evolutionary logic that benefits those who make the laws, would create legal systems that so often defined adultery as a crime for women, not for men—a social world that makes men freer than women. (Wrangham and Peterson, 1996:242).

Law represents the regulation of violence, as well as the means through which violence is authorized or inflicted by the state on the individual (Cover 1986). It is a set of rules by the state that regulate under what circumstances violence may be used, as well as the punishment for using violence outside of permitted circumstances. Judicial decisions in themselves authorize the use of violence to punish an offender, or may inflict violence on the victim of a crime by not punishing the offender (Cover 1986). Law enforcement represents the state's authorized use of violence to enforce the laws according to the discretion and priorities of political leaders.

Critically, laws have historically and traditionally been the province of men. Law reflects agreements and understandings between men, and laws regulating sexual violence reflect men's agreements about access to women (MacKinnon 2000:175). These laws form the basis of our patriarchal society and reflect an unwillingness of men to intervene in other men's coercive tactics. For example, before women had representation in the legal system, laws were forthright about the status of women as the property of men. Now that women have acquired the right to vote and to participate fully as citizens, laws reflect this more equal status. However, the prevalence of violence against women belies the idea that women are truly equal and that all citizens are treated the same under the law.

But even if women enjoyed equal status under the law, law enforcement remains a critical juncture at which many legal protections intended to protect women from male violence fail. As Wrangham and Peterson (1996) noted, political power is built on physical power, which is ultimately the power of violence or its threat. It is also being able to "count on someone coming to their aid—the police or the military or the mob or the family, or the royal guard" (Wrangham and Peterson 1996:242). Law enforcement, however, more often reflects male-male alliances and policies of nonintervention than it does straightforward implementation of protective laws. For example, some studies have estimated that 40% of police officers have been involved in abusing their own intimate partners (Johnson 1991, Neidig et al. 1992), suggesting that police officers might be inclined to empathize with, rationalize, or ignore the severity of a batterer's actions. This kind of problem is aggravated by the ability of batterers to bond with police officers over sports or common friends in the military. Indeed, these tactics are so prevalent that officers are specifically trained to recognize such attempts as manipulation (O'Dell 1997).

In order to understand how male alliances visibly contribute to violence against women, one might consider the claim of a woman in the United States trying to escape a polygamist cult in which sexual abuse was rampant. Carol Jessup, in her recent memoir, described how the police were of the same cult in the area and would return the runaway wives to their husbands instead of aiding them in escaping or addressing the abuse (Jessop 2007). She also noted the non-interventionist agreement among males when she stated that she could not go to the hospital without her husband's permission because "the volunteer ambulance drivers . . . were all members of the FLDS. Because of this they were under enormous pressure not to interfere with another man's family" (Jessop 2007:214). Women's attempts to resist male sexual coercion can thus easily be undermined by informal male-male alliances, in particular among law enforcement.

A major source of the problem is that because men tend to be bigger and stronger than women, women most often cannot defend themselves directly against threats of men's violence. In addition to the sexual dimorphism, however, there is a significant difference in the proclivity of men and women to use violence. In the United States, for example, men commit 85% of all violence and 85% of intimate partner violence (Bureau of Justice Statistics 2001). Women could theoretically use weapons to compensate for the inequality in body strength, but generally do not. Women's two main sources of protection from violence thus appear to be a male guardian, such as an intimate partner or family member, or the state. This gender disparity in personal strength and in the use of violence leads to what has been called a "male protection racket"—a phenomenon in which females learn to rely on males for protection from other males (Griffin 1970:10–11). A similar dynamic is in evidence in primates where females form friendships with males in order to gain protection from other males (Smuts 1985; Palombit, Chapter 15 in this volume). The unfortunate irony in this arrangement is that women are at a much greater risk of male sexual violence at the hands of their intimate partners than from strangers (Bureau of Justice Statistics 2001).

The male protection racket has particular salience in the U.S. Constitution's promise of "equal protection under the laws" (U.S. Constitution, Fourteenth Amendment, Section One). Equal protection under the laws has a unique meaning as applied to questions of sexual violence. If women generally do not use violence, but must rely on state intervention for protection, then perhaps equal protection means something different for women citizens than for men. For the state to provide meaningful equal protection, it must take into account the gendered disparity in the use of violence, and the need to protect women from male sexual coercion. An analysis of the laws concerning rape and domestic violence illustrates the problems we must overcome to effectively address male sexual coercion through the law.

Rape

Rape was initially a crime of trespass to property, and the right to bring an action used to lay in the hands of the offended property owner—either the husband or the father, depending on the woman's marital status. Therefore, when a woman was raped, the offense was deemed to be committed against a man, not against her own person.

According to William Blackstone, whose *Commentaries on Law from 1765–1769* are regarded as the most authoritative British legal text, rape was known as "carnal knowledge of a woman forcibly and against her will." Blackstone cited the injunction made famous by British Chief Justice Lord Hale in 1680 that rape "is an accusation easily to be made and hard to be proved, and harder to be defended by the party accused, tho never so innocent." If a rape was proven, the punishment was quite severe and sometimes resulted in capital punishment for the offender. The severity of the punishment was given as the reason for the law's encoding of suspicion of a woman's word that she had been raped. Blackstone states: "in order to prevent malicious accusations, it was then the law, (and, it seems, still continues to be so in appeals of rape) that the woman should immediately after, go to the next town, and there make discovery to some credible persons of the injury she has suffered; and afterwards should acquaint the high constable of the hundred, the coroners, and the sheriff with the outrage" (Blackstone, Book I, 1765–1769).

Of all regulations of gender-motivated violence, rape law has proven to be the most resistant to legal reform, and the historical notions from the seventeenth century are still very much in evidence today. "Although Hale's warnings are no longer a part of the court process, rape proceedings are still conducted as if all women tend to lie, or at least become confused about violations of their own body" (Madigan and Gamble 1991:15). Even today in law schools across the country, students are taught this maxim underlying rape law. Society routinely questions the victim for her lack of control or veracity and fails to accurately question the perpetrator for his lack of control or veracity. Juries are still instructed with Hale's centuries-old warnings. Formally, this principle has been codified in court rules specific to proving rape cases. These rules included (and sometimes still include—formally and informally) fresh complaint rules and the necessity for corroboration through witnesses or physical evidence (Schulhofer 1998:18–19). The fresh complaint rules—which act as a reminder of Hale's expectation of running to the next village—provide that a rape victim is expected to report her rape promptly after the incident. The requirement for prompt reporting assumes that the victim, having undergone the trauma of rape, will be willing and able to report her rape, yet this notion is in tension with the reality of rape victimization. Research has shown that some victims never report their rape; some report only upon being re-victimized, some report months later, some immediately (Herman 1992, 1997).

The FBI, which records the nation's rape statistics, currently defines rape as "carnal knowledge of a female forcibly and against her will" (FBI Uniform Crime Reports 2004), a definition quite similar to that found in Blackstone's *Commentaries* hundreds of years earlier. Most state rape statutes reflect some aspect of these rules. In the recent past, rules that would require the victim to prove her "utmost resistance" have been removed from formal statutory requirements, but still persist in society's expectations of how a rape victim should act. Yet, such an attitude fails to account for the reality of rape perpetration. Given the grave bodily danger that a victim faces during rape, and the fear of being beaten or killed in addition to being raped, many women resist fighting back as a matter of survival. The sexual dimorphism in body size between men and women also makes it unreasonable in many cases to expect a woman to fight back. Further angering a sexually aggressive male can make him increasingly violent and deter a woman from "forcibly" resisting rape to the utmost (Hartman 2008).

The force requirement coupled with the "against her will" language meant that a rape victim had to prove that she did not want to have sexual intercourse with the perpetrator at the time in question. This notion presumes that, unless proven otherwise, she *did* want to have sex with him (MacKinnon 1989:175). This presumption has been criticized by a number of diverse scholars, including one who proposes that exactly the opposite presumption should be held. Stephen Schulhofer compares rape to the following situation: "when a doctor asks if a patient wants a probe inserted into his rectum to check for tumors, we do not assume that the patient's silence indicates consent. The patient's willingness must be made explicit. And even then, because the doctor has a special duty of disclosure, the patient's permission is not sufficient until he considers the risks and gives 'informed consent.'" When considering the importance of consent to any invasion to a human's bodily integrity, the presumption should be, as we recognize in other cases, that the intrusion was not wanted. By assuming the opposite, that consent is given unless explicitly proven otherwise, the law suggests that women's bodies are readily available for the use and abuse of others.

Unlike any other crimes, victims of rape who decide to report the attacks that they have suffered are forced to go through a series of police, judicial, medical, and mental health proceedings, which are often described as being worse than the rape itself. Because of the humiliation and degradation that rape victims experience in these proceedings, it is not surprising that they have become known as the "second rape" (Madigan and Gamble 1991). Madigan and Gamble have described this "second rape" as "the act of violation, alienation, and disparagement a survivor receives when she turns to others for help and support. It can occur only if she has been brave enough to tell someone of her assault." They note that the only way to prevent the second rape is by "[k]eeping the [initial] rape a secret" (Madigan and Gamble 1991). Rates of rape underreporting indicate that this is exactly what most rape victims do.

Domestic Violence and Marriage Laws

According to Angela Browne, the first law of marriage (credited to Rome's mythical founder, Romulus) required married women "as having no other refuge, to conform themselves entirely to the temper of their husbands and the husbands to rule their wives as necessary and inseparable possessions" (Browne 2000). Although centuries old, this statement has resonance in today's society. The idea of women being required "to conform themselves entirely to the temper of their husbands" is evident in the current dynamics of domestic violence, where the violence serves as the punishment for women's failure to obey the behavioral dictates of their intimate partners.

"Statutes of chastisement" indicate patriarchal support for the acceptability of wife-beating. According to Blackstone's *Law Commentaries*, chastising one's wife was not only regarded as a right, but as a husband's duty, for "for as the husband is to answer for her misbehavior, the law thought it reasonable to entrust him with this power of chastisement, in the same moderation that a man is allowed to correct his apprentices or children. . . . The civil law gave the husband the same, or a larger, authority over his wife: allowing him for some misdemeanors, to beat his wife severely with scourges (whips) and cudgels (stout sticks with rounded heads) . . . for others only moderate chastisement" (Blackstone's *Commentaries,* Book I).

The most significant lesson from legal history is the difference between a husband killing his wife and a wife killing her husband. Blackstone describes this legal difference as follows:

> Husband and wife, in the language of the law, are styled baron and femme . . . if the baron kills his femme, it is the same as if he had killed a stranger or any other person, but if the femme kills her baron, it is regarded by the laws as much more atrocious crime, as she not only breaks through the restraints of humanity and conjugal affection, but throws off all subjection to the authority of her husband. And therefore the law denominates her crime a species of treason, and condemns her to the same punishment as if she had killed the king. And for every species of treason . . . the sentence of the woman was to be drawn and burnt alive. (Blackstone's *Commentaries,* Book I)

The disproportionate punishment meted to a woman for rising up against her king reflects the patriarchal agreement that the man is "king of his castle" and that the state will most severely punish a challenge to his authority. The statutes of chastisement indicate that a man's reign has been specifically delegated to him by the state. This delegation of regulatory responsibility to the husband leaves unfettered his discretion to rule his wife as he sees fit, with checks and balances only for extreme violence. As such, violence and rule of law produce what can be described as "controlled violence"—much like sports such as boxing, where violence is permitted and encouraged, provided it is not excessive or results in death (Balfour 2008).

It is against this background that wife-beating has gone from a legal entitlement or duty in the eighteenth century as described by Blackstone to a legal prohibition in the late nineteenth century. As late as 1824, the Mississippi Supreme Court upheld the right of the husband to chastise his wife, but only to a "reasonable" degree. The Court in that case invoked a deep reluctance to engage itself in matters of domestic relations ". . . in order to prevent the deplorable spectacle of the exhibition of similar cases in our courts of justice. . . . To screen from public reproach those who may be thus unhappily situated, let the husband be permitted to exercise the salutary restraints in every case of misbehaviour, without being subjected to vexatious prosecutions, resulting in the mutual discredit and shame of all parties concerned" (*Bradley v. State,* 1 Miss. 156, Supreme Court of Mississippi, 1824).

The nineteenth-century feminist movement brought about legal reforms that challenged the chastisement prerogative of the husband. Although these reforms resulted in some measure of status improvement for wives, they still did not achieve equality with their husbands (Siegel 1996). Instead, the challenge led to a change in justificatory rhetoric, in which judges adopted a privacy consideration and asserted that the law should not interfere in cases of wife-beating to protect marital harmony (Siegel 1996). Siegel refers to this process as "preservation through transformation" referring to preservation of the underlying discriminatory behavior through a transformation of the public rationale given to explain its existence or tolerate it under the law (Siegel 1996). Privacy rhetoric became the cloak behind which patriarchal violence continued at home unchecked by the law.

Thus, in both rape and domestic violence (including homicide), our legal rubric is based on centuries-old notions of male entitlement to control female sexuality. Women are still disbelieved when they bring forth rape charges (Torrey 1991), are overwhelmingly victims of domestic violence (Bureau of Justice Statistics 2001), and are punished disproportionately for killing their abusive spouses (Jones 1996).

Third Proposal: The theory of "patriarchal violence" explains the variability in forms and severity of male sexual coercion as that which is necessary to preserve a patriarchal social order.

Several chapters in this volume demonstrate the variability of human male sexual coercion and show that one function is to promote a male's reproductive success. However, consistent with Smuts's prediction that humans "exhibit more extensive male dominance and male control of female sexuality than is shown by most other primates" (Smuts 1995:1), men also exhibit more extreme forms of sexual violence against women; some of these do not appear to be functional with respect to reproductive success. Here I discuss two types of male sexual violence that seem to be unique to the human species: gang rape and domestic violence homicide. Both were at issue in two landmark Supreme Court cases that addressed women's rights to challenge gender-based violence. These cases demonstrate how weak the American legal system can be in protecting women from sexual violence, and support my proposal that sexual coercion promotes the patriarchal order. I consider the two types of violence in turn.

Gang Rape

> Cross-cultural research demonstrates that whenever men build and give allegiance to a mystical, enduring, all-male social group, the disparagement of women is, invariably, an important ingredient of the mystical bond, and sexual aggression the means by which the bond is renewed. As long as exclusive male clubs exist in a society that privileges men as a social category, we must recognize that collective sexual aggression provides a ready stage on which some men represent their social privilege and introduce adolescent boys to their future place in the status hierarchy. (Peggy Reeves Sanday, *Fraternity Gang Rape*, 19–20 [1990])

Gang rape presents the most obvious, and the only commonly recognized, example of men conspiring to violently harm women (Schaffer 2006). Gang rape is a crime in which the importance of male bonds is clearly visible. It is the apotheosis of the male-male alliances that form the basis of patriarchy in that the men demonstrate and enact male supremacy over women. An analysis of gang rape cases that were heard in the appellate courts in 2005 and the first half of 2006 revealed that "every single one of the rapes featured groups of men bonding with one another by committing violence against women. While the twisted details of these appellate gang rape cases differ, the cases' basic contours are identical every time. All of the cases center on all-male groups of perpetrators, whether formal gangs or informally affiliated friends, who consciously conspire to harm female victims as a unifying amusement" (Schaffer 2006:1–6).

The South American Mehinaku tribe, where women are systematically punished by gang rape for simply looking at flutes used in certain male rituals, provides a poignant example of a patriarchal society that uses gang rape to communicate loyalties between males at the expense of fracturing alliances males have with females (Smuts 1996, citing Gregor 1990:493). These alliances begin to form at an early age when boys and girls mimic the gang rape scenario in games of pretend (Smuts 1996 citing Gregor 1980: 114). For the Mehinaku the gang rape has become a methodical reaction to female insubordination.

Although different theories might explain the cultural significance of gang rapes, bonding between the males is widely concluded to be a major driving force. In gang rape cases in the United States, one man often has sex with his girlfriend and then offers to share her with his friends, as if she is a tradable commodity owned by the male. Pornography, commonly used as a male bonding medium among fraternity members (Sanday 1990:34–35), has reflected a significant increase of gang rapes in the past few years (Jensen 2004). This may well contribute to the normalization of gang rapes and the proliferation of such cases in recent years (Schaffer 2006:1–6).

Researcher Robert Jensen has studied the increasing violence in mainstream pornography, as well as the increase in the demand for "double penetration" scenes in which two men simultaneously anally and vaginally penetrate a woman (Jensen 2004:9). The degradation of the female and her reduction to "three holes and two hands" in pornography is a recurrent and necessary theme (Jensen 2004:12). In his study of mainstream pornographic videos, Jensen found the "Gang Bang Girl" video, which involved a football coach chiding his team that he did not want to lose with a bunch of "fags," so the men were to prove their masculinity by all having sex with Kimberly, one of the cheerleaders. This genre was so popular that at the time he published his findings, there were 34 videos made in the series (Jensen 2004). Multi-perpetrator sexual assaults on campus have become so prevalent that educational institutions receiving federal funding under the Violence against Women Act have been advised by the Department of Justice to institute specific training on prevention (Rosenfeld 2007).

Domestic Violence Homicide

Evolutionary perspectives on domestic violence homicide, or uxoricide (wifekilling), as well as studies in the domestic violence field, converge on the finding that sexual jealousy and possessiveness are central motivating factors present in these cases. But killing one's intimate partner seems counterintuitive to reproductive fitness. Killing a female intimate prevents her from being either a reproductive or a labor resource for the household. However, the theory of patriarchal violence contends that even though domestic violence homicide

might seem counterproductive, it contributes to the overall strength of the patriarchal order. It considers the male-male alliance implications of domestic violence homicide outside of the implications for the individual couple.

Similar but distinguishable from the male entitlement continuum is the
theory that male sexual proprietariness is a driving force behind coercive violence against wives (Wilson and Daly, Chapter 11 in the present volume).
Male sexual jealousy and proprietariness motivate over 80% of wife-killing
cases (Wilson and Daly 1998). This is a key insight and comports well with
other work in the field of domestic violence homicide. Jacqueline Campbell's
studies on domestic violence homicides and attempted homicides enabled her
to develop a danger assessment tool for use in domestic violence cases. The
danger assessment is a list of lethality indicators that were commonly present
in these cases. Sexual jealousy, possessiveness, and threats to kill a woman if she
leaves the abuser are key indicators of a potential homicide (Campbell 2003;
Adams 2007).

According to Wilson and Daly (1992a):

> Because human male fitness is and always has been limited primarily by ac-
> cess to the reproductive efforts of fertile women, we have proposed that male
> sexual proprietariness is an evolved motivational/cognitive subsystem of the
> human brain/mind. . . . In proposing that men take a proprietary view of
> women's sexuality and reproductive capacity, we mean that men are motivated
> to lay claim to particular women as songbirds lay claim to territories. . . . But
> proprietariness has further implication of a sense of entitlement, which if not
> unique to our species, is at least uniquely highly developed. Trespass against
> one's property provokes not only a hostile response to the trespasser, but
> moralistic feelings of aggrievedness and indignation as well . . . moral indig-
> nation against those who violate one's property rights is, arguably, a human
> universal and an aspect of our evolved psychology that would be pointless
> without the social complexity entailed by coalitions, reputations and politics.

The concept of male sexual proprietariness usefully focuses on the idea of
women as property, reflected in the legal history of male sexual coercion. It also
acknowledges the possibility of the larger social implications of uxoricide, beyond the question of individuals being motivated by reproductive concerns.
Thus Wilson and Daly (Chapter 11 in this volume) mention the "social complexity entailed by coalitions, reputations and politics," while Rodseth and Novak (Chapter 12) observe that "wife abuse became more of an arena for men to

demonstrate or jockey for political power (citation omitted) . . . once again, male aggression against women seems to involve, if not require, a masculine audience."

A critical political and social context of domestic violence homicide is that it sends a clear message to all victims of domestic violence: it tends to further silence and subordinate battered women. Women who are in abusive relationships get the message that murder is a possibility; batterers may use the information intentionally to intimidate and threaten their victims. Batterers are known to use news of domestic homicides as specific examples of what could happen to their victims.

For example, when O. J. Simpson was accused of murdering his wife Nicole Brown Simpson and her friend Ron Goldman, men around the country threatened to "O. J" their female partners. When Carol Cross was murdered by her longtime abusive partner from whom she had recently obtained an order of protection from court, men in Lewiston (Maine) used the murder as a threat to their own victims: "newspaper articles about the murder . . . leaving them around the house threatening that the women could end up like Cross if they dare to leave." (*Boston Globe,* September 1999; Abused Women's Advocacy Program 1999). Although the "official" number of honor killings in Turkey was 43 victims in 2004, one survey indicates that fully 66% of female respondents in eastern Turkey fear that they could become victims of honor killings (Women Living Under Muslim Laws 2001:11–12). The symbolic meaning of femicide is not lost on either other batterers or their victims.

The Supreme Court Rulings on Gang Rape and Domestic Violence Homicide

Out of several thousand petitions, the Supreme Court selects less than 100 cases for review each year. Although I am not claiming any conspiratorial action on the part of the highest court in the United States, it is remarkable that the two most important cases concerning the rights of women to challenge male sexual violence involved a gang rape and domestic violence homicide.

The first case, *United States v. Morrison,* (120 S. Ct. 1740 (2000)) concerned the constitutionality of a provision of the federal Violence against Women Act (VAWA), which created a right to be free from gender-motivated violence. The Civil Rights Remedy, as it became known, was the product of four years of extensive congressional testimony that included unanimous approval from the National Association of Attorneys General and testimony from 38 state attorneys general stating that "the 'current system for dealing with violence against women is inadequate,' . . . it was against this record of failure at the state level that the Act was passed."

Very briefly, the following facts were set forth in *Morrison*. Christy Brzonkala alleged that she was gang raped by two football players at the Virginia Technical Institute University when she was a freshman. The attack occurred in her dorm room about a half hour after she met them. The two assailants took turns raping her, stating that she "better not have any diseases" because they refused to use condoms. In the months following the assault, one of the assailants stated publicly in the cafeteria that "he liked to get girls drunk and f—— the s—— out of them." Although the Court recognized that the rape would qualify as "gender-motivated violence," it held that the federal courts were not the proper forum to address the problem of such violence. Echoing nineteenth-century jurisprudence around domestic violence, the late Chief Justice William Rehnquist had publicly objected to the Civil Rights Remedy before it was passed, expressing concern that the statute "could involve the federal courts in a whole host of domestic relations disputes" (Siegel 1996). In a sense, the decision represents federal nonintervention in the matters of the state in a homologous way to male judges not wanting to tell another man what he can do or not do in his own home. Indeed, "one of the primary goals of supporters of the Act was to overcome centuries of assumptions about the public and private spheres that have operated to deny women full equality under the law" (Goldfarb 2000).

Although a full discussion of this decision is beyond the scope of the current chapter, the salient point is that the Supreme Court, without overtly recognizing the plight of abused women, struck down the right to be free from gender-motivated violence. It found that the remedy exceeded the limits of congressional power to regulate under the Commerce Clause. To regulate under this Clause, Congress must find a substantial effect on interstate commerce, which it explicitly did in this case. Finding violence against women to be insufficiently related to interstate commerce, the Court found no federal jurisdiction. Instead, it found that the appropriate forum for relief would be the state court.

The second case, *Town of Castle Rock v. Gonzalez* (2005: 125 S. Ct. 2796), arose after Simon Gonzales, the estranged husband of Jessica Gonzales, kidnapped and murdered their three young daughters. Ms. Gonzales called the police upon learning that her daughters were missing. Despite having an order of protection against him and a history of threatening behavior that was known to the police, the police ignored multiple calls from Jessica over the course of the evening, repeatedly telling her to call back if Simon did not return the children. After midnight when Simon had still not returned the children, she went to the police station and filed a report. The officer took the report and then went out to dinner. In the early hours of the next day, Simon opened fire on the Castle Rock Police station. The officers returned fire and killed him. The three girls were found already murdered in Simon's truck, with a gun he had purchased that night.

The Court ruled that Jessica Gonzales had no right to enforcement of her order of protection. Justice Scalia, writing for the majority, noted that while he usually defers to the state's judgment in passing legislation, Colorado could not have meant to pass a statute mandating police to enforce orders of protection. The decision overrode the legislative history and intent of the statute, which provided specifically for mandatory action to counteract the traditional inaction of police departments in response to domestic violence. The decision restored police discretion to do nothing in response to pleas from an abused woman to enforce a state-granted order of protection.

The jurisdictional questions in both cases are important. In the case involving federal legislation, the Court said that women should go to the state for relief. In the case involving the state statute, the Court overrode the state legislature and invalidated the statute. In both *Morrison* and *Castle Rock,* the Court professed to care about women's rights but struck down the very basis on which women could assert them. The result is the upholding of a patriarchal right to rule at home and to use male sexual coercion and violence. Catharine MacKinnon (MacKinnon 2000:175) best describes this patriarchal agreement and its translation into law:

> [W]omen are the least equal at home, in private; they have had the most equality in public, far from home. It is in the private, man's sovereign castle, where most women remain for a lifetime, where women are most likely to be battered and sexually assaulted, and where they have no recourse because the private, by definition, is inviolable and recourse means intervention. For physically and sexually violated women, going public with their injuries has meant seeking accountability and relief from higher sovereigns, men who have power over the men who abused them because they are above, removed

from, hence less likely to be controlled by those abusers. . . . One way to describe this dynamic is to observe that men often respect other men's terrain
as sovereign in exchange for those other men's respect for their own sovereignty on their own terrain. As a result of such balances that men with power
strike among themselves, represented in the shape of public institutions,
men have the most freedom at home, and women gain correspondingly
greater equality, hence freedom, the further from home they go.

*Fourth Proposal: The potential of female-female alliances to overcome male sexual
aggression would be strengthened by confronting directly the ways in which male-
male alliances fracture female bonding.*

In both rape and domestic violence, male-male alliances function to fracture
the formation of female-female alliances in profoundly important ways. Because of the prevalence of male sexual violence, women learn to seek male protectors as the first line of defense. Such learned dependence might blind women
from recognizing their own collective power to defend against male aggression.
In rape cases, women divide against themselves by subscribing to rape myths
that blame the victim rather than the perpetrator. In domestic violence cases,
abusers typically isolate their victims as a tactic to increase control over them
and limit their ability and willingness to seek help outside the home.

Rape myths are powerful forces that keep victims from obtaining justice
and redress for rape. These include the idea that rape is sex, that a woman
changed her mind, and that she was "asking for it" by sexually provocative behavior or by being in the wrong place. For women, to blame another woman
for provoking her own attack preserves the idea that rape is something within
her control rather than something for which she is at risk. This may explain
why rape myths are still so prevalent today, as they "protect non-victims from
feeling vulnerable" (Benedict 1992:18). Victim blaming represents an important fracture of female-female alliances, as women try to distance themselves
from other women who are raped in this attempt at self-protection. Furthermore, the belief that a woman can provoke her own rape assumes that men
cannot control—or are not responsible for controlling—their own sex drives.
Focusing on the rape victim rather than the perpetrator prevents women from
recognizing the systemic and pervasive occurrence of rape.

In domestic violence cases, isolation is a well-known tactic of batterers. This
tactic is described as "Controlling what she does, who she sees and talks to,

what she reads, where she goes; limiting her outside involvement; using jealousy to justify actions" (Domestic Abuse Intervention Project: Duluth:1989). Isolating a victim from her female kin and other alliances further increases the batterer's power over her. An example of this isolationist behavior by male chimpanzees was described by Goodall (1986). Males sometimes attempt to isolate a female from other males by forcing her, through violence or threats of violence, into accompanying him to a remote part of their community range. If she tries to call out or flee, he may beat her. On these "consortships" the male has exclusive sexual access to the female and may impregnate her (Tracy and Crawford 1999:35).

This chimpanzee behavior is strikingly similar to the behavior of batterers who commonly rip the telephone out of the wall to prevent the woman from calling for help. This tactic is so widespread that some order of protection forms now provide a box to check for restitution if the batterer has damaged the victim's cell phone.

Preventing the female from seeking help in this way punishes the victim both for seeking assistance from the legal system and for trying to access or form female alliances that might provide protection from further abuse. Forms of male sexual coercion that prevent communication of the abuse to others can be seen to directly fracture the formation of female alliances.

However, bonobos offer an inspiring model of strong female-female alliances (Wrangham and Peterson 1996). The lesson of bonobos—that male sexual coercion can be eliminated through collective physical resistance—could inspire women to form coalitions that engage in some element of physical resistance to male domination. For example, these could take the form of women-led law enforcement agencies created and staffed primarily by women.

Bonobo females seem to operate from an understanding that if one can be aggressed upon, than all or any could be as well. This lesson could be particularly applicable to prostitution. The high levels of posttraumatic stress and sexual violence suffered by prostituted women are devastating (Farley 2003). Moreover, in the United States the vast majority (around 75%) of women who enter prostitution were victims of child sexual abuse and became prostituted between the ages of 14 and 17 (Center for Prostitution Alternatives 2003). Female-female alliances that recognize female vulnerability to male sexual coercion and the collective ability to physically resist should therefore be encouraged. Such alliances could alter the patriarchal structure of our society.

Conclusions

Evolutionary perspectives deepen our understanding of the dynamics of male sexual coercion and thus our ability to understand the law's potential to disrupt it. Such critical knowledge informs this chapter in important ways, as described in this section.

The proposals set forth in this article suggest several related conclusions.

1. Law should address male sexual coercion as a prevalent, rather than an aberrant, behavior, whatever its functions and origins might be.
2. Law enforcement should recognize the male-male alliances that perpetuate male dominance if it is serious about disrupting the occurrence of such violence.
3. Understanding the function of patriarchal violence can help legal policymakers develop more meaningful responses to male sexual coercion and promote the rights of women to be free from such violence, promoting the thriving of all.
4. The potential of female-female alliances to stop male sexual coercion could be more fully realized by understanding the way male alliances fracture female coalitions; addressing the prostitution of women by men is recommended as an effective campaign to attack the root of patriarchal violence.

Acknowledgments

This chapter would not have been possible without the extraordinary assistance of Adam D. J. Balfour, LL.B, LLM, who helped bring the arguments to fruition in earlier drafts. Richard Wrangham was also exceptionally helpful as a co-professor and the source of inspiration for my exploration into evolutionary biology as a field that might inform my legal work on violence against women. I appreciate the comments and feedback from many of my students in the Theories of Violence: Gender and Sexuality Reading Group and in Professor Wrangham's and my seminar on Theories of Sexual Coercion at Harvard. I am especially grateful for the editorial assistance of Harvey Berkman, Lisa Cloutier, and Caitlin Hartman. Other students, colleagues, and former students who provided valuable help include Alisha Bhagat, Mark Egerman, Terry Fisher, Mary Anne Franks, Kathryn Knisely, Maura Klugman, Margaret Lazarus, Brenda

Lee, Rebecca Leventhal, Sandra Pullman, Roberta Oster Sachs, Sarah Schaffer, Leah Satine, Shauna Shames, and Karen Techepin.

References

2003 Massachusetts Domestic Violence Homicide Report, Jane Doe Inc., 2006; available at http://www.janedoe.org/know/2003%20MA%20DV%20Homicide %20Report.pdf; accessed November 17, 2008.

Adams, D. *Why Do They Kill?: Men Who Murder Their Intimate Partners.* Nashville, Tenn.: Vanderbilt University Press, 2007, pp. 1–23, 163–219.

Benedict, H. *Virgin or Vamp: How the Press Cover Sex Crimes.* New York: Oxford University Press, 1992, pp. 13–24.

Blackstone's *Commentaries on the Laws of England,* Book I, Chapter 15; available at: http://www.avalon.law.yale.edu; accessed November 17, 2008.

Campbell, A. *A Mind of Her Own: The Evolutionary Psychology of Women.* Oxford: Oxford University Press, 2002, pp. 116–120.

Campbell, J. C., D. Webster, J. Koziol-McLain, C. R. Block, D. Campbell, M. A. Curry, F. Gary, J. McFarlane, C. Sachs, P. Sharps, Y. Ulrich, and S. A. Wilt. "Assessing Risk Factors for Intimate Partner Homicide." *National Institute of Justice Journal* 250 (2003):14–19.

deWaal, F. *Chimpanzee Politics: Power and Sex among Apes.* Baltimore, Md.: Johns Hopkins University Press, 2000.

deWaal, F., and F. Lanting. *Bonobo: The Forgotten Ape.* Berkeley: University of California Press, 1997.

Domestic Abuse Intervention Project, Duluth Project, Minnesota; available at http://www.duluth-model.org/; accessed November 17, 2008.

Farley, M. *Prostitution, Trafficking and Traumatic Stress.* New York: Haworth Press, 2003.

Fenno, C., Executive Director, Abused Women's Advocacy Program. Personal communication, September 1999.

Goldfarb, S. "Violence against Women and the Persistence of Privacy." *Ohio State Law Journal* 61 (2000): 1.

Griffin, S. *Rape: The Politics of Consciousness.* San Francisco: Harper and Row, 1986, pp. 10–11.

Jensen, R. "Cruel to Be Hard: Men and Pornography." *Sexual Assault Report,* January/February (2004): 33.

Jessop, C. *Escape.* New York: Broadway Books, 2007, pp. 211–217.

Johnson, L. B. *On The Front Lines: Police Stress and Family Well-Being.* Hearing before the Select Committee on Children, Youth, and Families. House of Representatives: 102 Congress, First Session, May 20 (pp. 32–48). Washington D.C.: U.S. Government Printing Office, 1991.

Jones, A. *Women Who Kill.* Boston: Beacon Press, 1996.

Karjane, H., B. Fisher, and F. Cullen. "Sexual Assault on Campus: What Colleges and Universities Are Doing about It." National Institute of Justice, Washington, D.C., 2005.

Katz, J. *The Macho Paradox: Why Some Men Hurt Women and How All Men Can Help.* Naperville, Ill.: Sourcebooks, 2006.

Kennedy, D., and W. W. Fisher III, eds. *The Canon of American Legal Thought.* Princeton, N.J.: Princeton University Press, 2006.

Kennedy, D. "Sexual Abuse, Sexy Dressing and the Eroticization of Domination." *New England Law Review* 26 (1992): 1309–1393.

Kummer, H. *In Quest of the Sacred Baboon: A Scientist's Journey.* Princeton, N.J.: Princeton University Press, 1995.

MacKinnon, C. A. *Sexual Harassment of Working Women.* New Haven, Conn.: Yale University Press, 1979.

———. *Feminism Unmodified: Discourses on Life and Law.* Cambridge, Mass.: Harvard University Press, 1982.

———. *Toward a Feminist Theory of the State.* Cambridge, Mass.: Harvard University Press, 1989.

———. "Disputing Male Sovereignty: On *United States v. Morrison.*" *Harvard Law Review* 114 (2000) 135, 175.

———. *Men's Laws, Women's Lives.* Cambridge, Mass.: Harvard University Press, 2006.

———. *Are Women Human? And Other International Dialogues.* Cambridge, Mass.: Harvard University Press, 2006.

Madigan, L., and N. C. Gamble. *The Second Rape: Society's Continued Betrayal of the Victim.* New York: Lexington Books, 1991.

Merrill, M. R. "Murder Has Abuse Victims Terrified." *Boston Globe,* September 12, 1999, C11–12.

Miccio, G. K. "Exiled from the Province of Care: Domestic Violence, Duty and Conceptions of State Accountability." *Rutgers Law Journal* 37 (2005): 111.

Minow, M., M. Ryan, and A. Serat, eds. *Narrative, Violence and the Law: The Essays of Robert Cover.* Ann Arbor: University of Michigan Press, 1995.

Neidig, P. H., H. E. Russell, and A. F. Seng. Interspousal aggression in law enforcement families: A preliminary investigation. *Police Studies* 15 (1992): 30–38.

O'Dell, A. "Domestic Violence Homicides." *The Police Chief,* February 1996, pp. 22–23.

Pateman, C. *The Sexual Contract.* Stanford, Calif.: Stanford University Press, 1988.

Pelka, F. "Raped: A Male Survivor Breaks His Silence." *On the Issues: A Progressive Woman's Quarterly* 22 (1992): 8–11.

Rosenfeld, D. "Law Enforcement Sends Mixed Signals." *Chicago Tribune,* July 31, 1994.

———. "Why Doesn't He Leave: Restoring Liberty and Equality to Battered Women." In *Directions in Sexual Harassment Law,* eds. Catharine A. MacKinnon and Reva Siegel, pp. 531–549. New Haven, Conn.: Yale University Press, 2004.

————."Doing the Right Thing: Using Title IX to Create Safe and Equal Campuses." Campus Training and Technical Assistance Institute, U.S. Department of Justice, Office on Violence Against Women, January 2007 (on file with author).

Sanday, P. R. *Fraternity Gang Rape: Sex, Brotherhood and Privilege on Campus.* New York: New York University Press, 1990.

Schaffer, S. "Closing Ranks: The Male Conspiracy of Violence against Women." Gender Violence, Law and Social Justice, Harvard Law School, May 2006 (on file with author), 1–6.

Schulhofer, S. *Unwanted Sex.* Cambridge, Mass.: Harvard University Press, 1998.

Siegel, R. B. "The Rule of Love: Wife Beating as Prerogative." *Yale Law Journal* 105 (1996): 2117.

————"Why Equal Protection No Longer Protects: The Evolving Forms of Status-Enforcing State Action." *Stanford Law Review* 49 (1997): 1111.

Smuts, B. B. *Sex and Friendship in Baboons.* Hawthorne, N.Y.: Aldine Publishing Company, 1985.

————. "The Evolutionary Origins of Patriarchy." *Human Nature: An Interdisciplinary Biosocial Perspective* 1 (1995): 1–32.

Smuts, B. B., and R. W. Smuts. "Male Aggression and Sexual Coercion of Females in Nonhuman Primates and Other Animals: Evidence and Theoretical Implications." *Advances in the Study of Behavior* 22 (1993):1–63.

————"Male Aggression against Women: An Evolutionary Perspective." *Human Nature* 1 (1992): 1–32.

Torrey, M. "When Will We Be Believed? Rape Myths and the Idea of a Fair Trial in Rape Prosecutions." *U.C. Davis Law Review* 24 (1991): 1013.

Tracy, K. K., and C. B. Crawford. "Wife Abuse: Does It Have an Evolutionary Origin?" In *To Have and to Hit,* eds. D.A. Counts, J.K. Brown, and J. C. Campbell, pp. 27–42. Urbana: University of Illinois Press, 1999.

Women Living Under Muslim Laws (WLUML). Roundtable on Strategies to Address 'Crimes of Honour' Summary Report. Occasional Paper No. 12, November 2001.

World Health Organization. "WHO Multi-country Study on Women's Health and Domestic Violence against Women: Summary Report of initial results on Prevalence, Health Outcomes and Women's Responses." Geneva: World Health Organization, 2005.

Wrangham, R., and D. Peterson. *Demonic Males: Apes and the Origins of Human Violence.* Boston: Houghton Mifflin Co., 1996.

V

SUMMARY AND CONCLUSIONS

18

Sexual Coercion in Humans and Other Primates: The Road Ahead

Richard W. Wrangham and Martin N. Muller

This volume is the first compilation of studies of male aggression toward females in species living in complex societies. It includes all the great apes (orangutans, gorillas, chimpanzees, and bonobos), and a selection of other cognitively complex primates and dolphins for which suitable data were available. Four analyses of human sexual coercion complement the nonhuman data. The book thereby offers new opportunities to understand the evolution and adaptive significance of a behavioral system that has considerable importance in the lives of animals and unpleasant relevance to our own species. In this chapter we discuss selected lessons arising from the new information that are relevant to understanding sexual coercion in humans, nonhuman primates, and other animals from an evolutionary perspective.

Relationship Violence: Sexual Coercion Is Often a Long-Term Strategy

The most detailed research on nonhuman sexual coercion has been on insects, including scorpion flies, fruit flies, water striders, and flour beetles (reviewed in Arnqvist and Rowe 2005). These studies have provided a critical theoretical foundation for the field by showing that sexual conflict, male coercion, and female counteradaptations are widespread and explicable in detail by evolutionary logic (Watson-Capps, Chapter 2). The sexually coercive tactics used by insects and most other animals are limited largely to "short-term coercion," that is, coercion concerned with the immediate act of mating, such as forced copulation, mate guarding, and punishment of females that mate with other males (for a possible exception in salamanders, see Jaeger et al. 2002). Short-term sexual coercion is also found in some primates, most notably orangutans in the form of forced copulation (Knott, Chapter 4). The most obvious form of sexual coercion in humans is rape, which has often been argued to be driven principally by the immediate goal of conception (reviewed by Emery Thompson, Chapter 14). For such reasons it might be thought that most sexual coercion in primates and humans would be short term.

This book suggests otherwise. In the nonhuman primate species reviewed here in which sexual coercion is most prominent (i.e., chimpanzees, hamadryas baboons, chacma baboons, and spider monkeys), sexual coercion is mostly long term because the aggression directed by males toward females achieves its goal by manipulating the future rather than the immediate behavior of the victim. Males in these species rarely practice forced copulation or forced consortships. Orangutans are a prominent exception to this trend, but our sample of primates suggests that sexual coercion more often involves a much longer temporal dimension than is implied by the classic notion of conceptive rape.

The theoretical basis for the evolution of long-term sexual coercion appears intuitively simple and has been elaborated in one instance by Clutton-Brock and Parker (1995), who showed that male punishment of female mating resistance can be an effective adaptive strategy if females modify their behavior in response to the actions of known males. Rather than requiring rare or complex circumstances to evolve, the major constraints on long-term sexual coercion are social and cognitive. Thus, species must live in stable social networks and have sufficient cognitive abilities to sustain social relationships. Such requisites include individual recognition, memory of specific events, and sophisticated learning.

There are two main examples of long-term sexual coercion. First, infanticide has long been recognized as generating a conflict of interest between the sexes and pervasively influencing sexual strategies and social relationships (Clarke et al., Chapter 3, and Palombit, Chapter 15), but it has not always been treated as a form of sexual coercion (Clutton-Brock and Parker 1995). However, infanticide tends to accelerate a female's return to a state of fecundability and therefore reduces the chance that the killer will already be dead or expelled from her group before she is ready to mate. For these reasons it conforms to Smuts and Smuts's (1993) definition of a sexually coercive behavior because it increases a male's chance of mating with the mother of the victim and/or decreases the chance that she will mate with other males (Watson-Capps, Chapter 2). Infanticide is therefore appropriately regarded as a form of long-term sexual coercion.

Second is a behavior that we call *conditioning aggression.* We restrict this term to cases in which females experience aggression from males who have no immediate interest in mating (e.g., because the females are not ovulating or have no sexual swelling). When such aggression is shown to lead to an increase in future mating opportunities for the aggressive male, while being costly for the female and having no other detectable function, we classify it as sexual coercion. Conditioning of females by males is clearly shown in chimpanzees (Muller et al., Chapter 8), and hamadryas baboons (Swedell and Schreier, Chapter 10). It is also suspected in spider monkeys (Link et al., Chapter 7), and may occur in mountain gorillas (Robbins, Chapter 5) and chacma baboons (Kitchen et al., Chapter 6). In addition, it appears to be an appropriate description of much human male aggression toward females, including wife-beating, acquaintance rape, and certain kinds of gang rape (e.g., Wilson and Daly, Chapter 11; Emery Thompson, Chapter 14; Rosenfeld, Chapter 17). Whether wife-beating qualifies as sexual coercion, however, depends on its leading to increased sexual access by the husband, a reduction in adultery by the wife, or both. Unless those effects are known, we cannot be certain that wife-beating is sexual coercion in the sense of Smuts and Smuts (1993).

The principal mechanism by which sexually selected infanticide benefits the killer is well established: it accelerates the return of the victim's mother to fecundability. By contrast, the mechanisms by which conditioning aggression leads to increased mating success are less well known. Conditioning aggression can have various kinds of effect. In hamadryas, the result of male aggression by a breeding male toward a new female in his group is to bring her closer, promoting their long-term social bond (Swedell and Schreier, Chapter 10). No such increase in proximity has been reported between a male chimpanzee and a female that he physically attacks, and our impression instead is that she routinely flees from him. Either way, whether aggression tends to increase or reduce immediate proximity, a longer-term increase in the aggressor's mating success can in theory occur because the female is more likely to respond to the advances of the aggressor (Clutton-Brock and Parker 1995), or because she is less likely to respond to the advances of other males in the presence of the aggressor, or both (Muller et al., Chapter 8). The relative importance of these two avenues has not yet been established for any of the primates in which conditioning aggression has been reported and is an important problem for the future.

The occurrence of long-term sexual coercion in nonhuman primates is illuminating because it warns against simplistic explanations. First, such "relationship violence" means that even when males are aggressive to females in nonmating contexts, they may be exhibiting sexual coercion. Second, it raises the possibility that coercive interactions that at first appear to be short term may in fact incorporate a long-term component. For example, whereas forced copulations represent only a conception attempt in scorpion flies, in species such as orangutans (Knott, Chapter 4) and humans (Emery Thompson, Chapter 14) they could in theory have social significance by conditioning a female to respond favorably in future interactions. Knott considers this possibility and concludes that it is unlikely for orangutans because males are aggressive to females only immediately prior to and during copulation, and male aggression rarely if ever leads to serious injury. However, whether females who have experienced successful forced copulation become less resistant to their aggressors in the future is not yet known. Even in the case of orangutans, therefore, the adaptive significance of forced copulation may hold surprises.

In sum, the occurrence of long-term relationships in certain large-brained animals requires us to extend our understanding of the adaptive significance of sexual coercion beyond the systems found in cognitively simpler species. This conclusion raises the possibility that across species, the nature of sexual coercion will tend to reflect the level of cognitive ability.

Species Differences in Sexual Coercion

Across primates there is indeed some indication of aspects of sexual coercion being associated with taxonomy, but there has been little suggestion that such clusterings are due to shared cognitive features. Thus Clarke et al. (Chapter 3) report that effective sexual harassment occurs primarily in the Catarrhines (Old World monkeys and apes), and they suggest that this is because males can most easily harass females by day, on the ground, and if females are relatively small, conditions commonly found among the Catarrhines. Similarly, Connor and Vollmer (Chapter 9) note that consortships are more common in dolphins than in chimpanzees, and they attribute the difference partly to features that differ widely in general between cetaceans and primates, such as the costs of locomotion. These proposals echo phylogenetic correlations found in some other animal groups. For example, in birds the incidence of forced copulation is strongly associated with the occurrence in males of an intromittent organ, which is generally present for reasons unrelated to sexual coercion (Low et al. 2005).

By contrast, among closely related species, patterns of sexual coercion do not fit into tidy taxonomic boxes. The diversity is well illustrated by the great apes. Bonobos show that male sexual coercion is not a biological imperative, since none has been found despite a careful search (Paoli, Chapter 16). However, in chimpanzees, the closest relative of bonobos, males exhibit frequent long-term sexual coercion, that is, harsh conditioning aggression and infanticide, as well as occasional short-term sexual coercion, primarily in the form of forced consortships (Muller et al., Chapter 8). Like chimpanzees, male mountain gorillas are also prone to be aggressive toward females. But aggression by male gorillas is so mild that it is not yet clear whether it imposes sufficient costs on females to qualify as sexual coercion (Robbins, Chapter 5). Orangutans, by contrast, commonly practice short-term (and possibly long-term) coercion in the form of forced copulations (Knott, Chapter 4). None of these great apes show the same patterns of male aggression toward females in humans, where many types of sexual violence can be found, but in which there are nevertheless some cross-cultural consistencies, such as a high frequency of husbands beating their wives (Campbell 1999).

Hamadryas and savanna baboons are even more closely related to each other than great apes are, yet they exhibit similarly diverse behaviors. Male hamadryas are generally less aggressive than other baboons, but they are nevertheless highly coercive to females with whom they are forming a new relationship (Swedell and Schreier, Chapter 10). Male aggression toward females is more frequent in various savanna baboons. The aggression includes infanticide that varies in frequency from low in olive baboons to high in chacma baboons, a species in which it has large demographic and social effects (Palombit, Chapter 15). Infanticide is not only a form of sexual coercion itself, but as Clarke et al. (Chapter 3) discuss, it also generates a conflict of interest between males and females that leads to further forms of coercion. Link et al. (Chapter 7) also report that among the South American ateline monkeys, patterns of sexual coercion range from absent (in muriqui) to infanticide and conditioning aggression in others.

Based on general principles, such variation among close relatives is expected to be explained by variation in social systems, largely owing to independent factors such as ecological pressures on grouping, locomotor style, and sex differences in body size and fighting ability (Clarke et al., Chapter 3). Paoli's analysis of the differences in sexual coercion between bonobos and chimpanzees illustrates the strengths and limitations of this approach (Chapter 16). He shows that, as expected, the greater importance of sexual coercion in chimpanzees is associated with their having smaller and less stable groups, less completely concealed ovulation, and more socially dominant males. Furthermore, differences in ecology provide suggestive explanations for the differences in grouping patterns, and hence possibly for the differences in social relationships. Yet as he notes, the causal pathways have not yet been established precisely.

Similar uncertainty applies to understanding the evolutionary ecology of all the primates in this book. Understanding the ultimate causes of social systems will be an important contribution to a more complete theory of primate and human sexual coercion because at present it is unclear how much sexual coercion is responsible for the social systems, versus being a response to them. Clearly, there must be some important relevant species differences in evolutionary ecology and biology: if there were not, every species would have similar patterns of sexual coercion.

This issue is particularly intriguing when we consider humans. Of course, for several reasons human sexual coercion presents a difficult analytical challenge. The quality of data is routinely poor, both because of the sensitivity of the topic and because sexual coercion occurs largely in private, so researchers must be alert to numerous biases (Wilson and Daly, Chapter 11; Emery Thompson, Chapter 14; Rosenfeld, Chapter 17). Substantial variation across cultures also complicates the problem, but there is an even more fundamental difficulty. Many nonbiologists hold the mistaken notion that an evolutionary explanation of behavior requires us to see it as inevitable.

As Emery Thompson (Chapter 14) explains, however, this characterization of biological determinism is misconceived. All the authors in this book share the biologists' conventional view that because phenotypes depend on interactions between genes and environments, individuals' behavioral strategies are not predictable from their genes alone. Furthermore, behavior responds sensitively to context. This means that if its costs or benefits change, whether because of changes in social relationships, ecology, or (in human cases) the law, biologists confidently expect the frequency and the nature of behavior to be flexible as well. Contrary to the fear of many nonbiologists, therefore, there is nothing about an evolutionary analysis that implies immutability. So we can search for a common basis of sexual coercion in humans and other animals undeterred by the false notion that an evolutionary analysis reduces our hope for more effective legal and societal responses.

We do not suggest that the common basis has yet been identified, however. For instance, Clarke et al. (Chapter 3) attribute the majority of sexual coercion in primates to the threat of infanticide. Since humans are vulnerable to infanticide (Daly and Wilson 1988), this raises the possibility that the fundamental conflicts that underlie much sexual coercion in human society emerge from the same intersexual conflict stimulated by the threat of infanticide. The problem with this hypothesis is that, in order to test it, we need to consider the influence of various sources of conflict between male and female interests that are unique to humans; and these have not yet been fully established.

Patriarchy Changes the Equation for Human Sexual Coercion

Biologists have responded to the challenge of explaining human sexual coercion mainly with the conventional evolutionary gambit of focusing on reproductive outcomes. As Smuts (1995:22) argued, "the ultimate goal of male control over females is reproduction: men coerce, constrain, and dominate women in order to maintain control over female sexuality and the offspring women produce." This perspective sometimes works well. The ability of men to coerce women is certainly partly due to factors that are found in other animals, such as sex differences in size, strength, aggressiveness, fearfulness, and libido (Wood and Eagly 2002); and male dominance over females is used partly to achieve conceptive goals. For example, male reproductive striving based on dominance over females provides the background for Wilson and Daly's claim that "male sexual proprietariness is an evolved motivational/cognitive subsystem of the human brain/mind," and hence for their understanding of wife-killing as a rare failure in a system of behavioral control that is generally well adapted for promoting conjugal fidelity (Chapter 11).

As Rodseth and Novak (Chapter 12) argue, however, a primatological view of sexual coercion in humans is incomplete because human cultural complexity creates new kinds of social logic—such as beating women in order to make peace among men. We suggest that two unique features of humans are particularly important in shifting the dynamics of sexual violence away from a purely reproductive logic.

First is patriarchy, a cultural system that can be operationalized as the extent to which women lack social power, control, and authority, and are regarded as inferior to men (Broude 1990). Patriarchy stems from male-male alliances that, partly through the use of language, lead to a collective ability by men to control women, including their sexuality (Smuts 1995). Patriarchy thus tends to amplify power differentials between the sexes. For this reason it is a critical part of the equation for understanding sexual coercion.

The sexual control that patriarchy gives men over women is sometimes exerted in the service of reproductive goals, such as in harems (Dickemann 1979). However, men can also use their patriarchal power to control women's sexuality for nonreproductive purposes, such as political goals. These practices are coercive in the ordinary sense of meaning that women are forced to behave sexually against their will, but they do not conform to Smuts and Smuts's (1993) definition of sexual coercion, which assumes that male dominance serves reproductive ends.

For example, among Australian hunter-gatherers, wives might be required to have sex with multiple men at special ceremonies (Elkin 1938; Berndt and Berndt 1988). They could also be lent to a visiting man, or be given sexually by a husband to a man with whom the husband had quarreled, in order to erase a debt or make peace. Women could also be sent on sexual missions into dangerous situations. When potential attackers were seen approaching a group, one response was to send women out to greet them. If the strange men were willing to forego their attack, they signaled their intent by having sexual intercourse with the female emissaries. If not, they sent the women back and then attacked. The final stage of peacemaking between two tribes almost always involved an exchange of wives. Elkin (1938) reported that women sometimes lived in terror of the use that was made of them at ceremonial times.

In theory, women in such societies could be so libidinous that what appears from the outside to be enforced polyandry might actually have been a culturally sanctioned erotic enjoyment of multiple partners. This argument has been made for the Canela, a Brazilian hunter-farmer society in which nulliparous females were similarly expected to have sex in various polyandrous arrangements, including being given to their husbands' friends, having sex with multiple males on hunting expeditions, and having sequential sex with the men's society. Crocker and Crocker (1994) reported that the young women who engaged in these practices were encouraged to do so by older female relatives who reported that they greatly enjoyed sex. Yet whether or not the young women found these interactions pleasurable, they had no choice in the matter. "Sexually stingy" girls were advised by older female kin that they "simply had to go through with this generosity to men, even if it were unpleasant" (Crocker and Crocker 1994:157). Girls were told that if they did not submit themselves to sequential sex, they would never win the ceremonial belts that were required to get a husband and become a mother. If they persisted in refusal, they were gang-raped by up to six youths.

These practices in Australia and Brazil illustrate that men's ability to control the sexuality of young women does not always serve the functions of increasing their mating access or preventing other males from mating. In the cases described, men used their power to promote alliances with other men, reduce the risk of male-male violence, or give sexual rewards to those whom they wished to please. Arranged marriages provide another familiar example common to many societies. Patriarchy enhances men's coercive power by strengthening their alliances. Human sexual coercion is therefore not always explicable by the same principles as among nonhumans.

The second unique feature of human society that causes sexual coercion to lead to nonreproductive (or indirectly reproductive) goals is that unlike nonhuman primates, in which pair bonds are based only on a sexual relationship, in human pair bonds, sexual and economic relationships are intertwined. This means that within the pair bond a man may have nonsexual reasons to be aggressive to a woman. For instance, Collier and Rosaldo (1981) show that men in prestate societies need wives in order to be able to have predictable meals, to be able to entertain other men, and thus to function as respected members of the community of dominant males. Much male aggression, jealousy, and competition occurs in the context of wives as domestic producers, such as beatings that result from a wife failing to produce a timely or adequate meal (Collier and Rosaldo 1981). The general lesson is that conflict between husbands and wives can arise not only over matters concerned with sexual fidelity, but also for economic or political reasons.

The greater range of functional possibilities for human sexual coercion means that, even when reproductive goals look important, care must be taken to check their significance compared to nonreproductive functions. Gang rape, for example, might superficially appear to serve reproductive ends by each male taking the opportunity for a possible conception. But it is clearly often far more than that. In state societies at war, gang rape is sometimes promoted as a tool of terror (Schwendinger and Schwendinger 1983). In prestate societies, gang rape was commonly ordained as a punishment for females who contravened societal norms, such as failing to have sex as demanded or viewing men's secret ceremonies (Meggit 1987). It could also be a component of ceremonies such as marriage. In one case, the lack of reproductive significance is clearly suggested by the aftermath: "In the Boulia district (Queensland) a pubescent girl is caught by a number of men; they forcibly enlarge the vaginal orifice by tearing it downward with their fingers, which have possum twine wound round them, then have sexual relations with her, collecting the semen and drinking it ritually" (Berndt and Berndt, 1988:181).

Gang rape in such cases seems to be employed primarily to teach women about the power of men and the wisdom of conforming to norms. It is thus functionally more similar to the "conditioning aggression" found in some nonhuman primates than to the forced copulation of orangutans. It illustrates that sexual coercion in humans includes a wide set of adaptive pathways because such behaviors sometimes have no direct reproductive significance, even though they may still be importantly advantageous to the rapists. For instance they might help to promote a man's integration into his social group, or they might reduce the chance that a woman's behavior will erode the symbolic but critically important power of cultural norms. Power can be used in multiple ways, not merely to improve sexual access.

An evolutionary analysis of human sexual coercion thus faces two important conceptual challenges that have not always been considered. First, when evaluating the functional significance of an aggressive act it must discriminate the relative importance of short-term and long-term consequences. In particular, it must be open to the idea that long-term consequences could be important because if so, the immediate results may be less significant than they initially appear.

Second, even where women's sexual behavior is being controlled, an evolutionary analysis must decide whether men's coercive power is used for reproductive or nonreproductive purposes. In most cases more data are needed. For instance, Emery Thompson (Chapter 14) makes a compelling case that acquaintance rape can be a form of conditioning aggression (as we call it here), although there seem to be no critical tests of this idea. Similarly, the obvious adaptive hypothesis for wife-beating (where it stops short of uxoricide) is that it serves the aggressor's interests by promoting her subordination. But even assuming that women are more easily controlled by their husbands as a result of being raped and beaten, the link to sexual coercion in Smuts's sense remains to be shown.

These considerations thus add to the observations by Rodseth and Novak (Chapter 12) and Rosenfeld (Chapter 17) that male aggression toward females does not necessarily lead to improved mating access. Rodseth and Novak cite evidence that males who are violent to females may benefit by status enhancement with respect to other males, whereas Rosenfeld suggests that legal tolerance of male violence toward females may serve to promote patriarchal power within society—rather as gang rape has been argued to do in a variety of small-scale societies. Rosenfeld (Chapter 17) cites two recent Supreme Court cases showing that in the United States the law is surprisingly tolerant of certain kinds of male aggression toward women. She suggests that this tolerance occurs because the law is made principally by men and that male violence toward women serves the interests of men by helping to maintain the patriarchy. (The men in question are presumably the dominant men, i.e., those for whom the control of women will provide the most important benefits.) The implication is that the aim of the judges who interpret the law is to use men's control of women to erode female power. Admittedly, there are also alternative possibilities. For instance, maintenance of the patriarchy could be an incidental consequence of efforts to protect the innocent by following the principle of accused individuals being presumed innocent until proven guilty. But either way, an important conclusion is that patriarchal violence does not function in the same way as male aggression toward females in nonhuman primates.

Incidents of men's aggression toward women thus present many obstacles to the inference that the behavior offers a simple parallel to nonhuman sexual coercion. Analysts must decide whether the aggressor benefits in terms of mating access at all, whether mating benefits occur in the short term or long term, and through what mechanisms any benefits occur. They must also recognize the possibility that economic or social benefits can in some circumstances be more important than mating benefits. By taking such problems into account, it will be increasingly possible to discriminate evolutionarily novel forms of sexual violence in humans from patterns more typical of nonhuman primates.

Female Counterstrategies Are Similar across Humans and Nonhuman Primates

The range of counterstrategies employed by females to respond to male aggression is remarkably similar among socially complex primates and dolphins, and can be summarized as (1) temporarily fleeing to a different part of the range, (2) submitting to and appeasing the abuser, (3) forming a bond with a nonabusive protector, and (4) fighting back with the support of female allies.

In some species, multimale mating represents an additional female counterstrategy to the specific tactic of male infanticide. Ironically, however, this behavior appears to exacerbate some of the other forms of male coercion discussed in this book (Clarke et al., Chapter 3).

Flight to a different part of the range is a common short-term strategy for evading male coercion, at least for females in fission-fusion species. Connor (Chapter 9) describes female dolphins bolting from consorting males in escape attempts that can result in long chases. Female chimpanzees can also run away from males who are attempting to form consortships, and may leave mixed-sex parties after being attacked by a male (Goodall 1986). An inherent risk of flight is future punitive aggression from the male, particularly if he engages in immediate pursuit.

The strategy of appeasement has been perfected by hamadryas females, who learn quickly how far they can stray from their leader male without incurring a neck-bite and how to employ grooming to soothe an irritated male (Swedell and Schreier, Chapter 10). Female hamadryas appear so compliant that they never voluntarily leave their leader males, all such transfers being initiated by male aggression (Swedell and Schreier, Chapter 10). Bachmann and Kummer (1980) found limited evidence for female agency through field experiments showing that males were more likely to respect the "ownership" of leaders whose females demonstrated affection for them and more likely to attempt to steal females who appeared to dislike their leader males. This subtle effect, however, was restricted to medium- and low-ranking males; high-ranking males disregarded female preferences. Further attesting to the submissive temperament of female hamadryas, Biquand et al. (1994) reported that vasectomizing a leader male was a wholly effective means of contracepting the females in his harem, surely a singularity for primates living in multimale, multifemale groups. When taken to such extremes, appeasement may inflict costs on females by restricting their ability to find a high-quality mate. However, this cost may be low if only high-quality males are capable of preserving a group of females in the face of extreme male-male competition.

Cases of near-total female submission to male coercion can be found in human societies exhibiting extreme patriarchal ideologies. Female genital mutilation, for example, including clitoridectomy, excision, and infibulation, is commonly practiced by Arab Muslims in Sudan's Nile Valley. The stated purpose of the procedure is explicitly coercive—to reduce women's sexual desire and thus diminish their interest in "illicit" sex (Hayes 1975). However, it is women rather than men who rigorously police, defend, and perpetuate the practice, especially grandmothers who want to protect their granddaughters from being viewed as sexually licentious (Hayes 1975). In such cases, male power is so pervasive that women preemptively facilitate men's coercive tactics.

Another prevalent female counterstrategy is to form a long-term bond with a noncoercive protector male, or "hired gun" (Smuts and Smuts 1993). Much evidence suggests that gorilla females form bonds with silverbacks at least in part to protect their offspring from infanticidal males (Wrangham 1986; Harcourt and Stewart 2007). Palombit (Chapter 15) shows that within chacma baboon groups, females form special affiliative relationships with particular males, the primary function of which appears to be infanticide protection. Similar friendships in olive baboons have more generalized benefits, including protection from male aggression in noninfanticidal contexts (Smuts 1985). Smuts (1992) reviews reports from a range of human societies showing that women who lack the protection of a mate are at particular risk for abduction and rape, and argues that the threat of male coercion played a major role in the evolution of human pair-bonding. Although protection of this kind is critically important in some human societies, the solution is imperfect, as several authors in this volume note that when women are abused, it tends to be at the hands of their primary partner.

Fighting back is often a difficult option for females because in many species males are larger. Thus, physical resistance by females is most practicable when they can count on the support of female allies. Smuts and Smuts (1993) argue that, distinct from female coalitions directed toward other females, coalitions against males in primates often involve unrelated females. They note reports of such coalitions successfully preventing males from joining social groups and driving males out of groups. However, few studies directly address the effectiveness of female coalitions at reducing male coercion.

In most of the species reviewed in this book, female coalitions against males are either absent (e.g., among spider monkeys, hamadryas baboons, gorillas, orangutans) or appear most effective against subordinate males. Among the relatively monomorphic dolphins of Shark Bay, cooperative female coalitions are employed only against juvenile males (Connor, Chapter 9; Connor et al. 1992). And Nishida (2003) reported that female chimpanzees at Mahale often successfully enlist the help of other females to retaliate against aggression from older adolescent males. Bonobos represent a prominent exception, as females do form aggressive coalitions against males, and these coalitions have been characterized as playing a critical role in establishing female dominance in this species (i.e., the *female-defense alliance hypothesis:* Parish 1996). Paoli (Chapter 16), however, argues that the importance of such alliances for female status is ambiguous, as the few data that exist from the wild suggest both that females can often dominate males without assistance and that males and females are equally likely to be female coalition partners (Hohmann and Fruth 2003). The female-defense alliance hypothesis is therefore not yet established.

Parallel caution is needed before female chimpanzees can be regarded as successfully countering male aggression with female-female alliances. For example, although Newton-Fisher (2006) reported such cooperation in Budongo, his definition of coalitionary aggression was not restricted to directed charges, physical attacks, or chases. It included vocalizations (waa-barks, generally taken to indicate a mixture of fear and hostile defiance) and unspecified "threat gestures." This means that while females may have responded in chorus to male aggression (as they commonly do at our site and others), it is unknown whether they formed true aggressive alliances that reduced male coercion. The circumstances in which females are able to form effective alliances among each other, and the frequency and effectiveness of this strategy, remain important problems for detailed examination in bonobos, chimpanzees, and other primates.

In some nonhuman primates, confusing paternity through promiscuous mating represents an additional female counterstrategy to the specific male tactic of infanticide (reviewed in Clarke et al., Chapter 3). Multimale mating is facilitated by a range of physiological adaptations, including multiple cycling, an extended follicular phase, and sexual swellings (Clarke et al., Chapter 3; Connor, Chapter 9). However, van Schaik et al. (2000) argue that females who subtly bias paternity toward high-ranking males can gain additional protection for their offspring, by inducing those males to defend their infants against potential aggressors. Thus, they predict that females can minimize their risk of infanticide by mating promiscuously early in the follicular phase (to give all males a nonzero probability of conception) and adopting a preference for high-ranking males when conception is most likely to occur (reviewed in Clarke et al., Chapter 3).

To date, the only direct claims for such biasing of paternity in a promiscuous primate come from studies of wild chimpanzees that remain problematic. At both Mahale (Matsumoto-Oda 1999) and Taï (Stumpf and Boesch 2005), females have been observed to mate more restrictively during periods when they were most fecund. At Taï, females were more likely to solicit matings from some males and to resist the advances of others, around ovulation, and this has been presented as evidence for female choice. As described in Muller et al. (Chapter 8), we believe that an important problem in this literature has been a failure to distinguish whether female approaches to males, during periods when the females are most attractive, are made out of free choice (as Stumpf and Boesch 2005 assume) or out of fear. The same holds true for female resistance. Our experience suggests that fear is important. For example, long-term data from Kanyawara indicate that during the periovulatory period, females are most likely to solicit copulations from the males who have been most aggressive toward them in both estrous and nonestrous periods (Muller et al., Chapter 8). We suggest that future studies of female "preference" must do more to control for the potential effects of male coercion and particularly this form of long-term "conditioning aggression." Although there is an understandable interest in the notion that females use their agency to advance their goals of finding high-quality males, we should be prepared for the possibility that in some species the coercive power of male aggression is much greater than has sometimes been considered.

Despite the broad similarities between many of the female counterstrategies to male coercion available to humans and nonhuman primates, human language and culture do, of course, allow for unique approaches, especially in the form of the law. Although Rosenfeld (Chapter 17) focuses on the U.S. legal system's limitations in this realm, and the ways in which male coalitions might subvert women's protections, that system arguably protects women from male violence more effectively than many. Rosenfeld rightly points out that large numbers of rapes go unreported in the United States, and rules of evidence sometimes appear biased against the victim. Nevertheless, in 1997 approximately 88,000 men were incarcerated in U.S. state prisons for violent sex offenses—approximately 9.7% of the state prison population (Greenfeld 1997). And between 1981 and 1999 the U.S. Department of Justice (Farrington et al. 2004) estimated the probability that a rape offender would be convicted was 15.2%, comparable to other nonlethal crimes in their survey: vehicle theft (11%), residential burglary (14%), robbery (19%), and assault (20%). Although these rape statistics are clearly problematic because of systematic underreporting, it is instructive to compare this situation with a country like Pakistan, where the legal system is more explicitly dominated by men's interests. Prior to 2006, when President Pervez Musharraf signed into law a bill curtailing Sharia directives on rape, establishing rape under Pakistani law required at least four male adult Muslim eyewitnesses to the act of penetration (Quraishi 2000). In cases where such witnesses were lacking, the Pakistani legal system sometimes concluded that intercourse must have been consensual and charged the rape victim with adultery, the prescribed punishment for which was public stoning or whipping (Quraishi 2000).

Worldwide surveys on the incidence of rape support the notion that women are significantly less likely to be victimized in the United States than in many countries. The World Health Organization, for example, estimates that 9.1% of women in the United States are forced into their first experience of sexual intercourse (Krug et al. 2002:153). Although this figure is appallingly high, it is low by global standards. Note the discouraging statistics from Cameroon (37.3%), Peru (40%), South Africa (28.4%), or a nine-country average from the Caribbean (47.6%). Although the extent to which these disparities reflect a lack of appropriate legislation or law enforcement is uncertain, we share Rosenfeld's (Chapter 17) optimism that a deeper understanding of the dynamics of men's sexual coercion will increase the potential for the law to disrupt it. Such an understanding will necessitate both an appreciation of how men's coercive strategies are similar to those employed by nonhuman primates and a realization that patriarchy creates novel forms of evolutionary logic in humans that can produce unique forms of coercion.

References

Arnqvist, G., and L. Rowe. *Sexual Conflict.* Princeton, N.J.: Princeton University Press, 2005.

Bachmann, C., and H. Kummer. "Male Assessment of Female Choice in Hamadryas Baboons." *Behavioral Ecology and Sociobiology* 6 (1980): 315–321.

Berndt, R. M., and C. H. Berndt. *The World of the First Australians.* Canberra: Aboriginal Studies Press, 1988.

Biquand, S., A. Boug, V. Biquand-Guyot, and J.-P. Gautier. "Management of Commensal Baboons in Saudi Arabia." *Revue d'Ecologie* 49 (1994): 213–222.

Broude, G. J. "The Division of Labor by Sex and Other Gender-related Variables: An Exploratory Study." *Behavior Science Research* 24 (1990): 29–49.

Campbell, J. C. "Sanctions and Sanctuary: Wife Battering within Cultural Contexts." In *To Have and to Hit: Cultural Perspectives on Wife Beating,* eds. D. A. Counts, J. K. Brown, and J. C. Campbell, pp. 261–285. Chicago: University of Illinois Press, 1999.

Clutton-Brock, T. H., and G. A. Parker. "Sexual Coercion in Animal Societies." *Animal Behaviour* 49 (1995): 1345–1365.

Collier, J. F., and M. Z. Rosaldo. "Politics and Gender in Simple Societies." In *Sexual Meanings: The Cultural Construction of Gender and Sexuality*, eds. S. B. Ortner and H. Whitehead, pp. 275–329. Cambridge: Cambridge University Press, 1981.

Connor, R. C., R. A. Smolker, and A. F. Richards. "Dolphin Alliances and Coalitions." In *Coalitions and Alliances in Animals and Humans*, eds. A. H. Harcourt, and F. B. M. de Waal, pp. 415–443. Oxford: Oxford University Press, 1992.

Crocker, W. H., and J. Crocker. *The Canela: Bonding through Kinship, Ritual, and Sex.* Fort Worth, Tex.: Harcourt Brace College Publishers, 1994.

Daly, M., and M. I. Wilson. *Homicide.* New York: Aldine de Gruyter, 1988.

Dickemann, M. "The Ecology of Mating Systems in Hypergynous Dowry Societies." *Social Science Information* 18 (1979): 163–195.

Elkin, A. P. *The Australian Aborigines: How to Understand Them.* Sydney: Angus and Robertson, 1938.

Farrington, D. P., P. A. Langan, and M. Tonry, eds. "Cross-National Studies in Crime and Justice." U.S. Department of Justice: Bureau of Justice Statistics, 2004.

Goodall, J. *The Chimpanzees of Gombe: Patterns of Behavior.* Cambridge, Mass.: Harvard University Press, 1986.

Greenfeld, L. A. "Sex Offenses and Offenders: An Analysis of Data on Rape and Sexual Assault." U.S. Department of Justice: Bureau of Justice Statistics, 1997.

Harcourt, S., and K. Stewart. *Gorilla Society: Conflict, Compromise, and Cooperation between the Sexes.* Chicago: University of Chicago Press, 2007.

Hayes, R. O. "Female Genital Mutilation, Fertility Control, Women's Roles, and the Patrilineage in Modern Sudan: A Functional Analysis." *American Ethnologist* 4 (1975): 617–633.

Hohmann, G., and B. Fruth. "Intra- and Inter-Sexual Aggression by Bonobos in the Context of Mating." *Behaviour* 140 (2003): 1389–1413.

Jaeger, R. G., J. R. Gillette, and R. C. Cooper. "Sexual Coercion in a Territorial Salamander: Males Punish Socially Polyandrous Female Partners." *Animal Behaviour* 63 (2002): 871–877.

Krug, E. G., L. L. Dahlberg, J. A. Mercy, A. B. Zwi, and R. Lozano, eds. *World Report on Violence and Health.* Geneva: World Health Organization, 2002.

Low, M., I. Castro, and A. Berggren. "Cloacal Erection Promotes Vent Apposition during Forced Copulation in the New Zealand Stitchbird (Hihi): Implications for Copulation Efficiency in Other Species." *Behavioral Ecology and Sociobiology* 58 (2005): 247–255.

Matsumoto-Oda, A. "Female Choice in the Opportunistic Mating of Wild Chimpanzees *(Pan troglodytes schweinfurthii)* at Mahale." *Behavioural Ecology and Sociobiology* 46 (1999): 258–266.

Meggitt, M. J. "Understanding Australian Aboriginal Society: Kinship Systems or Cultural Categories?" In *Traditional Aboriginal Society: A Reader*, ed. W. H. Edwards, pp. 113–137. Melbourne, Australia: Macmillan, 1987.

Newton-Fisher, N. "Female Coalitions against Male Aggression in Wild Chimpanzees of the Budongo Forest." *International Journal of Primatology* 27 (2006): 1589–1599.

Nishida, T. "Harassment of Mature Female Chimpanzees by Young Males in the Mahale Mountains." *International Journal of Primatology* 24 (2003): 503–514.

Parish, A. R. "Female Relationships in Bonobos *(Pan paniscus):* Evidence for Bonding, Cooperation, and Female Dominance in a Male-Philopatric Species." *Human Nature* 7 (1996): 61–69.

Quraishi, A. "Her Honor: An Islamic Critique of the Rape Laws of Pakistan from a Woman-Sensitive Perspective." In *Windows of Faith: Muslim Women Scholar-Activists in North America,* ed. G. Webb, pp. 102–135. Syracuse, N.Y.: Syracuse University Press, 2000.

Schwendinger, J. R., and H. Schwendinger. *Rape and Inequality.* Beverly Hills, Calif.: Sage, 1983.

Smuts, B. B., and R. W. Smuts. "Male Aggression and Sexual Coercion of Females in Nonhuman Primates and Other Mammals: Evidence and Theoretical Implications." *Advances in the Study of Behavior* 22 (1993): 1–63.

Smuts, B. B. *Sex and Friendship in Baboons.* New York: Aldine, 1985.

———. "Male Aggression against Women: An Evolutionary Perspective." *Human Nature* 3 (1992): 1–44.

———. "The Evolutionary Origins of Patriarchy." *Human Nature* 6 (1995): 1–32.

Stumpf, R. M., and C. Boesch. "Does Promiscuous Mating Preclude Female Choice? Female Sexual Strategies in Chimpanzees *(Pan troglodytes verus)* of the Taï National Park, Côte d'Ivoire." *Behavioral Ecology and Sociobiology* 57 (2005): 511–524.

Wood, W., and A. H. Eagly. "A Cross-Cultural Analysis of the Behavior of Women and Men: Implications for the Origin of Sex Differences." *Psychological Bulletin* 128 (2002): 699–727.

CONTRIBUTORS

INDEX

Contributors

JACINTA C. BEEHNER, Departments of Anthropology and Psychology, University of Michigan, Ann Arbor

THORE J. BERGMAN, Departments of Psychology and Ecology and Evolutionary Biology, University of Michigan, Ann Arbor

DOROTHY L. CHENEY, Department of Biology, University of Pennsylvania, Philadelphia

PARRY CLARKE, Department of Anthropology, University of Utah, Salt Lake City

RICHARD C. CONNOR, Department of Biology, University of Massachusetts, Dartmouth, North Dartmouth

CATHERINE CROCKFORD, Department of Psychology, University of Pennsylvania, Philadelphia

MARTIN DALY, Department of Psychology, McMaster University, Hamilton, Ontario, Canada

ANTHONY DI FIORE, Center for the Study of Human Origins, Department of Anthropology, New York University; New York Consortium in Evolutionary Primatology

MELISSA EMERY THOMPSON, Department of Anthropology, University of New Mexico, Albuquerque

ANNE L. ENGH, Department of Biology, University of Pennsylvania, Philadelphia

JULIA FISCHER, Department of Psychology, University of Pennsylvania, Philadelphia

MALLORIE A. HATCH, School of Human Evolution and Social Change, Arizona State University, Tempe

SONYA M. KAHLENBERG, Department of Anthropology, Harvard University, Cambridge, Massachusetts

DAWN M. KITCHEN, Department of Anthropology, The Ohio State University, Columbus

CHERYL D. KNOTT, Department of Anthropology, Harvard University, Cambridge, Massachusetts

ANDRES LINK, Center for the Study of Human Origins, Department of Anthropology, New York University; New York Consortium in Evolutionary Primatology

MARTIN N. MULLER, Department of Anthropology, University of New Mexico, Albuquerque

SHANNON A. NOVAK, Department of Anthropology, Maxwell School of Citizenship and Public Affairs, Syracuse University, Syracuse, New York

RYNE PALOMBIT, Department of Anthropology, Rutgers University, New Brunswick, New Jersey

TOMMASO PAOLI, Centro Interdipartimentale Museo di Storia Naturale e del Territorio, University of Pisa, Pisa, Italy

GAURI PRADHAN, Anthropologisches Institut and Museum, University of Zurich, Switzerland

MARTHA M. ROBBINS, Max Planck Institute for Evolutionary Anthropology, Leipzig, Germany

LARS RODSETH, Department of Anthropology, University of Utah, Salt Lake City

DIANE L. ROSENFELD, Harvard Law School, Cambridge, Massachusetts

AMY SCHREIER, City University of New York, New York Consortium in Evolutionary Primatology

ROBERT M. SEYFARTH, Department of Psychology, University of Pennsylvania, Philadelphia

STEPHANIE N. SPEHAR, Center for the Study of Human Origins, Department of Anthropology, New York University; New York Consortium in Evolutionary Primatology

LARISSA SWEDELL, Department of Anthropology, Queens College, City University of New York; New York Consortium in Evolutionary Primatology

CAREL VAN SCHAIK, Anthropologisches Institut and Museum, University of Zurich, Switzerland

NICOLE L. VOLLMER, Department of Biology, University of Massachusetts, Dartmouth, North Dartmouth

JANA J. WATSON-CAPPS, Department of Biology, Georgetown University, Washington, D.C.

KYLEB WILD, Department of Anthropology, University of California San Diego, La Jolla

MARGO WILSON, Department of Psychology, McMaster University, Hamilton, Ontario, Canada

ROMAN M. WITTIG, Department of Biology, University of Pennsylvania, Philadelphia

RICHARD W. WRANGHAM, Department of Anthropology, Harvard University, Cambridge, Massachusetts

Index

Milton Keynes UK
Ingram Content Group UK Ltd.
UKHW050609160424
441078UK00002B/28/J